Studies in Systems, Decision and Control

Volume 2

For further volumes:
http://www.springer.com/series/13304

Aleksander M. Nawrat
Editor

Innovative Control Systems for Tracked Vehicle Platforms

Springer

Editor
Aleksander M. Nawrat
Institute of Automatic Control
Silesian University of Technology
Gliwice
Poland

ISSN 2198-4182 ISSN 2198-4190 (electronic)
ISBN 978-3-319-04623-5 ISBN 978-3-319-04624-2 (eBook)
DOI 10.1007/978-3-319-04624-2
Springer Cham Heidelberg New York Dordrecht London

Library of Congress Control Number: 2014930024

Printed on acid-free paper

Springer is part of Springer Science+Business Media (www.springer.com)

"First they ignore you, then they laugh at you, then they fight you, then you win."

Mahatma Gandhi

Preface

The book was motivated by an urgent need for designing and implementation of innovative control algorithms and systems for tracked vehicles. Nowadays the unmanned vehicles are becoming more and more common. Therefore there is a need for innovative mechanical constructions capable of adapting to various applications regardless the ground, air or water/underwater environment. Military operations in order to complete successfully require a vast amount of information and high quality equipment and vehicles capable of movement both in a city and terrain. An example of such are tracked vehicles. As well as in today's and tomorrow's army, tracked vehicles will be one of the most important equipment. There are multiple various activities connected with tracked vehicles. They can be distributed among three main groups: design and control algorithms, sensoric and vision based information, construction and testing mechanical parts of unmanned vehicles.

Many research groups works on design and implementation of control systems for unmanned vehicles. One of the most challenging tasks is to design robust control system that allows human operator stay away from operation zone which can be very dangerous. There are few approaches to this problem. First proposition of control system is remote control. Human operator is away from vehicle and has full control on the vehicle. Operator controls signals which are send directly to vehicle actuators. Control signals are transferred from ground station to vehicle by wireless link. Another approach is tele-operation where human operator has visual feedback from cameras mounted on vehicle. Final approach is to implement autonomy functions on vehicle which is able to do tasks without human intervention. That option requires full sensor system with control processor unit on the vehicle.

As well as in today's and tomorrow's army, tracked vehicles will be one of the most important equipment. As it is known this equipment is expensive, and extremely stressful experiments on this equipment should be conducted carefully. An answer for this problem is to conduct experiments on a computer model. Design, construction of such models require research in multiple fields. One of the main threats to tracked vehicles are mines and IEDs. Therefore it is extremely important to research the impact of explosive on the tracked vehicles and especially the crew inside. It is known that the impact of explosion on the crust (hull) of a vehicle creates a secondary wave phenomenon. The phenomenon rests on the interaction of the bottom and sides of the vehicle as a membrane.The membrane moves the pressure pulse acting on the armour inside the vehicle. High mobility in all terrain and weather conditions is one of the most significant features of the light armoured vehicles. As well as that, such vehicles have good independence, setting them as a primary military equipment to be used by stabilization forces of many countries. However, they are exposed to many threats among which are improvised explosive device (IED). As a consequence, contemporary technical and tactical requirements put strong emphasis on proper layout of the hull in order to provide a good cover for the crew.

This book is not only intended for mechanical and sensory officers and engineers. It can be treated as a valuable source of information for constructors and developers of tracked vehicles or accompanying systems. Scientists and researchers involved in mechanics, control algorithms, image processing, computer vision, data fusion, or IC can find many valuable suggestions and solutions. We took an effort to present multiple solutions and challenges in the fields above. Moreover students of higher years of study in the fields of mechanics, automation, computer science, robotics or IC can benefit greatly. The authors encourage to actively explore the following chapters of the book and, in case of questions or concerns, invite you to an open and frank discussion of these matters.

Gliwice
September 2013

Aleksander M. Nawrat

Contents

Part I: Design and Control Algorithms for Unmanned Vehicles

Part II: Sensoric and Vision Based Information for UGVs

Part I
Design and Control Algorithms for Unmanned Vehicles

Many research groups works on design and implementation of control systems for unmanned vehicles. One of the most challenging tasks is to design robust control system that allows human operator stay away from operation zone which can be very dangerous. There are few approaches to this problem. One common concept of those approaches is structure of unmanned system. In all described in details in this chapter approaches there are two main elements: ground control station and vehicle control system. Those two element communicate to each other via link. In most cases it is wireless link. In underwater systems communication is performed by wired connection.

First proposition of control system is remote control. Human operator is away from vehicle and has full control on the vehicle. Operator controls signals which are send directly to vehicle actuators. Control signals are transferred from ground station to vehicle by wireless link. System on board of unmanned vehicles set control values on actuators. This approach is to replace the human factor in danger zone with unmanned robotized vehicles. One of the most important tasks is to design control system is presented, which allows remote manual control of a number of robots by means of chosen input devices. Main focus is placed on the problem of user friendly control, reducing operator workload and computer controlled intervention increasing safety of subject vehicle.

In tele-operation approach human operator has visual feedback from cameras mounted on vehicle. Operator Doesn't need direct visual contact with vehicle and he is limited only by communication link power. This approach guarantees security for human operator.

Final approach is to implement autonomy functions on vehicle which is able to do tasks without human intervention. That option requires full sensor system with control processor unit on the vehicle. It's also required advanced data processing and control algorithms. Role of human operator is limited to sets mission goals and observe how vehicle achieving them.

One of the most important component of the unmanned vehicles are control base stations. They are necessary for all of mentioned approaches. Most of them uses rugged computers but some are designed directly for selected type of unmanned vehicle. Hardware of base control station should have two parts. First allows for basic control of vehicle and should be safe, redundant and robust. Second part can have advanced futures. In most cases its computer program with modules such as map module, sensor visualization module.

This chapter includes a number of important challenges in the fields mentioned above. At the same time valuable suggestions and conclusions from authors are presented and discussed in detail.

Control System of Mobile Robot Group

Artur Babiarz, Radosław Zawiski, Michał Skrzypek, and Aleksander M. Nawrat

Silesian University of Technology, Institute of Automatic Control, Akademicka 16, 44-100 Gliwice, Poland
{artur.babiarz,radoslaw.zawiski,aleksander.nawrat}@polsl.pl, avenicus@gmail.com

Abstract. In this paper the Master Control System is presented, which allows remote manual control of a number of robots by means of chosen input devices. Main focus is placed on the problem of user friendly control, reducing operator workload and computer controlled intervention increasing safety of subject vehicle.

Keywords: mobile robots, vision system, Master Controller.

1 Introduction

Remote manual control ability of unmanned vehicles or robots is often underestimated. But in fact it needs no less effort than the automatic control. This is mainly due to the fact, that the input device signals have to be properly transmitted and converted in order to have the movement smooth and controllable by human. Setting speed of a robot to maximum value by pushing a given button will result in nothing else than a crash and damage of equipment. Proper control algorithms have to be created and tuned.

When we think about manual control we often think about things we need to do by hand in our everyday life. Although there are more complicated systems we use, like riding a bicycle or driving a car. Some of us has learned skill to operate those vehicles. After gaining proficiency in such a skill, it is natural for us, we do it automatically because we have the feeling of controls. Such manual control is a mechanical one. It utilizes a collection of mechanical parts, such as rods or tension cables. Steering wheel in a car directly changes the orientation of wheels, change of gear at the cars gearbox directly changes used gear, etc. This is the most basic method of control, widely used in vehicles around the world, as it successfully accomplishes its goal. As more complicated vehicles and planes were built, and as rapid growth of technology has allowed, more of the intelligent operations had relied on electronic devices. It has become possible to apply machine intelligence to control. That idea was mainly utilized in flight control, as it has always been a very complicated problem, consisting of vast number of variables [1,2,3]. Vision systems are used mainly in mobile robotics for determining the position robot. A very simple way to generate a path with the vision presented in [14]. This method is called *funnel lane*. The coordinates of feature points during the replay phase are compared with those obtained during the teaching phase in order to determine the turning direction. Another problem is the determination of the actual speed of the

A.M. Nawrat (ed.), *Innovative Control Systems for Tracked Vehicle Platforms*, Studies in Systems, Decision and Control 2,
DOI: 10.1007/978-3-319-04624-2_1, © Springer International Publishing Switzerland 2015

mobile robot. The authors [12] propose the use of a Kalman filter (KF)-integrated optical flow method to measure the velocity of mobile robots using a downward-looking camera. In [9] the authors present a group of robot motion planning using distributed control and vision system. The control laws are distributed in the sense that they require information only from neighboring robots. Examples of the application of the vision for mobile robot control in a limited area can be found in [7,10]. The use of vision and Q-learning controller to robust grasping by wheeled mobile robots was presented in [13]. Control law based on sliding-mode theory in order to drive mobile robots to a target location, which is specified by a previously acquired reference image is presented [6].

2 Environment

Base for construction of the control system is the RoboSoccer project [11], referred to as an Environment. Its main modules are used in order to reuse the work already done, making possible focusing on the real control algorithms, in turn giving back a great testing and learning facility. Figure 1 shows the general overview of the RoboSoccer project system. Its divided into three main parts [5]:

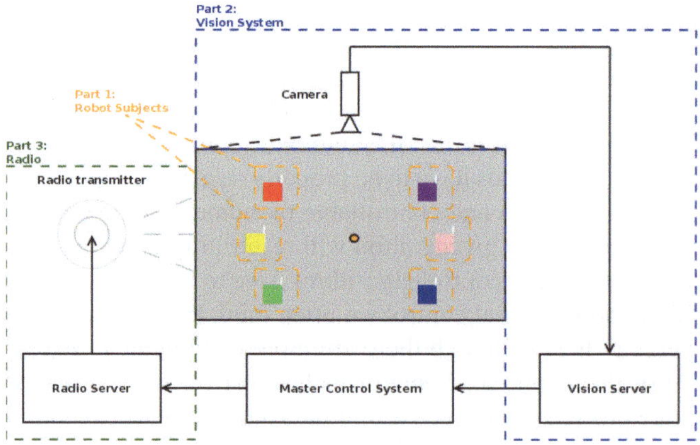

Fig. 1. RoboSoccer system overview

1. Robot subject.
2. Vision system.
3. Radio system.

Each of the parts is described in [4] in details.

The subject to control is a robot, shown at figure 2 with a two-wheeled differential drive, operated by two DC motors. Each of motor has an incremental magnetic encoder, that counts how much a wheel has rotated, and it is the only sensor available at the vessel. It has an on-board micro controller constructed, based on ATMEL microprocessor. It is able to realize simple control procedures, making it possible for Master Control

Fig. 2. Robot

System to focus on more sophisticated algorithms [8]. The wireless connection, serving as input for commands, is provided by the Xbee module, properly interfaced by the software in the on-board controller. The same Xbee module exists at the other end, in Radio system of the project. The robot provides interface for following actions, which can be commanded through the Radio:

- Change the speed of wheels i.e. set linear velocity of a robot.
- Rotate by given angle, at given angular velocity or at given constant radius.
- Move forward/backward by given distance.
- Go to a given location.

Important fact is ensuring that the robot does not slide, as it would input errors to the the incremental magnetic encoder, which runs vital part of every algorithm at the robot. That is why the on-board controller avoids skid of wheels, by limiting linear and angular acceleration. It has to be taken into account by the Master Control System, so that the system is cohesive. The limits are calculated in work [8] and are listed in Table 1. The a_{max} was obtained experimentally by checking whether a set acceleration causes any sliding and repeating the experiment successively. The value that is used was additionally lowered by 50% in order to minimize the risk of skid, while the α_{max} was calculated from the interaction of wheels with the ground [8].

Table 1. Limits

Limit	Value	Unit
Acceleration a_{max}	1	$\frac{m}{s^2}$
Angular acceleration α_{max}	0.5	$\frac{rad}{s^2}$

3 Master Control System Structure

The Master Controller is implemented in Java programming language as an application with graphical user interface. It is called **qView** and on Figure 3 can be seen in the middle. Control algorithms are constructed in its basic layer and graphical user interface utilizes them on application layer allowing user interaction and gives the visual feedback. Additional application is the **qPhone**, that is a midlet application for cellphones.

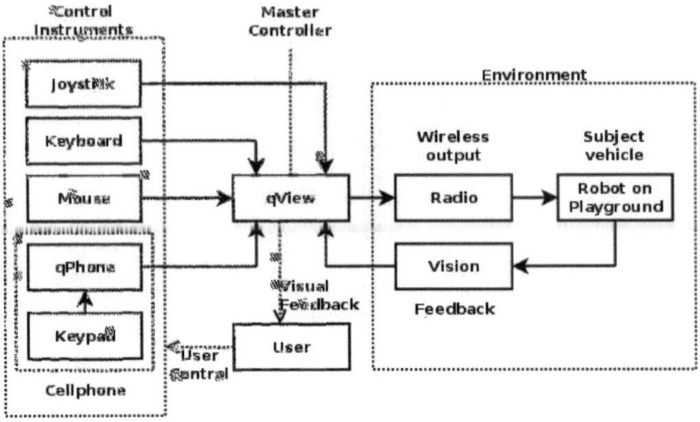

Fig. 3. Master Control System structure

Utilizing a bluetooth connection between cellphone and PC computer with qView application installed, it transmits the cellphones keypad events to qView control algorithms in order to fulfill one of primary goals - possibility to control a robot with a cellphone. Such construction - only transmission of events - makes it able to reuse algorithms, without need of implementing whole new separate controller. Additional application is the qPhone, that is a midlet application for cellphones. Utilizing a bluetooth connection between cellphone and PC com- puter with qView application installed, it transmits the cellphone's keypad events to qView control algorithms in order to fulfill one of primary goals - possibility to control a robot with a cellphone. Such construction - only transmission of events - makes it able to reuse algorithms, without need of implementing whole new separate controller.

3.1 Speed and Acceleration

Basing on two coordinates, speed is calculated. Coordinates for the calculation are chosen every third coordinate, in order to minimize the vision system's variations. Time between them is recorded. Thus we have all the data we need:

$$x_t - \text{x coordinate at time t} \tag{1}$$

$$y_t - \text{y coordinate at time t} \tag{2}$$

$$x_{t-p} - \text{x coordinate at time t-p} \tag{3}$$

$$y_{t-p} - \text{y coordinate at time t-p} \tag{4}$$

$$p - \text{time between two successive coordinates} \tag{5}$$

Velocity is the rate of change of position. The average velocity v of an object moving through a displacement during a time interval (Δt) is described by the formula:

$$v = \frac{\Delta x}{\Delta t} \tag{6}$$

Velocity in coordinate system is a composition of velocities through displacements along axes x and y, as shown on Figure 4 and is a directed vector.

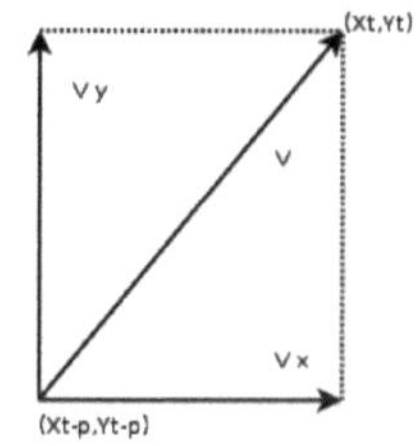

Fig. 4. Velocity

Speed is the magnitude of velocity, that we want to display for the user. Velocities along axes are calculated in following way:

$$v_x = \frac{|x_t - x_{t-p}|}{p} \tag{7}$$

$$v_y = \frac{|y_t - y_{t-p}|}{p} \tag{8}$$

Using the well known Pythagorean theorem magnitude of object velocity is calculated:

$$v = \sqrt{v_x^2 + v_y^2} \tag{9}$$

Acceleration is a rate of change of speed. Average acceleration is the change in velocity (Δv) divided by the change in time (Δt):

$$a = \frac{\Delta v}{\Delta t} \tag{10}$$

Having following data:

$$v_t - \text{speed at time t} \tag{11}$$

$$v_{t-p} - \text{speed at time t-p} \tag{12}$$

$$p - \text{time between two speed calculations} \tag{13}$$

We can calculate acceleration with following equation:

$$a = \frac{v_t - v_{t-p}}{p} \tag{14}$$

In order to decrease variance and have the value more readable for the user, an average of two consecutive acceleration values is calculated. Both speed and acceleration are shown in the user interface for each robot in a way presented in Figure 5.

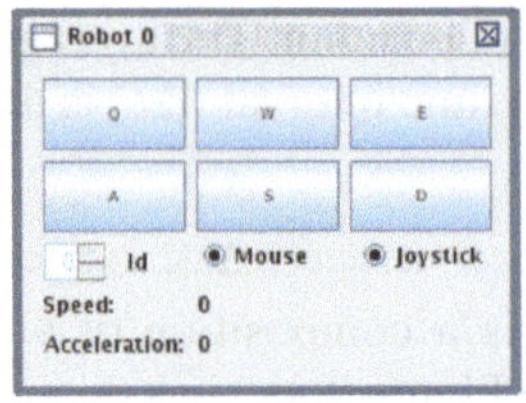

Fig. 5. qView speed and acceleration display

4 Control Algorithms

Control instrument signals have to be translated in order to give the user a way to control a given object. Input devices differ from each other in some ways, but also some things may be in common. Best practice is to join the common things together and create one joint controller, then creating more advanced controllers on its basis for each of control instrument. As there can be numerous control instruments connected to qView, in order to provide proper control, only one input device can be allowed to control a given robot. This is done by simple locking mechanism, in which input devices ask for permission to control a robot, and if one is not controlled at a time, permission is granted and lock is activated. In this time no other control instrument can get access to that robot.

Control Variables
Two things in the robot can be controlled - linear and angular speed. Both are accessible directly through functions and this direct approach is most frequently used:

1. Robot::setSpeed(speed).
2. Robot::setTurningSpeed(angSpeed).

Although the on-board controller on the robot also gives another way to control the angular speed, through the Robot::setTurningAngle(radAtan) function. It is a turning radius control algorithm implemented in the robot software and detail description is presented in work [8]. It allows to control turning of the robot at constant radius. The parameter given to this function - $atan(\frac{1}{r})$ - can be understood as an angle at which robot would turn if it had third leading wheel. Although in order to keep the turning radius constant, the algorithm has to limit the linear speed accordingly,which could cause that the actions of operator are not natural and not as intended.

4.1 Keyboard Control

Biggest issue with keyboard control is that buttons are on/off switches. Transition between one to another is virtually instantaneous, while in order to give fine ability to control a robot, smooth transition should be introduced. Additional variable providing this feature is time. Having a button event coming as an input from the user is one variable. If we count time for how long user has the button pushed, we get another variable, thus we can create a function of time, that controls linear and angular speeds of the robot. Few different functions were considered, starting from the simplest linear function, through exponential function ending on logarithmic one. Eventu- ally exponential function was chosen because of the fact it is easily tuned, is bounded, is not complicated computationally and fits human operator, as is perceived as natural reaction. Therefore following functions are used for control:

$$v = v_{max}\left(1 - e^{\frac{-t}{T_v}}\right) \tag{15}$$

$$\omega = \omega_{max}\left(1 - e^{\frac{-t}{T\omega}}\right) \tag{16}$$

$$r = r_{max}\left(1 - e^{\frac{-t}{T_r}}\right) \tag{17}$$

where:

$$v - \text{speed}$$

$$\omega - \text{angular speed}$$

$$r - \text{turning radius angle}$$

$$v_{max}, \omega_{max}, r_{max} - \text{maximum values}$$

$$T_v, T_\omega, T_r - \text{time constants}$$

$$t - \text{is time since button click}$$

4.2 Cellphone Control

Cellphone keypad has much in common with computer keyboard. Push events are created in the same way, which leads to best solution for cellphone controller: events are not resolved on the cellular phone, but are transmitted to master controller, which does the main job. It is a good practice to reuse work already done, and in this problem we can reuse keyboard controller for cellphone keypad. That leads to conclusion that the best solution for the cellphone appli- cation is to transmit events to the PC computer, not control robot directly. This causes that the control algorithms and computer controlled intervention algorithms are the same for cellphone controller. The protocol for wireless transmission chosen is Bluetooth, as it is most available in public cellphones.

4.3 Mouse Control

In mouse control two techniques were proposed and implemented:

1. Mouse events connected to the field graphical representation.
2. Virtual mouse pad.

It is most natural for a user to directly interact with feedback. That is why there is possibility to interact with the graphical representation of the field with use of mouse. User is allowed to click the field with his mouse, issuing commands. Figure 6 visualizes the mouse click with respect to robot position. Three different commands are implemented using this technique:

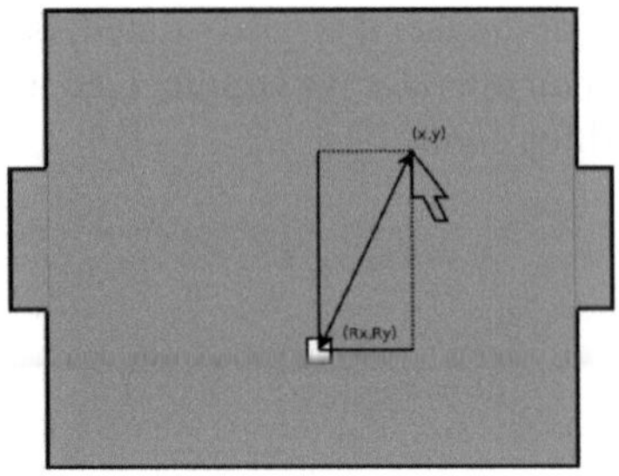

Fig. 6. Mouse click distance calculation

Left mouse button click Move to
Middle mouse button click Go distance
Right mouse button click Look at **Move to** command directly sends a $Robot :: gotoXY(x,y)$ with proper coordinate arguments x and y, that have to be inverted from graphical user interface (GUI) coordinates in order to comply with global positioning system.

Go distance sends a $Robot :: move(d)$, with distance calcu- lated in following way:

$$d = \sqrt{(x - r_x)^2 + (y - r_y)^2} \tag{18}$$

where:

$$x, y - \text{x and y coordinates of mouse click}$$

$$r_x, r_y - \text{x and y coordinates of robot's current position}$$

Look at command sends a $Robot :: lookAt(x,y)$. The coordinates x and y are in global positioning system of the system, so they only need to be inverted from GUI.

Virtual Mouse Pad
Figure 7 presents the virtual mouse pad available in qView. Mouse click or drag on its area causes an event, which coordinates are translated from GUI coordinate system (red) to magnitude of linear and angular speed (green) in following way:

$$v = -v_{max}\left(\frac{y}{\frac{padheight}{2}} - 1\right) \tag{19}$$

$$\omega = -\omega_{max}\left(\frac{x}{\frac{padwidth}{2}} - 1\right) \tag{20}$$

where:

$$x, y - \text{coordinates of mouse click}$$

$$v_{max}, \omega_{max} - \text{maximum values of speeds}$$

$$padheight, padwidth - \text{height and width of the virtual pad}$$

And these magnitudes are then sent to chosen robot using commands $Robot :: setSpeed(speed)$ and $Robot :: setTurningSpeed(angSpeed)$. Additionally there are safety procedures implemented, that when mouse leaves the virtual pad or mouse click is released, robot is commanded to stop by sending linear and angular speed equal to 0.

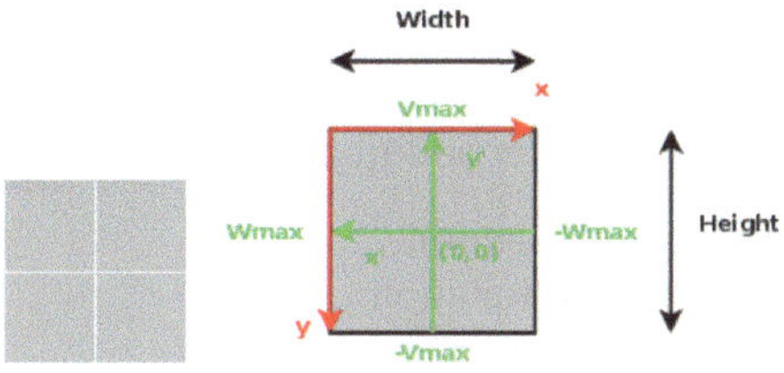

(a) in GUI (b) in theory

Fig. 7. Virtual mouse pad

4.4 Joystick Control

Joystick control is similar to the mouse virtual pad control, as the joystick variations are also contained in a rectangle. Main difference is that joystick needs a lot more of service and support in order to capture the events. After successful implementation of gathering data from joystick control instrument, it is basically the same translation as in case of mouse virtual pad shown in Figure 8. I.e. the red system of joystick deviation in axes x and y is translated to axes x' and y', which correspond to angular and linear speed respectively, as presented in following equations:

$$v = -v_{max} \cdot y \tag{21}$$

$$\omega = -\omega_{max} \cdot x \tag{22}$$

where:

$$x, y - \text{magnitudes of joystick deviation in x and y axes}$$

$$v_{max}, \omega_{max} - \text{maximum values of speeds}$$

Skid limits for mouse virtual pad and joystick are neglected, as the intervention on control of on-board controller does not influence the feeling of control using these control instruments.

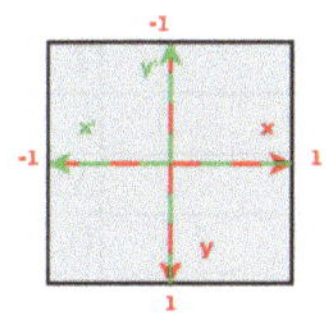

Fig. 8. Joystick translation

4.5 Tuning for Finer Control

Each of the control instruments has its own specific features, so the user has different feeling of control with different device. That is why each of the control instruments has to be in fact tuned separately. The parameters available for tuning of every instrument are:

1. Maximum linear speed.
2. Maximum angular speed.
3. Maximum radial turning magnitude.

And for keyboard/cellphone controller additionally:

1. Time constant of linear speed function.
2. Time constant of angular speed function.
3. Time constant of radial turning function.

User has the ability to change these values at his willing in GUI. Although the default values in qView are already optimized and ready "out of the box" and equal to the values in Table 2. Such values were obtained by experimental work. Higher values of mouse and joystick parameters come from the fact that these control instruments are very precise an user can control himself the maximum value, while the restriction is only for safety reasons in order not to exceed some critical values. On the other hand, keyboard control has the limitation, that the button is on/off state. We can vary the magnitude in time, but user has limited control over that, and maximal value of speeds have to be properly set in order to give fine control. Above parameters are changeable by the user, although the change is lim- ited in some strict boundaries that would break the safety of vehicle.

Table 2. Tuning parameters

Keyboard		
Parameter	Value	Unit
speed Max	0,6	$\frac{m}{s}$
speed time constant	2	s
turning Max	1,5	$\frac{rad}{s}$
turning time constant	2	s
radius Max	2	$degree$
radius time constant	2	s
Mouse		
Parameter	Value	Unit
speed Max	0,8	$\frac{m}{s}$
turning Max	2	$\frac{rad}{s}$
Joystick		
Parameter	Value	Unit
speed Max	0,8	$\frac{m}{s}$
turning Max	2	$\frac{rad}{s}$

5 Experimental Results

Trajectory Control

Speed limitation algorithm has a serious drawback. The distance measure- ment forward and backward is calculated from the coordinate center of the robot, while it is a rectangle from the top, so at low angles (see Figure 9(a)) to the edges speed is not limited, while robot scratches the edge with its side. This situation has to be avoided and this is done by the trajectory control algorithm. Idea of the trajectory control is to, at one hand: induce angular speed preventing the collision, and at the other: disallow human turning control at the dangerous side. All that only in a case when the robot is going to touch the edge. As the speed limiter works perfectly at high angles (see Figure 9(b)), we do not need to change trajectory and intervene human control at those high angles. It was decided that the trajectory control works only at angles below 60 to the edge. The trajectory change algorithm includes few parameters, all of which are discussed in following sections.

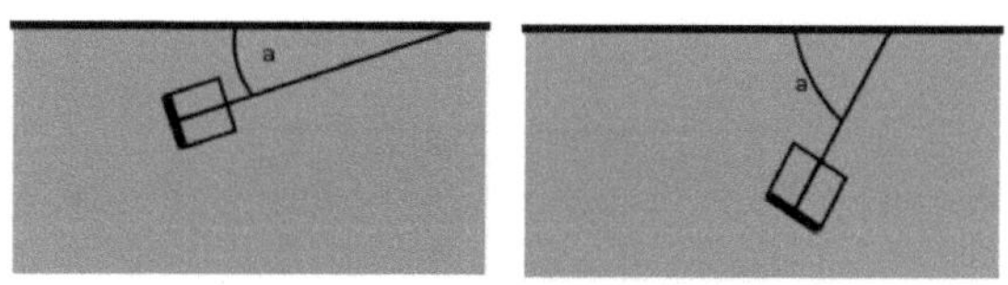

(a) Low angle (b) High angle

Fig. 9. Robot orientation to the edge

Clockwise / Counter-Clockwise

In order to properly change the trajectory, we need to know at which side the turn should be induced. That is why a clockwise χ value is introduced:

$$\chi = \text{-1clockwise direction}$$

$$\chi = 1\text{counter-clockwise direction}$$

The value corresponds and complies with the argument of function:

$$Robot :: setTurningSpeed(angSpeed)$$

This value has to be properly calculated from the coordinate system and playground. It is calculated for each of edges separately:

$$\chi_{leftedge} = -sin(orientation)$$

$$\chi_{rightedge} = sin(orientation)$$

$$\chi_{topedge} = \quad cos(orientation)$$

$$\chi_{bottomedge} = cos(orientation)$$

Above simple equations utilize the change of sign of the trigonometric functions. The value from the closest edge - the one of minimum 5,9 to 5,12 is taken.

Rate of Change
The turning speed should be faster at high angles, and slower at low angles (see Figure 9). In order to achieve a smooth function, also trigonometric functions are used. This parameter is marked as γ. Each of the edges has it own trigonometric equation, and the closest edge is chosen:

$$\gamma_{leftedge} = -cos(orientation)$$

$$\gamma_{rightedge} = cos(orientation)$$

$$\gamma_{topedge} = sin(orientation)$$

$$\gamma_{bottomedge} = -sin(orientation)$$

This can be graphically interpreted (for top edge) in way shown in Figure 10. The parameter γ is the blue-green plot.

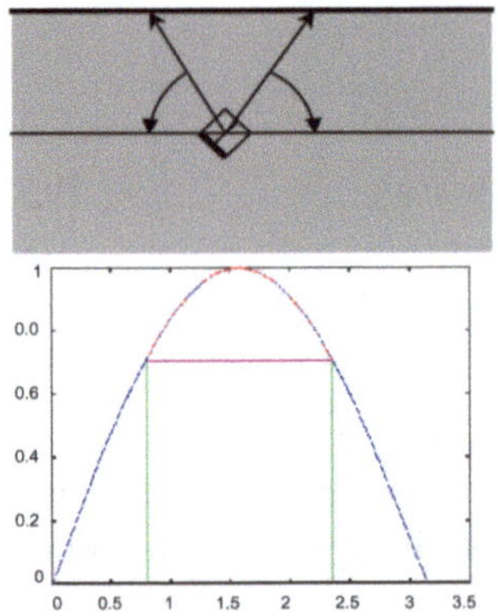

Fig. 10. Rate of change graphical interpretation

Distance from the Edge
The intervention should be induced only in direct neighborhood of the edge. Thus distance at which the trajectory algorithm is working is limited. The minimal distance from the edge is used for this purpose (d). The limit is marked as parameter:

$$d_{limit} = 0,15 \quad [m] \tag{23}$$

Additionally the magnitude of trajectory change should also be higher closer to the edge, and smaller farther from it. The parameter δ that achieves that is defined in following way:

$$\delta = \frac{d_{limit}}{d} \tag{24}$$

Plot of Equation 24 is shown at Figure 11.

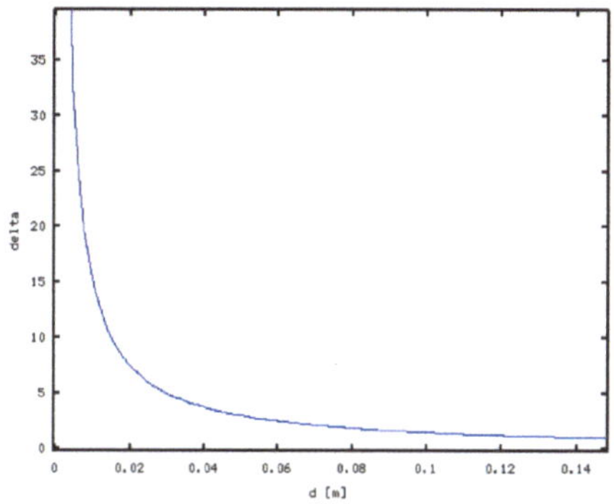

Fig. 11. Delta parameter

Velocity
The fourth and last parameter is simply the linear speed of the robot as the angular speed should be proportional to it.

The Trajectory
Parameters discussed in previous sections construct a final function that gives the angular velocity that has to be sent to robot in order to change its trajectory properly:

$$\omega = \chi \cdot \gamma \cdot \delta \cdot v \tag{25}$$

It is sent with the following function $Robot :: setTurningSpeed(angSpeed)$ to the robot. The master controller checks whether signal coming from the human operator contradicts this calculated value. If it is, it overrides the human signal and sends the intervening signal. If it is not - i.e. human signal will not cause an accident with an edge - no intervening action is taken, as it would unnecessarily influence the manual controls.

Human Control Override

The human control override can be seen on Figure 12. The human op- erator controls the robot near the edge of the field, moving from the right to the left. He issues commands to move forward and turn right, which would cause a collision. This action is being overridden by computer controlled in- tervention by disallowing such an action, and additionally leading the robot back to right track to collision free trajectory. This override can be seen on Figure 12(a) - the high peaks to the negative values are human controlled commands, and the "gaps" between them are override by the algorithm, that cause a shallow turn to the other side.

(a) First

(b) Second

Fig. 12. Photos from the camera

6 Conclusions

The Master Control System achieved all the stated goals. It is a reliable controller for manual control using various control instruments, allows tuning of the parameters of its control algorithms, default values of these are optimized, so that a learning curve of the controls is very steep. Feedback of current state of playground is simple and informative, and the graphical user interface is user friendly - the operator can start controlling robots right away even without reading the manual. Numerous functions can be checked, their effects on each other read and concluded from.

References

1. Babiarz, A., Jaskot, K.: Experimental Construction of Walking Robot. Mechanics 27(3), 91–95 (2008)
2. Babiarz, A., Jaskot, K.: Układ inercyjny do pomiaru orientacji obiektów. Przegld Elektrotechniczny 86(11a), 323–333 (2010)
3. Babiarz, A., Jaskot, K.: The Concept of Collision-Free Path Planning of UAV Objects. In: Nawrat, A., Simek, K., Świerniak, A. (eds.) Advanced Technologies for Intelligent Systems. SCI, vol. 440, pp. 81–94. Springer, Heidelberg (2013)
4. Babiarz, A., Bieda, R., Jaskot, K.: A Distributed Control Group of Mobile Robots in a Limited Area with a Vision System. In: Nawrat, A., Kuś, Z. (eds.) Vision Based Systems for UAV Applications. SCI, vol. 481, pp. 157–175. Springer, Heidelberg (2013)
5. Skrzypek, M.: Master Control System of Mobile Robot Group. Masters thesis, Silesian University of Technology, Gliwice (2010)
6. Becerra, H.M., Lopez-Nicolas, G., Sagues, C.: A Sliding-Mode-Control Law for Mobile Robots Based on Epipolar Visual Servoing From Three Views. IEEE Transactions on Robotics 27(1), 175–183 (2011)
7. Livatino, S., Muscato, G., Privitera, F.: Stereo Viewing and Virtual Reality Technologies in Mobile Robot Teleguide. IEEE Transactions on Robotics 25(6), 1343–1355 (2009)
8. Łakota, T.: Design, implementation and programming of mobile robot controller. Masters thesis, Silesian University of Technology, Gliwice (2010)
9. Moshtagh, N., Michael, N., Jadbabaie, A., Daniilidis, K.: Vision-Based, Distributed Control Laws for Motion Coordination of Nonholonomic Robots. IEEE Transactions on Robotics 25(4), 851–860 (2009)
10. Riggs, T.A., Inanc, T., Weizhong, Z.: An Autonomous Mobile Robotics Testbed: Construction, Validation, and Experiments. IEEE Transactions on Control Systems Technology 18(3), 757–766 (2010)

11. Smoliński, M.: Vision system for group of mobile robots. Masters thesis, Silesian University of Technology, Gliwice (2006)
12. Xiaojing, S., Seneviratne, L.D., Althoefer, K.: A Kalman Filter-Integrated Optical Flow Method for Velocity Sensing of Mobile Robots. IEEE/ASME Transactions on Mechatronics 16(3), 551–563 (2011)
13. Ying, W., Haoxiang, L., de Silva, C.W.: A Hybrid Visual Servo Controller for Robust Grasping by Wheeled Mobile Robots. IEEE/ASME Transactions on Mechatronics 15(5), 757–769 (2010)
14. Zhichao, C., Birchfield, S.T.: Qualitative Vision-Based Path Following. IEEE Transactions on Robotics 25(3), 749–754 (2009)
15. Bieda, R., Jaskot, K., Jędrasiak, K., Nawrat, A.: Recognition and Location of Objects in the Visual Field of a UAV Vision System. In: Nawrat, A., Kuś, Z. (eds.) Vision Based Systems for UAV Applications. SCI, vol. 481, pp. 27–45. Springer, Heidelberg (2013)
16. Mezyk, A., Switonski, E., Kciuk, S., Klein, W.: Modelling and investigation of dynamic parameters of tracked vehicles. Mechanics and Mechanical Engineering 15(4), 115–130 (2011)
17. Kciuk, S., Kciuk, M., Turczyn, R., Martynowicz, P.: An Investigation of the Behaviour of Magnetorheological Fluids in The Rotary Shock-Absorber. Advanced Materials Research 628, 512–517 (2013)
18. Barnat, W., Panowicz, R., Niezgoda, T.: Numerical and Experimental Comparison of Combined Multilayer Protective Panels. Acta Mechanica et Automatica 6(1), 13–16 (2012)
19. Barnat, W., Panowicz, R., Niezgoda, T., Dybcio, P.: Numerical analysis of IED detonation effect on steel plate. Acta Mechanica et Automatica 6(1), 10–12 (2012)
20. Panowicz, R., Barnat, W., Niezgoda, T., Szymanczyk, L., Grzymkowski, J.: Numerical-Experimental Investigation of Squared-based Metal Pyramids Loaded With a Blast Wave From a Small Explosives Charge. Acta Mechanica et Automatica 6(1), 49–52 (2012)
21. Barnat, W., Niezgoda, T., Panowicz, R., Sybilski, K.: The Influence of Conical Composite Filling on Energy Absorbtion During The Progressive Fracture Process. WIT Transactions on Modelling and Simulation 51, 625–633 (2011)
22. Niezgoda, T., Panowicz, R., Sybilski, K., Barnat, W.: Numerical Analysis of Missile Impact Being Shot By Rocket Propelled Grenades with Rod Armour. WIT Transactions on Modelling and Simulation 51, 625–633 (2011)
23. Barnat, W., Dziewulski, P., Niezgoda, T., Panowicz, R.: Application of Composites to Impact Energy Absorbtion. Computational Materials Science 50(4), 1233–1237 (2011)
24. Niezgoda, T., Barnat, W.: Analysis of Protective Structures Made of Various Composite Materials Subjucted to Imapct. Materials Science and Engineering A 483-484(1-2 C), 705–707 (2008)
25. Daniec, K., Jedrasiak, K., Koteras, R., Nawrat, A.: Embedded Micro Inertial Navigation System. Applied Mechanics and Materials 249, 1234–1246 (2013)
26. Jędrasiak, K., Bereska, D., Nawrat, A.: The Prototype of Gyro-Stabilized UAV Gimbal for Day-Night Surveillance. In: Nawrat, A., Simek, K., Świerniak, A. (eds.) Advanced Technologies for Intelligent Systems. SCI, vol. 440, pp. 107–116. Springer, Heidelberg (2013)
27. Jędrasiak, K., Nawrat, A., Wydmańska, K.: SETh-link The Distributed Management System for Unmanned Mobile Vehicles. In: Nawrat, A., Simek, K., Świerniak, A. (eds.) Advanced Technologies for Intelligent Systems. SCI, vol. 440, pp. 247–256. Springer, Heidelberg (2013)
28. Jędrasiak, K., Nawrat, A.: Image Recognition Technique for Unmanned Aerial Vehicles. In: Bolc, L., Kulikowski, J.L., Wojciechowski, K. (eds.) ICCVG 2008. LNCS, vol. 5337, pp. 391–399. Springer, Heidelberg (2009)
29. Nawrat, A., Jedrasiak, K.: SETh System Spatio-Temporal Object Tracking Using Combined Color and Motion Features. In: WSEAS International Conference, Proceedings, Mathematics and Computers in Science and Engineering, vol. 9 (2009)

30. Jedrasiak, K., Daniec, K., Nawrat, A.: The Low Cost Micro Inertial Measurement Unit. In: Industrial Electronics and Applications (ICIEA), pp. 403–408 (2013)
31. Bereska, D., Jędrasiak, K., Daniec, K., Nawrat, A.: Gyro-Stabilized Platform for Multispectral Image Acquisition. In: Nawrat, A., Kuś, Z. (eds.) Vision Based Systems for UAV Applications. SCI, vol. 481, pp. 115–121. Springer, Heidelberg (2013)
32. Daniec, K., Iwaneczko, P., Jędrasiak, K., Nawrat, A.: Prototyping the Autonomous Flight Algorithms Using the Prepar3D Simulator. In: Nawrat, A., Kuś, Z. (eds.) Vision Based Systems for UAV Applications. SCI, vol. 481, pp. 219–232. Springer, Heidelberg (2013)
33. Babiarz, A., Bieda, R., Jędrasiak, K., Nawrat, A.: Machine Vision in Autonomous Systems of Detection and Location of Objects in Digital Images. In: Nawrat, A., Kuś, Z. (eds.) Vision Based Systems for UAV Applications. SCI, vol. 481, pp. 3–25. Springer, Heidelberg (2013)
34. Bereska, D., Daniec, K., Fraś, S., Jędrasiak, K., Malinowski, M., Nawrat, A.: System for Multi-Axial Mechanical Stabilization of Digital Camera. In: Nawrat, A., Kuś, Z. (eds.) Vision Based Systems for UAV Applications. SCI, vol. 481, pp. 177–189. Springer, Heidelberg (2013)

Concept of Implementation of Remote Control Systems into Manned Armoured Ground Tracked Vehicles

Adam Bartnicki[1], Marian Janusz Łopatka[1], Lucjan Śnieżek[1],
Józef Wrona[1,2] and Aleksander M. Nawrat[3]

[1] Military University of Technology, ul. gen. Sylwestra Kaliskiego 2, 00-908 Warsaw, Poland
{abartnicki,mlopatka,lsniezek,jwrona}@wat.edu.pl
http://www.wat.edu.pl/
[2] Industrial Research Institute for Automation and Measurements,
Aleje Jerozolimskie 202, 02-486 Warsaw, Poland
jwrona@piap.pl
http://www.piap.pl/
[3] Institute of Automatic Control, Silesian University of Technology,
Akademicka 16, 44-100, Gliwice, Poland

Abstract. The first idea to replace the human factor in danger zone with unmanned robotized vehicles that was defined and implemented at Military University of Technology has been presented in this article. The concept of implementation of components of the remote control system to transform the main battle tank T-72 into Unmanned Ground Vehicle to perform mine clearance tasks and cross-bedding the river has been described. Some results of research of the system have been presented.

Keywords: remote control systems, teleoperations, UCEV, UGV, ARV.

1 Introduction

Last ten years transformation of battlefield has been resulted by asymmetric warfare, so the nature of military conflicts is being changed. It creates more and more complex but after all dangerous tasks that warfighters are faced now. Asymmetric character of contemporary warfare causes imbalance in the means of power, resources and costs resulted from this asymmetry. In consequence, armed forces need to develop complex and expensive operational capabilities and their application methods in order to fight against relatively simple and inexpensive techniques used by irregular enemy troops or terrorists (Czapla, Wrona, 2013).

Lessons Learned Experiences of Polish Armed Forces from Iraq, Afghanistan, Kosovo, Lebanon, Syria, and Chad indicate the most dangerous threats and challenges of contemporary battlefield create the needs for new military capabilities and show that existing vehicles and military equipment very often do not fulfill Operational Requirements (OR). Development of techniques and technologies but after all emerging and disruptive technologies help to achieve these capabilities, allow to think

A.M. Nawrat (ed.), *Innovative Control Systems for Tracked Vehicle Platforms*, Studies in Systems, Decision and Control 2,
DOI: 10.1007/978-3-319-04624-2_2,

about new Concepts of Operations, (CONOPs) and then OR for new military machines and equipments. The main focus of them is on safety of soldiers. That is why they should be able to fight against most dangerous threats at this very complex asymmetric battlefield. One of them are mines and Improvised Explosive Devices. Considering fighting against these threats some CONOPs and then OR have been worked out. One of the results of such a considerations is that warfighter should be replaced with machines that could carry out the complex tasks in dangerous environment that soldiers could be exposed to lost of health and even life. To do that in countermine operations there is a necessity to use remote controlled or even autonomous vehicles.

Operational possibilities and effectiveness of Unmanned Combat Engineer Vehicles (UCEV), Unmanned Ground Vehicles (UGV), Armoured Recovery Vehicle (ARV) depend on their design and possibilities of control systems – fig.1. Potentially the highest level of possibilities present the autonomous systems, because:

- lower price than platform operating in teleoperation system – they are not equipped with advanced system of transmission of all collected information concerning actual state of platform, environment, terrain and tactical imagery, and concerning executed tasks;
- multiplication of capabilities – one operator are able to monitor the operation a few or even dozen of robots that autonomously take necessary decisions while in teleoperation system the single platform are able to require to be controlled by even 3 operators responsible for maneuver, nawigation and on-board systems (reconnaissance, armaments, etc.);
- faster operation – work out decision and take operation in time of 1ns while operator's reaction in teleoperation system is 0,2-1s (depending on operator's concentration level and delay time in data transmission system); stress-resistant, while man is yield to stress being in danger – always, independently of threat, fulfill procedures being define by application developer;
- impartiality of operation – they have no bias against anything and fulfill the standard processes without discrimination in favor of or against any person by dint of gender, attire, religion, skin color, behaviour, etc. (Konopka S., Łopatka M. J., Muszyński T., Typiak A., 2008).

However, they require advanced systems of environment recognition, situation assessment and decision-making. It necessitates development of artificial intelligence. Conducted research (van den Heuvel Johan C.,2008), (Lundber Carl, 2008), (Schnaider F., 2009) indicate that the current level of technology development is too low to make possible execution of very complex missions on the contemporary battlefield autonomously.

Analysis that were done at Military University of Technology (Bartnicki A., Łopatka M.J., Typiak A., 2012), (Bartnicki A., Łopatka M.J., Muszyński T., Sprawka P., 2012) indicated complex missions on the contemporary battlefield autonomously, necessity and possibilities of usage of available technologies to develop platforms operating in teleoperation mode. By dint of limitations of such a solution, relatively low speed of performance of remote controlled platforms equipped with teleoperation system can be achieved. Taking into consideration these limitations the most needed applications of unmanned systems were pointed out:

- performance of mine clearance tasks;
- obstacle breaching performance;
- cross-bedding the river.

The execution of these tasks is possible thanks to robotization of existing Manned Armored Tracked Vehicles or designing new special systems. Their wide and fast implementation demands of elaboration of universal control kits that allow to control of existing manned vehicles remotely.

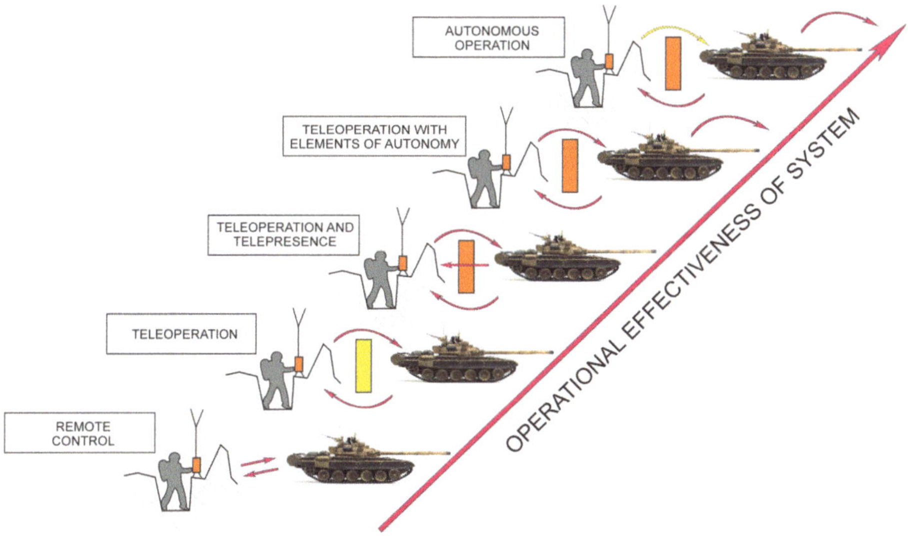

Fig. 1. The levels of robotization of mobile platforms with aspect of performance possibilities

Moreover, in designing of manned ground vehicles the simulation models play very important role. Such an approach creates synergy and make some availabilities for remote control systems. The efforts focused on this issue are described in the paper (Arkadiusz Mężyk, Eugeniusz Świtoński, Sławomir Kciuk, Wojciech Klein, 2011). There are several projects on military unmanned vehicles also performed in Poland (Bartnicki A., Dąbrowska A., Rubiec A., 2011).

2 Assumptions for Development of the Concept of Remote Controlled Systems of Chosen Manned Tracked Ground Vehicles

In the designing process of universal control system for Manned Tracked Ground Vehicles the interactions in the “Operator – Vehicle – Environment” system (fig.2).

The vehicles based on the tracked system and power transmission unit of the main battle tank T-72 were chosen to figure out the concept of remote control kit. The study of different scenarios by dint of threat to life the scenarios of cross-bedding the river by Armored Recovery Vehicle (ARV) and performance of mine clearance tasks by main battle tank T-72 fitted with certain mine-clearing devices (Combat Engineer Vehicle, CEV) were chosen.

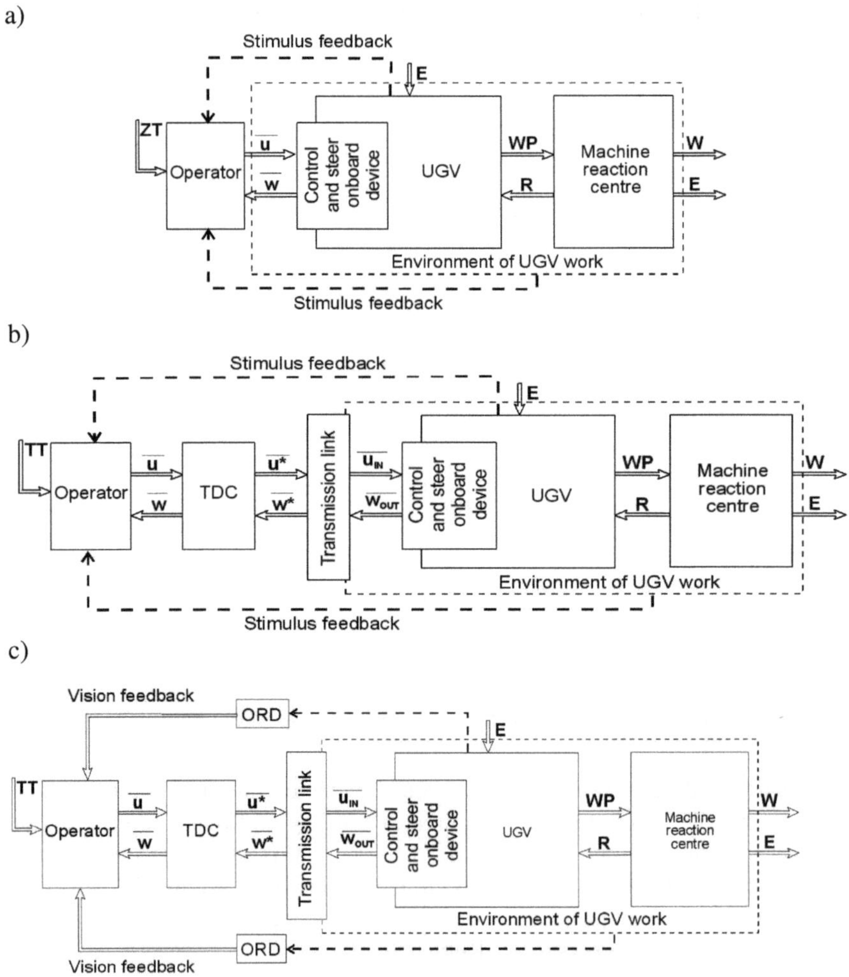

Fig. 2. Structures of operator-UGV-environment system: a) when direct control, b) when remoter control with stimulus feedback usage, c) when control in tele-operator system: TT technological tasks: $\bar{u}$- control signals generated by operator; $\overline{w}$- information for operator; $\overline{u^*}$, $\overline{w^*}$ -accordingly sent control signals and received information signals; $\overline{u_{IN}}$ - signals controlling UGV onboard control system; $\overline{w_{OUT}}$ - output information signals; E- outside energy; WP- working processes; R- machine reactions centre; E,W- efficiency and UGV work quality; TDC- technical devices of path control; ORD- observation-recognition

The issue is particularly essentials because the assumed goal of control depends on not only monitoring of movements of the UGV but after all on information about results of its impact on environment. It is crucial when the vehicle is fitted with certain special devices as it is in case of Combat Engineer Vehicle (CEV).

The appreciation of performance specific of UGV/UCEV allows to appropriate design of its control system, particularly when the teleoperation system is used. Advancement of on-board systems technology is extraordinary crucial for designing process of their control systems. If they are based mechanical solutions that operator's abilities requiring long term training are used – there is necessary to figure out the computer-added operator systems to minimize the quality requirements in the field of usage of remote control systems.

The next very important issue in designing process of control systems is to assign priorities – necessary information that are needed to effective control of UGV/UCEV.

The system showing the interconnections among components of UGV/UCEV during its performance that is presented in fig.2 allows to:

- separation of necessary factors and information being crucial to design automation and autonomous systems;
- separation of information needed to identify vehicle surroundings;
- separation of information coming from on-board and from remote operator stand that are necessary to control UGV/UCEV;
- allowing to assess the influence of noise signals on the elements of remote control system and assess their influence on final results of UGV/UCEV performance.

To work out the general design of control system of Unmanned CEV the following preliminary assumptions were taken:

- separated (mobile) operator's stand, moved on safe distance, were fitted with, apart from radio-TV link, high specialized visual system to monitor operating area and vehicle surroundings;
- high visibility of operating area assurance was fulfilled by application of necessary specialized equipment that remote operator's stand were fitted with;
- there was expected, in particular cases, the usage of independent unmanned vehicles to better monitor vehicle surroundings to improve situational awareness;
- UGV/UCEV were fitted with basic automation systems with measurement and decision-making microprocessor modules;
- there was possibility to link the remote control system with existing vehicle control system;
- the usage of technologically advanced elements of automation system and detectors that monitored vehicle surroundings were minimized because of high costs. There were a high risk to damage or even destroy them during the mission.

The proposed assumptions concerning the control systems gave some possibilities for flexible adaptation to unpredicted situation, keeping the necessary economical constraints. The elaborated system was designed for main battle tank T-72 to perform following missions: mine-clearance (fig.6) and cross-bedding the river (ARV - WZT-3 – fig.7).

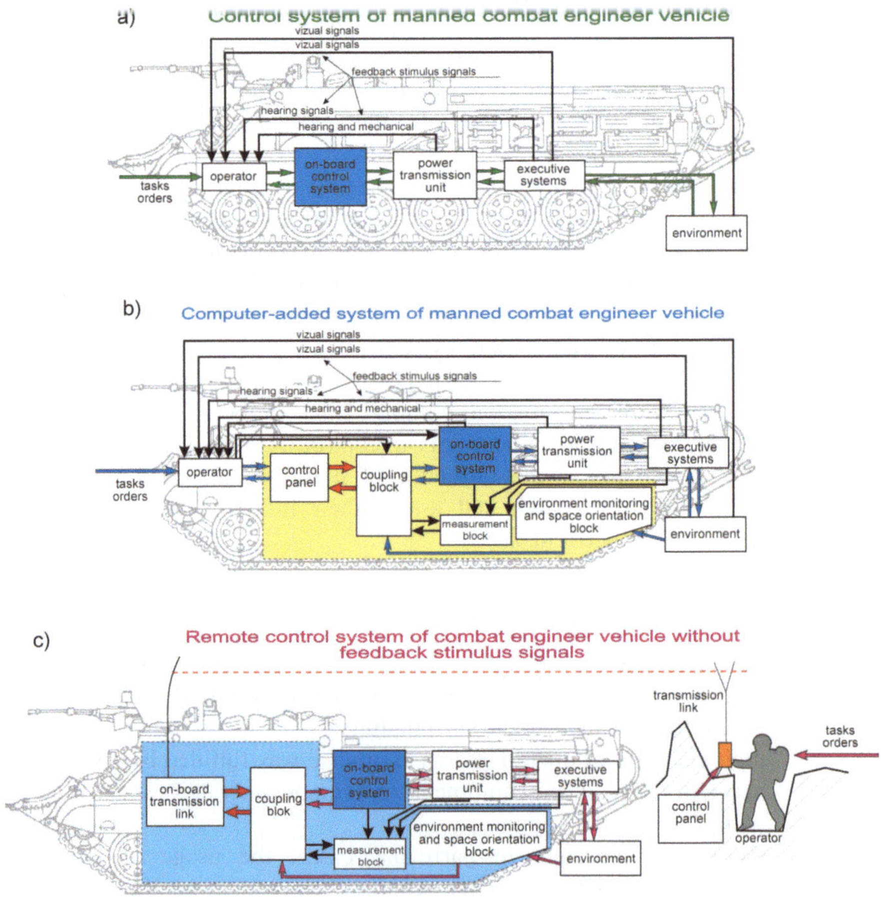

Fig. 3. The system "Operator – Vehicle – Environment": a) scheme of regular system "Operator – Vehicle – Environment" that operator directly realizes the control processes, b) scheme of system "Operator – Vehicle – Environment" with the computer-added system of operator work, c) scheme of system "Operator – Vehicle – Environment" that allows to remote control ARV/CEV

The next necessary stages of control systems development to ensure capabilities to remote control in teleoperation system is presented in fig.3. In case of regular systems of interconnections – in the system "Operator – Vehicle – Environment" (fig.3a), that operator directly control the vehicle, the consideration by him the stimulus feedback signals is essential. This system demanded a high qualifications and experiences of skilled operator and appropriate reaction on detected environmental stimulations. The lower requirements for operator generate the system presented in fig.3b – it was fitted with computer-added system of operator work – making the role of experiences and stimulus feedback signals limited. Extension of this system with next necessary modules of automatic realization of chosen functions in the system "Operator – Vehicle – Environment" (fig.3c) allows to remote control ARV/CEV without stimulus feedback.

Achievement of remote control capabilities (fig.3) must be preceded development of control system design by:

- displacement mechanical control by electric, electro-hydraulic and electro-mechanical executive systems;
- application of the computer-added system of operator work;
- application of continuous diagnostic and reaction systems to monitor and react on unacceptable working parameters;
- application the tools for vehicle executive devices monitoring to control the parameters of position, stability, slipperiness, etc.;
- application the tools for environment monitoring and space orientation.

3 Design of Vehicles' Control System with Teleoperator Function

The features of CEV's control systems that perform their tasks with teleoperator function imply following requirements and limitations responding to system design, information flow and control signals:

a) elimination of direct stimulus feedback signals between operator and vehicle requires development of systems that generate and send information to the operator and support the process of signals generation controlling the energetic path of vehicle;
b) elaboration of emergency procedures of control system performance in case of noise in transmission path to make possible (in emergency case) the autonomous operation of the vehicle
c) feedback signals generation for operator that their nature and range are very similar to standard signals that operator receives being in the vehicle cabin;
d) implementation of teleoperation to the standard CEVs demands development of diagnostic-measurement unit. It sends information to the operator stand in case of autonomous operation;
e) design of proposed remote control system should make possible its application in any vehicle equipped with power transmission unit of main battle tank T-72;
f) elaborated system should be placed in the operator's space, in the operator's compartment.

The above mentioned requirements and limitations were base to work out the functional scheme of UGV/UCEV remote control system (fig.4). As opposed to control system of standard CEV, in the system presented on the scheme the interconnections are shown. They result from the essence of remote control. The remote operator position makes necessary development of information flow channel. There is a need to add the channels of visual signals of surroundings – the operation space images. On the scheme, two independent channels from where the operator receives the images of CEV operation space are presented:

- from camera or cameras CEV is fitted with that send the visual signals of surroundings - in front of the vehicle, or they allow to monitor the operation space;
- external monitoring system that supported vehicle is fitted with or special monitoring system together with methods and algorithms of environment recognition, measurement distance to determined, characteristic elements of surroundings. It allows to more effective control of CEV, its devices and attachments. There are some research on methodology being focus on creation of a tool which would be able to construct a system for obtaining information about objects in the surroundings of the vision system (Bieda, R., Jaskot, K., Jedrasiak, K., Nawrat, A, 2013) and on support of land vehicle vision system configuration by serious gaming (SG) solution (Koźlak M., Kurzeja A. Nawrat A., 2013).

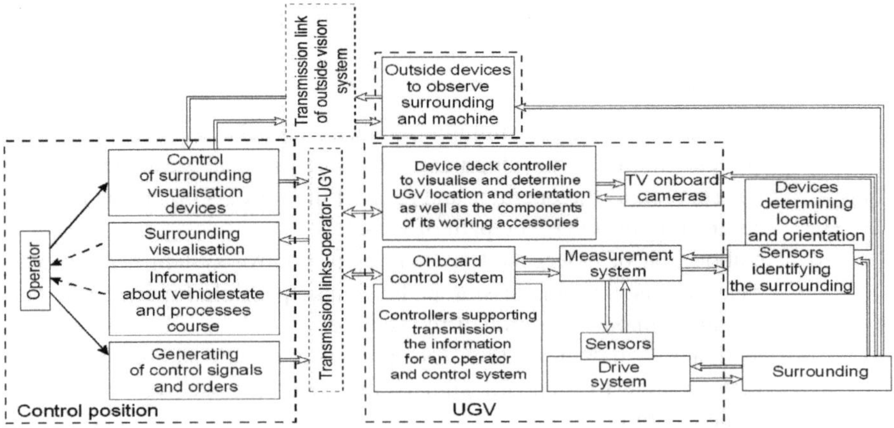

Fig. 4. Functional scheme of the remote control system of unmanned ground vehicle

The results of analysis of the above scheme are that development of on-board automation units is needed. There two basic units should be added. One that is responsible for receiving, converting operator's orders and sending them to the on-board control system and also generating information for operator. The second one, on the base of signals coming from the sensors, in case of lost of communication with operator, is responsible for execution of emergency procedures.

Assumed that, above mentioned characteristic features of the system should allow to control of CEV driving system and its devices and attachments in the real time. The exemplary variations of control systems due to assumed concept are presented in fig.5.

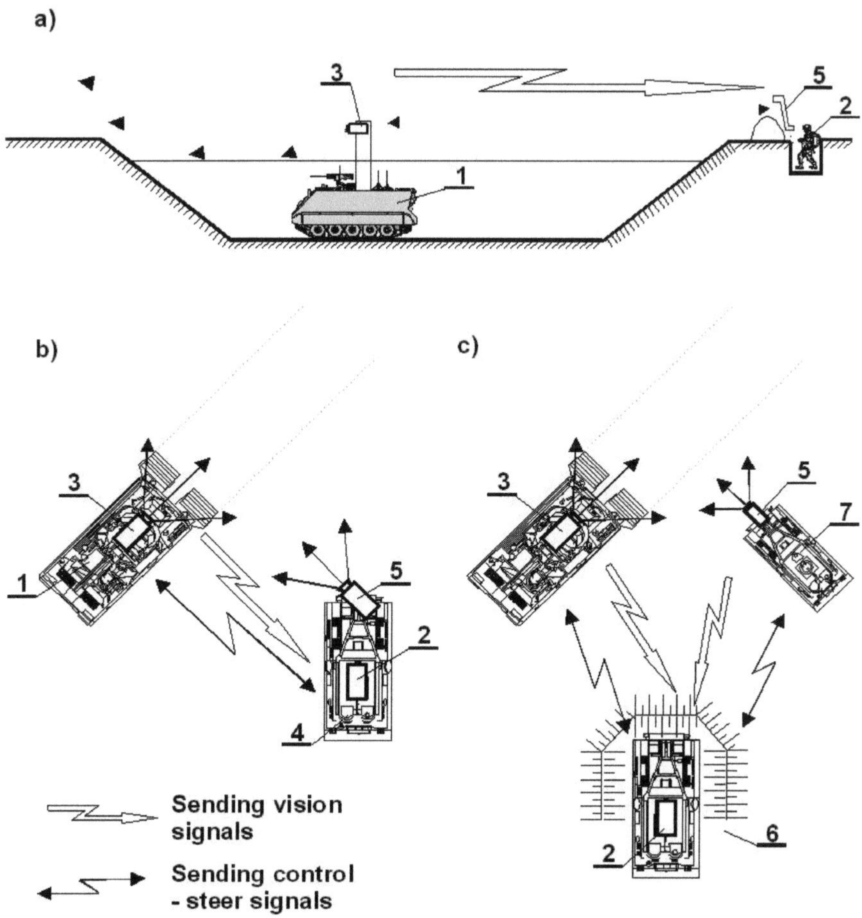

Fig. 5. Exemplary variants of unmanned ground vehicles remote control: a) from the immobile control – observation position (e.g. when vehicles crossing the bottom of the water obstacle), b) from the mobile control – observation position (when crossing narrow mine field, neutralisation and removing hazardous loads), c) with usage of mobile observation vehicle (crossing wide mine field): 1 – UGV; 2 – operator ; 3 – observation cameras, 4 – mobile control position, 5 – outside cameras or observation devices; 6 – immobile control position, 7 – unmanned observation vehicle

4 General Scheme of Remote Control Systems of Manned Armoured Ground Tracked Vehicles Designed on Base of Power Transmission Unit of Main Battle Tank T-72

To work out the concept of remote control systems of manned armoured ground tracked vehicles designed on base of power transmission unit of main battle tank T-72 assumed that system will consist of control unit with transmission link and executive unit. The system was tasked to assume operator's responsibility during mine clearance missions and cross-bedding the river and simultaneously monitor boundary values of power transmission unit to protect it. The system did the supervision functions in limited range.

a)

b)

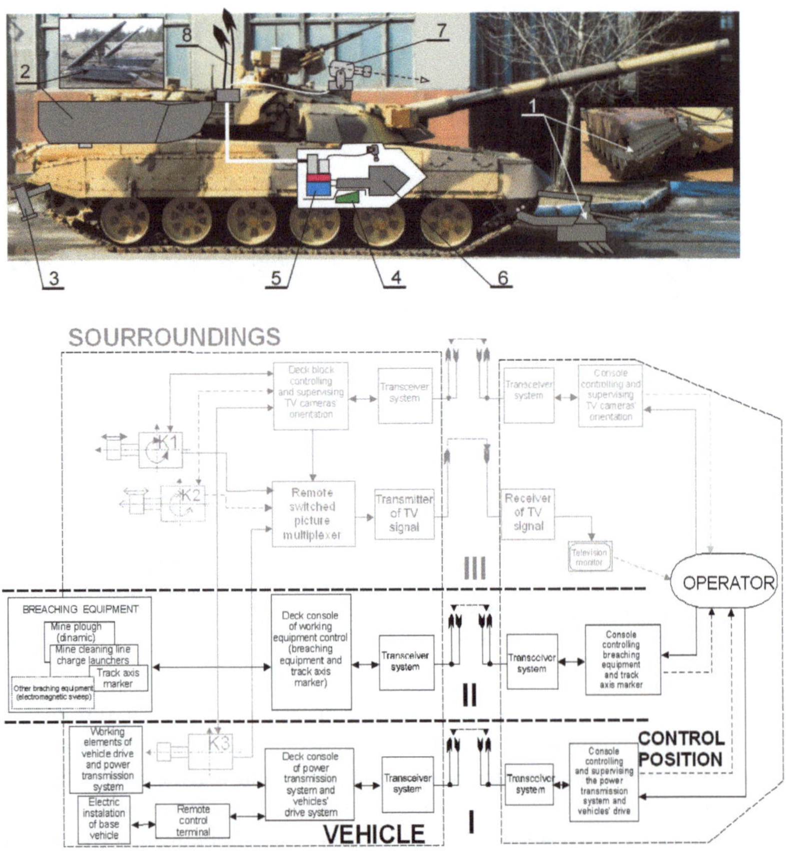

Fig. 6. Remote controlled tank T-72 with breaching equipment: a) general elements layout, b) the scheme of remote control system: 1 – mine plough, 2 - mine cleaning line charge launchers, 3 – track axis marker, 4 - deck fastening set of a part of remote control system, 5- electric and electronic modes, 6 – actuators, 7 – observation camera, 8 – a set of transceiving antennas

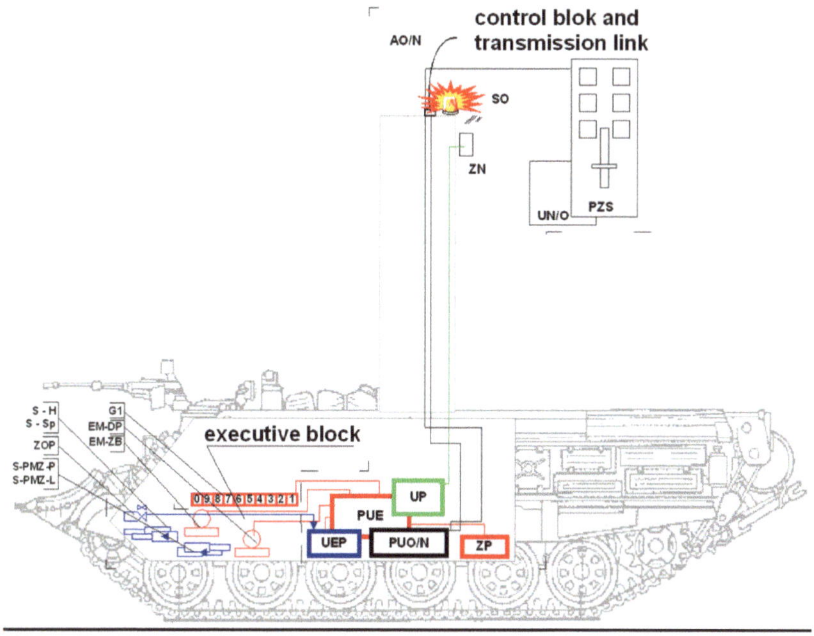

Fig. 7. The elements of remote control system of Armoured Recovery Vehicle (ARV – WZT-3) for cross-bedding the river application: ZOP – air valve, S-PMZ-P – right turn lever, S-PMZ-L – left turn lever, S-H – break pedal pneumatic actuator, S-Sp - clutch pedal pneumatic actuator, EM-ZB – change gear electro mechanism, EM-DP – accelerator electro mechanism, G1 – set of on-board outlets, AO/N – transmitting-receiving antenna, PZS – remote control panel, ZP – on-board power supply, UP – cameras switch system, PUE – on-board electronic system, UEP – electropneumatic system, PUO/N – on-board transceiver system, SSO – luminous caution signal, DSO – sonic caution signal

Following, necessary gauges and control mechanisms that operator could used during cross-bedding the river mission, controlling it remotely were:

- push-buttons „START" and „STOP" making turn-on and turn-off the engine possible;
- impeller pump switch to pump water out of engine compartment;
- push-buttons of engine rotational speed change;
- vehicle direction change levers;
- push-button of automatic opening of inlet and outlet covers;
- oil pressure in the gear box indicator;
- oil pressure in engine lubrication system indicator;
- coolant temperature indicator;
- engine oil temperature indicator;
- tachometer;
- push-button of sonic signal.

The experimental remote control system (fig.7) in case of cross-bedding the river mission consisted of following elements:

- remote control panel (PZS) with transceiver system (UN/O);
- on-board transceiver system (PUO/N);
- on-board electronic system (PUE);
- electro pneumatic system (UEP);
- executive unit.

The remote control panel, placed in close housing, makes possible the performance all control procedures (actions) anticipated for the needs of remote control of Armored Recovery Vehicle (ARV) – WZT-3 during the mission of cross-bedding the river. At the panel were placed:

- ignition switch;
- push-buttons „START" and „STOP" of engine;
- brake and clutch turn-on and off push-buttons;
- change of direction of travelling controllers;
- gears changes and accelerator levers;
- sonic signal and emergency stop engine push-buttons;
- necessary caution - signaling indicators.

The remote control panel was designed reflecting way of deployment and operation of main battle tank T-72 control elements to operator could remote controlled the vehicle without any problems.

The control signals were transmitted from remote control panel through transceiver system, where they were encoded and transmitted to electronic system where the controller was the main element of it. There the signals were converted into appropriate current and voltage pulses and through the electro pneumatic system sent to the executive unit of vehicle.

Application of controller allowed to automatic performs some early programmed control functions. It fulfilled also the tasks of automatic protection and diagnostics.

Moreover, there was worked out additional panel for cord control of electro pneumatic system and electro mechanisms of gears changes and accelerator levers for the need of tests and to control the vehicle in case of transmission link breakdown.

The executive unit of ARV (WZT-3) remote control system (fig.8,9,12) consisted of following control mechanisms:

- change of direction of travelling levers;
- brake and clutch pedals;
- accelerator pedal;
- gear box lever;
- and mechanisms that controlled them. Executive unit was the last element of remote control system of ARV (WZT-3) responsible for execution of operator orders that controlled vehicle remotely.

In the executive unit the control of break, clutch and change of direction of travelling levers were executed with the usage of pneumatic actuators. Gear box lever and accelerator pedal were controlled by electro mechanisms.

Compressed air were delivered through pressure reduction valve from the pneumatic system of ARV (WZT-3). There was installed additional valve on the air power take-off valve to make possible connection of vehicle pneumatic system with pneumatic mechanisms of executive unit.

Pneumatic drive was chosen because in such a case the connection of vehicle pneumatic system (ARV - WZT-3) with executive unit mechanisms was possible. There was also possible to assembly and disassembly thanks to the usage of fast hook up tools and relatively small size of pneumatic actuator and sufficient force on its piston rode.

The pneumatic actuators PT-90 that are used in the Engineering Mine Plough TW-92 (fig.10) were applied to control brake and clutch pedals. The actuators that control of change of direction of travelling levers assured their displacement from neutral position to the fully forward position whereas the actuators that controlled brake and clutch pedals assured full operational needs. Gear box lever and accelerator pedal were electric energy-powered. There were used the up-graded electro mechanisms UT-6D (fig.11). Thanks to such a solution there was very easy to choose particular gear. Electric system of control of gear box lever and accelerator pedal has an elements that allow to copy the their position. Thanks to that the transmitted to controller signals allowed to automatic execution some operations or their blocking in case of wrong arguments of operator.

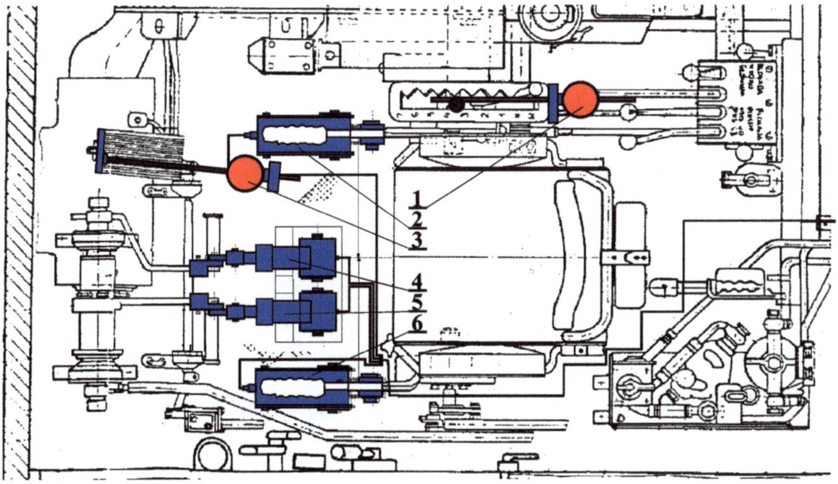

Fig. 8. Bird's-eye view on operator compartment of ARV (WZT-3) with remote control system executive unit elements: 1 – upgraded electromechanism UT-6D to drive gear box lever; 2, 6 - change of travelling direction levers pneumatic mechanisms; 3 - upgraded electromechanism UT-6D to drive the accelerator mechanism; 4 – break pedal pneumatic mechanism; 5 – clutch pedal pneumatic mechanism

In order to simplify assembly of electro pneumatic executive unit the special assembly elements were welded on to the bottom of the operator compartment. They have no impact on standard usage of vehicle in case of disassembly of remote control system on-board elements. Making displacement of electropneumatic executive unit the available space and elements of operator compartment were used. Rest of elements of remote control system were placed out of the operator compartment. Vehicle could be controlled manually after partially disassembly and disconnection of remote control system executive unit elements (fig.9).

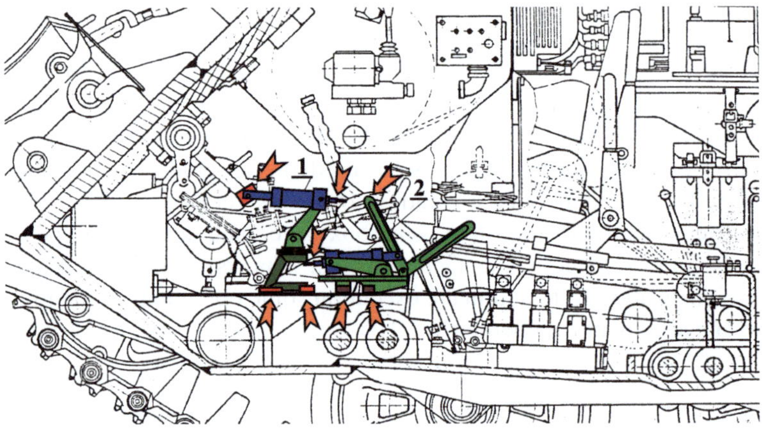

Fig. 9. End elevation of operator compartment of ARV (WZT-3) with remote control system executive unit elements: 1 – clutch pedal drive mechanism; 2 - change of travelling direction levers pneumatic mechanisms

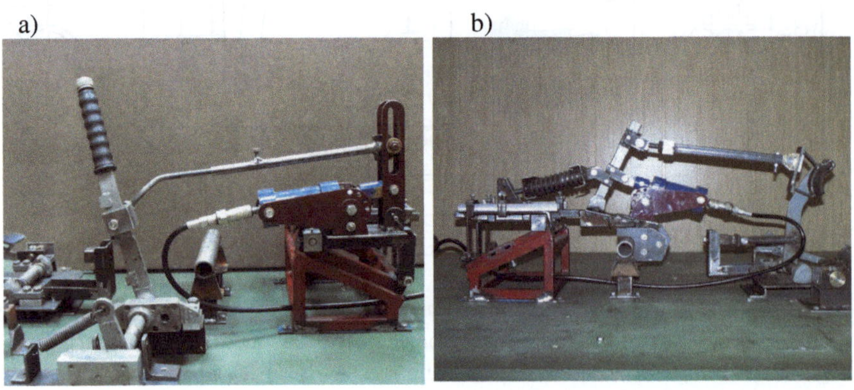

Fig. 10. Remote control system executive unit elements driven pneumatically: a) change of travelling direction levers drive mechanism, b) break pedal drive mechanism

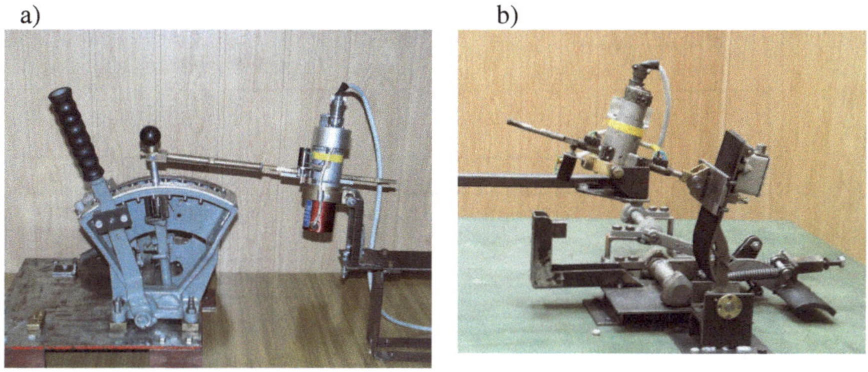

Fig. 11. Remote control system executive unit elements driver electrically: a) gear box lever drive mechanism, b) accelerator drive mechanism

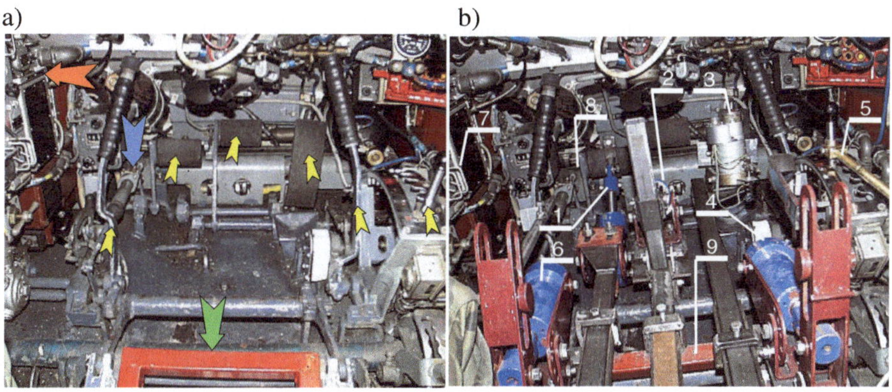

Fig. 12. The general view of operator compartment of main battle tank T-72 with remote control system executive unit elements T-72: a) to point the fixing points of remote control system executive unit elements out, b) fitted with z remote control system executive unit elements: 1 - clutch pedal pneumatic mechanism 2 - break pedal pneumatic mechanism; 3 - upgraded electromechanism UT-6D to drive the accelerator mechanism; 4, 6 - change of travelling direction levers pneumatic mechanisms; 5 - upgraded electromechanism UT-6D to drive gear box lever; 7 – board of measurement-test devices; 8 – air valve; 9 – special frame slipped in place of operator's seat

5 Results of Functional Research of Remote Control System Executive Unit

Correctness of adopted assumptions and design solutions of the remote control system with teleoperator function were verified as a consequence of conducted research.

Testing functionality of the system, the input and output signals of executive unit elements were recorded. In fig.13 the exemplary results of research of time of change of position of vehicle direction change lever are presented.

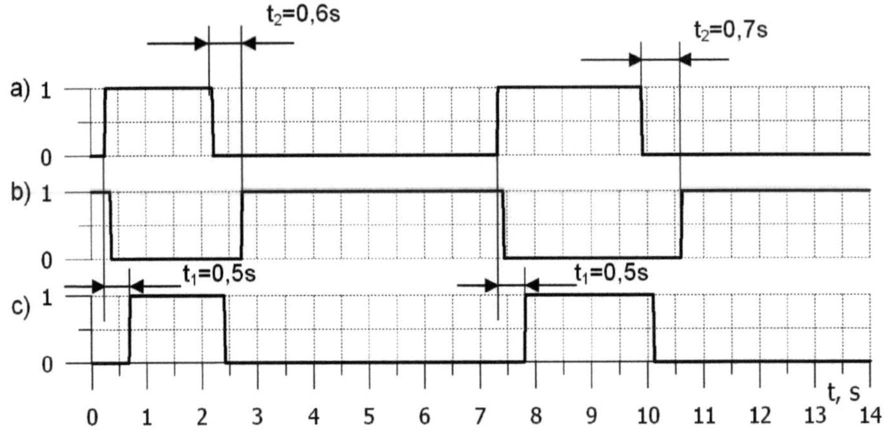

Fig. 13. The exemplary results of research of time of change of position of vehicle direction change lever: a) flow of control signal (square wave), b) vehicle direction change lever in "0" position, c) vehicle direction change lever in "1" position

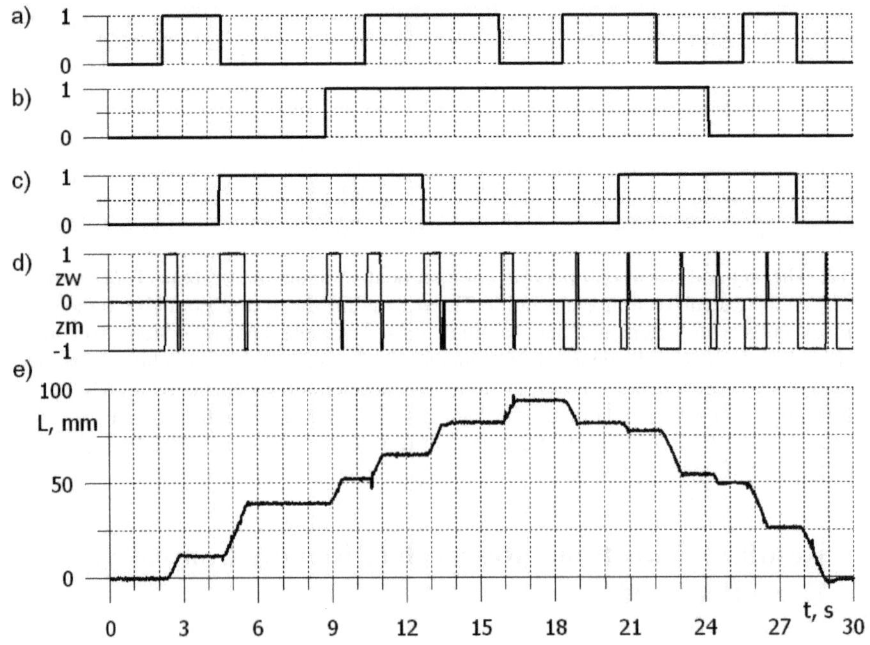

Fig. 14. The exemplary results of research of time of change of position of accelerator lever: a) the status of particular binary unit "0", b) the status of particular binary unit "1", c) the status of particular binary unit "2", d) flow of signal that control the engine of electromechanism UT-6D, e) diagram of positions of accelerator pedal

In fig.14 the results of research of time of change of position of accelerator pedal are submitted. There was used 3-binary unit Gray's code as a control signals. In such a way eight different positions of accelerator pedal were tested. In fig.14a,b,c the status of particular binary units are presented, whereas in fig.14d – the flow of control signals in case of increasing (upper part of diagram – zw) and decreasing (lower part of diagram – zm) fuel charge. To shorten time of change of position of accelerator pedal fast inversion of control signals was applied, to achieve swift stop.

In fig.15 the exemplary results of research of time of change of position of gear box lever for first gear, back gear and neutral position are presented.

In fig. 15c the flow of control signal during shift the gear lever from lower to upper gear (upper part of diagram) and during shift the gear lever from upper to lower gear (lower part of diagram) are presented. Similarly as in case of accelerator pedal control, to shorten time of change of position of gear box lever fast inversion of control signals was applied, to achieve swift stop.

In fig. 15d the diagram of position changes of gear box lever (Lb) in time is presented.

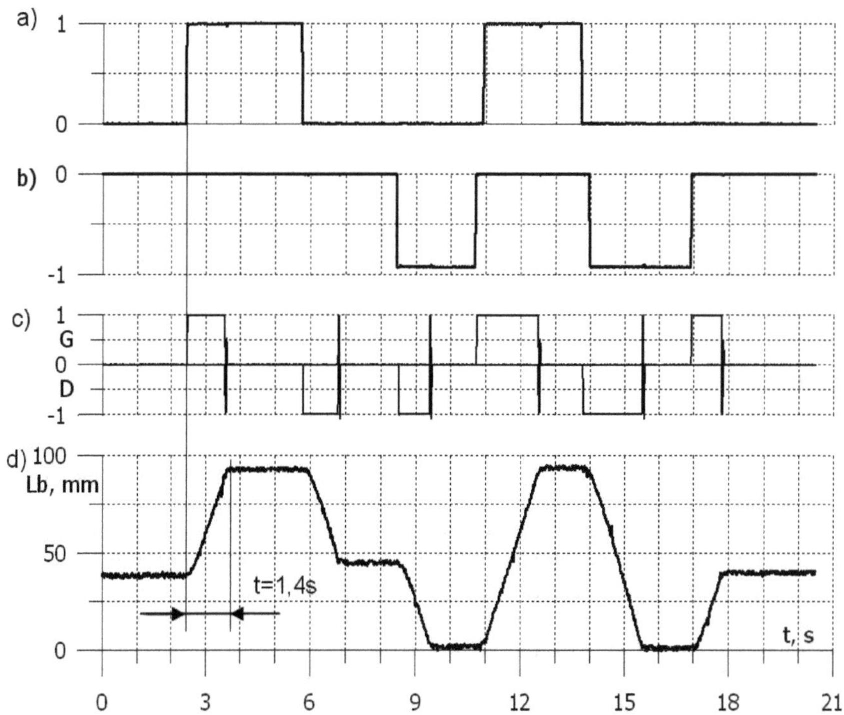

Fig. 15. The exemplary results of research of time of change of position of gear box lever: a) position of gear box lever on the control panel for the first gear, b) position of gear box lever on the control panel for the back gear, c) flow of control signal, d) position of the gear box lever

The collected results confirmed functional correctness of elaborated system. Tests that were done on the real object proved its suitability for remote controlled vehicles based on power transmission unit of main battle tank T-72. Achieved time of change of position of control elements of remote control system was comparable with the time that is achieved by operator in case of manned control.

6 Conclusions

It is necessary to study all main possible solutions for UGTV configurations. The lack of standard references for some technologies and protocols highlights the needs at national and international level of defining standards and possible regulation for unmanned systems, mainly at operational, interoperability and technological level to fulfil requirements for modular and open architecture such a systems. UGTV modules should be plugged into any vehicle with minimum effort, at the same time preserving the integration into the superior command and control network structure. In order to be useful, a robotic system should be technically reliable and not require upsetting changes into the military applications (e.g. using standard interfaces and procedures).

New generations of Ground Tactical Vehicles should result from Capabilities Needs then should be elaborated Concept of Operations and then Operational Requirements, mixing technical and operational (military) point of views. That is why this processes should be supported by research on new technologies that are critical to achieve new capabilities. These technologies should be developed in S&D works, implemented in the process of prototyping and production phase covering all Life Cycle Management including education and training. To unify such a system the methodology of qualification tests should be agreed with end users and among producers.

References

[1] Czapla, T., Wrona, J.: Technology development of military applications of unmanned ground vehicles. In: Nawrat, A., Kuś, Z. (eds.) Vision Based Systems for UAV Applications. SCI, vol. 481, pp. 293–309. Springer, Heidelberg (2013)

[2] Konopka, S., Łopatka, M.J., Muszyński, T., Typiak, A.: Bezzałogowe platformy lądowe, Nowoczesne technologie systemów uzbrojenia. Rozdział 4.1 monografii pod redakcją naukową Zygmunta Mierczyka; WAT Warszawa (2008)

[3] van den Heuvel Johan, C.: Unmanned Ground Vehicles: Autonomous or Teleoperated. Military Robotics. Cafe Royal, London (2008)

[4] Carl, L.: Swedish Armed Forces Unmanned Ground Vehicle (UGV) Projects. Military Robotics. Cafe Royal, London (2008)

[5] Schnaider, F.: International Robotics Competition from a Military View. Military Robotics. Le Meridien Piccadily, London (2009)

[6] Bartnicki, A., Łopatka, M.J., Typiak, A.: Problemy teleoperacji w sterowaniu bezzałogowymi platformami lądowymi; "Technologie podwójnego zastosowania"; Rozdział 5.1 monografii pod redakcją naukową Andrzeja Najgebauera; WAT Warszawa (2012)

[7] Bartnicki, A., Łopatka, M.J., Muszyński, T., Sprawka, P.: Bezzałogowe platformy lądowe w zadaniach ochrony konwojów I utrzymania dróg przemarszu; "Technologie podwójnego zastosowania"; Rozdział 5.1 monografii pod redakcją naukową Andrzeja Najgebauera; WAT Warszawa (2012)

[8] Mężyk, A., Świtoński, E., Kciuk, S., Klein, W.: Modelling and investigation of dynamic parameters of tracked vehicles. Mechanics and Mechanical Engineering 15(4), 115–130 (2011)

[9] Bartnicki, A., Dąbrowska, A., Rubiec, A.: Mobility problems of remote-controlled ground vehicles. Polish Journal of Environmental Studies 20(5A) (2011)

[10] Bieda, R., Jaskot, K., Jędrasiak, K., Nawrat, A.: Recognition and Location of Objects in the Visual Field of a UAV Vision System. In: Nawrat, A., Kuś, Z. (eds.) Vision Based Systems for UAV Applications. SCI, vol. 481, pp. 27–45. Springer, Heidelberg (2013)

[11] Koźlak, M., Kurzeja, A., Nawrat, A.: Verifying Unmanned Land Vehicle Vision System Configuration Using Serious Gaming Solution. In: Nawrat, A., Kuś, Z. (eds.) Vision Based Systems for UAV Applications. SCI, vol. 481, pp. 335–342. Springer, Heidelberg (2013)

[12] Kuś, Z., Nawrat, A.: Object Tracking for Rapid Camera Movements in 3D Space. In: Nawrat, A., Kuś, Z. (eds.) Vision Based Systems for UAV Applications. SCI, vol. 481, pp. 57–76. Springer, Heidelberg (2013)

[13] Kuś, Z., Nawrat, A.: Object Tracking for Rapid Camera Movements in 3D Space. In: Nawrat, A., Kuś, Z. (eds.) Vision Based Systems for UAV Applications. SCI, vol. 481, pp. 57–76. Springer, Heidelberg (2013)

The Selected Innovative Solutions in UAV Control Systems Technologies

Dariusz Nowak, Grzegorz Kopecki, Marek Orkisz, Tomasz Rogalski, and Paweł Rzucidło

Rzeszow University of Technology, Rzeszow, Poland
{darnow,gkopecki,mareko,orakl,pawelrz}@prz.edu.pl

Abstract. The paper presents selected original solutions concerning UAV control systems technologies. The paper does not describe general theories and global technical solutions used in UAV's technologies. It presents some tricks created and developed at Rzeszow University of Technology. Presented solutions have been developed and implemented into real control systems. They main goal is to improve control precision and increase system functionality. Results presented in this paper have been collected during real flight tests of different classes of UAVs.

Keywords: control systems, fly tests, practical solutions, UAV.

1 Introduction

Unmanned Area Vehicles (UAV) have become a popular working tool for many civilian and military users. They are currently used in more and more complex actions taken by army, police, fire department, municipal police and other to detect and observe different kinds of threats. But they are also used by, still rising, group of people for fun. It is result of the technological development during last years which has caused both availability and usefulness of UAV have become significantly bigger than in the past.

The full picture of the technological development of UAVs consists of some elements. There are drone structures technologies - leading to mass reduction, modern control systems - improving flight precision, electronics technologies - providing more efficient electrical drives and other science and technical disciplines among them.

The great development of many UAV technologies is possible because lot of science and technical centers over the World work under different, sometimes very significant and sometimes maybe negligible, problems related to UAV. Finally, they introduce to UAV systems many innovations. Therefore, innovative technologies and intensive works drive UAVs forward.

The paper intends to present selected results of research and technical works under UAV systems taken by Rzeszow University of Technology (RUT). The paper doesn't widely describe general solutions used in UAV systems. It focuses on some objectives

A.M. Nawrat (ed.), *Innovative Control Systems for Tracked Vehicle Platforms*,
Studies in Systems, Decision and Control 2,
DOI: 10.1007/978-3-319-04624-2_3,

strictly related to UAV's control systems which have been projected to improve selected UAV's characteristics. They had been implemented to real systems and have been tested in laboratories and real flight trials.

2 General Structure of the Control System

In spite of the paper presents selected detailed solutions improving UAV's features implemented into real systems, the general structure of the Control System (CS) they are implemented in should be described. The system consists of two parts: an on-ground control station and an airborne part. The ground station realizes typical for such systems tasks. It communicates with the airborne part by a radio link. It is not the subject of discussion taken later in this paper and is not to be described in details.

Navigation and control calculations are performed by the on-board part of the system. A logical structure of the airborne part of the CS consists of measurements modules (including ADC, AHRS, GPS), a radio link and actuators. All elements of the system are connected to a digital databus CAN 2B working in CAN-Aerospace standard (Fig. 1).

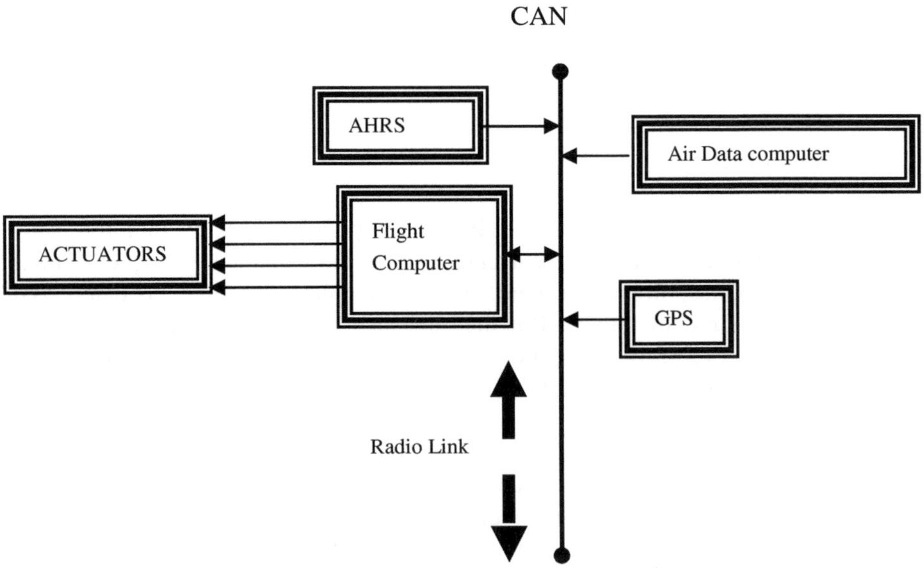

Fig. 1. Logical structure of the control system

3 Attitude Stabilization

The Control System of UAV must perform several different tasks related to flight stabilization, mission's tasks and other. One of them is maintaining desired (demanded) pitch and roll angles, what is the basis of the space attitude stabilization process, in fact. This level of the control is often the most inner control loop in the structure of control algorithms. Its precision of the work determines precision of the work of remaining outer, supervising control loops.

The paper presents modification of the typical PID controller which implemented into roll and pitch stabilization channel can improve controller's performance and control precision.

The idea of presented algorithms arises from wishes to project a controller which works like pilot. The discussion with pilots let formulate the rule which says "Pilot doesn't try to force the plane and doesn't correct its flight parameters unless it is absolutely necessary". So it can be said that, pilot never corrects plane's flight when it goes toward the demanded state itself. The pitch angle stabilization process has been chosen as a sample illustrating this rule (Fig. 2).

According to this idea, controller's channel using different rules in different states of aircraft states of flight, presented in Fig. 2, has been created (1).

$$action(t) = \begin{cases} g(e(t), \dot{e}(t)) & if \ \mathrm{sgn}(e(t)) \neq \mathrm{sgn}(\dot{e}(t)) \\ f(e(t), \dot{e}(t)) & if \ \mathrm{sgn}(e(t)) = \mathrm{sgn}(\dot{e}(t)) \end{cases}, \qquad (1)$$

where:

e - deviation of flight parameter from its demanded value.

The function $g(e(t), \dot{e}(t))$ in formula (1) can be equal zero in special cases. It means no correction action is taken in these cases.

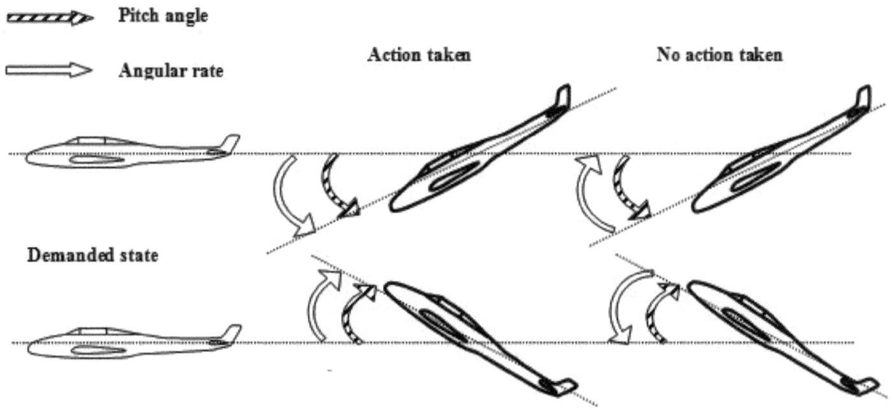

Fig. 2. When pilot correct flight - as a sample pitch angle channel chosen

To prepare the modified PID controller with the supplementary control channel (1), the classical PID controller as a starting point for calculations and analyses is set. The process of the creations of the modified controller runs in three steps. The typical PID controller is projected in the first step. Several nonlinear elements including (saturations, death zones etc.) are introduced to the structure of the controller in the step two. The function $g(e(t),\dot{e}(t))$ from equation (1) is implemented and tested in the third step. Finally the extended controller signed PID^2 is completed (2).

$$\delta S_{PID^2}(t) = \delta S_{PID}(t) + \delta S_{D'}(t)\,, \tag{2}$$

where:

δS_{PID^2} - complete control signal calculated by the controller.

δS_{PID} - control signal calculated by the PID root of the controller.

δS_{D} - control signal calculated by the supplementary channel of the controller.

$$\delta S_{D'}(t) = \begin{cases} 0 & \mathit{if}\ \mathrm{sgn}(e(t)) \neq \mathrm{sgn}(\dot{e}(t)) \\ f(e(t),\dot{e}(t)) & \mathit{if}\ \mathrm{sgn}(e(t)) = \mathrm{sgn}(\dot{e}(t)) \end{cases}, \tag{3}$$

where

e(t) - deviation of pitch angle in pitch channel or deviation of roll angle in roll channel.

Formula (3) informs that no correcting action is taken if aircraft's flight parameters vary leading the plane to the demanded state themselves.

The process of selection of controller's coefficients must be strictly connected with physical rules governing planes dynamics. The control signal $\delta S_{D'}(t)$ can be created according to different rules including linear and nonlinear formulas, expert knowledge etc.

One of methods of PID^2 controller's parameters tuning assumes that control signal $\delta S_{D'}(t)$ should be intense enough to stop rotation of the plane around X-axis or Y-axis respectively. Taking into consideration the dynamics of the plane, control signals generated by extra controller's channels have following forms (4), (5).

$$f(e(t),\dot{e}(t))_{roll}(t) = \frac{p(t)}{K_p}, \tag{4}$$

$$f(e(t),\dot{e}(t))_{pitch}(t) = \frac{q(t)}{K_q}, \tag{5}$$

where

Kp - gain in the transfer function defining plane dynamics from aileron deflection to roll rate [6],

Kq - gain in the transfer function defining plane dynamics from elevator deflection to pitch rate[6].

The theoretical research were verified in real flight tests. Tests included flights when the classical PID controller controlled pitch and roll, and flights when the PID2 controller was used. The experimental drone was a small 2.5 m wingspan airplane. Its maximal take-off mass was 24 kg and cruise speed was 45m/s. Approximated transfer functions defining its dynamics in lateral (pure roll mode) and longitudinal (short period mode only) channels, prepared on the basics of acquired real flight data are as follows (6),(7) - more details in [6].

$$G_{roll}(s) = \frac{p(s)}{\delta_a(s)} = \frac{K_p}{(T_{roll}s+1)} = \frac{-2}{1.1s+1}, \tag{6}$$

where:

p - roll rate deg/s,

δ_a - deflection of ailerons ±100%.

$$f(e(t), \dot{e}(t))_{pitch}(t) = \frac{q(t)}{K_q}, \tag{7}$$

where:

q - pitch rate deg/s,

δ_e - deflection of elevator ±100%.

Sample results are collected and presented on Fig. 3. The history of the pitch angle stabilization process during the flight at constant altitude with the use of two kinds of controllers is plotted.

Experiments were conducted in similar atmospheric conditions. Flight with the use of extended PID controller was characterized with less deviation of pitch angle. Oscillations of pitch angle were scientifically reduced. Less variations of pitch rate when extended PID was used testified the flight were less dynamic (flight was calmer).

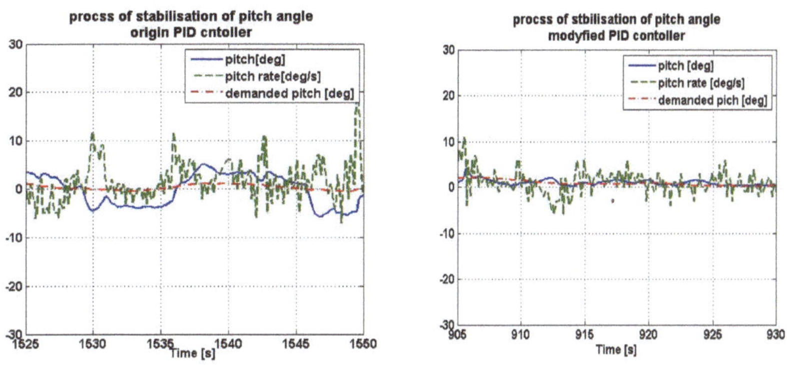

Fig. 3. The process of pitch angle stabilization by pure PID controller (left) and extended PID controller (right)

4 Precise Trajectory Tracking

The control system of the aircraft trajectory is usually based on classical control laws, using PID -based control [8][11][13][15][16]. In some solutions, quality requirements for the estimation of the trajectory are very high. The quality of control can be improved by means of additional correction signals, added to the signals calculated by the PID regulator. For example, during an ILS approach in lateral channel control, information about the track angle can be used, as shown in [10]. In this solution the track angle is treated as additional correction in the regulator. Due to such a solution control quality is significantly improved; however, it requires an additional measurement, which is not always possible. A small aircraft, both manned or unmanned, is very rarely equipped with radio navigation systems. Frequently, especially in low-cost solutions, the GPS receiver is the only system providing navigation data. Then, the desired trajectory is a line defined as two points with known geodic latitude and longitude. Method described in [10] for the case of ILS approach and generalized in [7] gives possibility of precise trajectory control. Let's introduce a model which generates an ideal trajectory. The difference between a trajectory generated by the model and a real trajectory and its integration can be used as correction signals. In order in ensure proper behavior of the system, the system is described in one state-space and the LQR method is implemented. The detailed description is shown in [7]:

The plant dynamics (aircraft + classical control system) is described in equations:

$$\dot{x} = Ax + Bu \ , \tag{8}$$

$$y = Cx \ , \tag{9}$$

where:

$$x(t) \in \Re^n \ , \ u(t) \in \Re^q \ , \ y(t) \in \Re^p \ , A \in \Re^{nxn} \ , \ B \in \Re^{nxq} \ , \ C \in \Re^{pxn}$$

Let's assume, we defined an ideal trajectory the aircraft should follow. The trajectory takes into account the system's dynamic. It is generated by a model with a structure analogical to the plant, but in the place of PID controllers, PD or P are used. The aircraft model and virtual control systems can be described as follows:

$$\dot{x}_m = A_m x_m + B_m u \ , \tag{10}$$

where:

$$x_m(t) \in \Re^n \ , \ u(t) \in \Re^q \ , \ A_m \in \Re^{nxn} \ , \ B_m \in \Re^{nxq}$$

As an additional signal, the difference between the real and calculated trajectory and integration of the difference mentioned can be used. The difference between the real, controlled trajectory and the trajectory generated by the plant and it's integration are additional states:

$$x_T(1) = x(n) - x_m(n)\,, \tag{11}$$

$$x_T(2) = \int_0^T x_T(1)dt\,, \tag{12}$$

Equations 8-12 can be described in one state-space:

$$\begin{bmatrix} \dot{x} \\ \dot{x}_m \\ \dot{x}_T(1) \\ \dot{x}_T(2) \end{bmatrix} = \begin{bmatrix} A & \mathbf{0} & 0 & 0 \\ \mathbf{0} & A_m & 0 & 0 \\ A(n,:) & -A_m(n,:) & 0 & 0 \\ \mathbf{0} & \mathbf{0} & 0 & 1 \end{bmatrix} \begin{bmatrix} x \\ x_m \\ x_T(1) \\ x_T(2) \end{bmatrix} + \begin{bmatrix} B \\ B_m \\ \mathbf{0} \\ \mathbf{0} \end{bmatrix} u\,, \tag{13}$$

where:

$A(n,:)$ – n-th row of matrix A,

$A_m(n,:)$ – n-th row of matrix A_m.

Equation 13 can be written as:

$$\dot{x}_s = A_s x_s + B_s u\,, \tag{14}$$

where:

$$x_s(t) \in \Re^{2n+2},\ u(t) \in \Re^q,\ A_s \in \Re^{(2n+2)x(2n+2)},\ B_m \in \Re^{(2n+2)xq}$$

Such state-space description is used for LQR control laws synthesis [1][14].

Let's assume the trajectory is defined by two points A (xA,yA), B(xB,yB) described in the Cartesian frame (Fig. 4). The aircraft position is also measured in the defined Cartesian frame. The aircraft position (xs,ys) is defined in state-space as a distance between the aircraft and point B and the angle ε shown in Fig. 4.

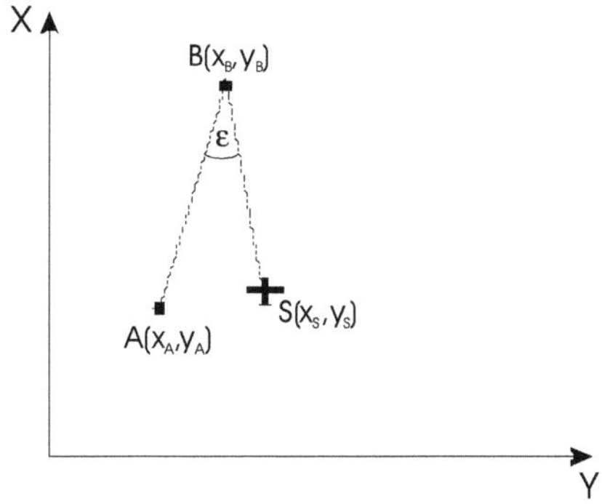

Fig. 4. Defined trajectory and geometrical dependencies

The aircraft classical control system stabilizing ε angle was defined as a plant. As a trajectory model was similar unit. All necessary limitations were introduced in the model. The role of limitations is explained in [7]. Fig. 5 and Fig. 6 show simulation results.

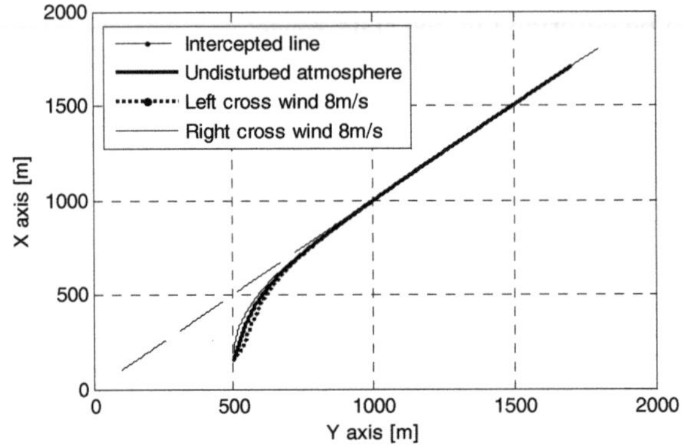

Fig. 5. Aircraft trajectory during interception of desired line. Simulation in steady atmosphere and during right cross wind.

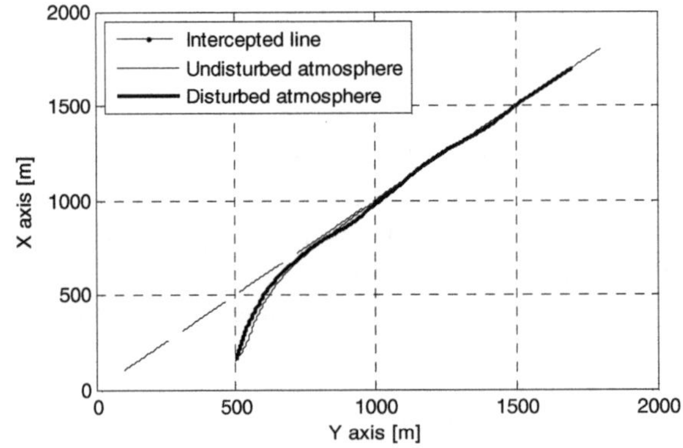

Fig. 6. Aircraft trajectory during interception of desired line. Simulation in steady and turbulent atmosphere.

As shown in Fig. 5 and Fig. 6, simulation results are good. The trajectory tracking is appropriate, and when compared to classical control system based on PID controllers, the system with the structure proposed has better control quality, which was shown in [10], on the example of LOC control. The structure of the control system analyzed in [7],[10] was analogical to the presented one.

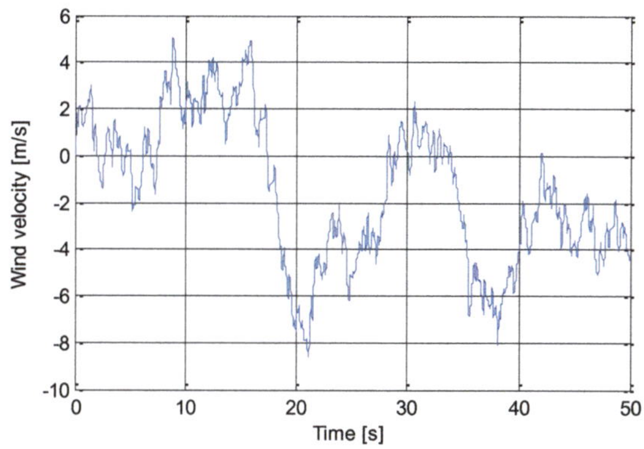

Fig. 7. Simulated cross-wind used in simulation presented in Fig. 5

5 The Concept of Using the ASG-EUPOS Corrections in the UAV Control and Navigation System

The process of controlling and navigating of an unmanned aerial vehicle (UAV) during a typical flight does not require accurate position information. The matter becomes much more complicated when the UAV makes a fully autonomous flight including take-off and landing stages. Augmenting of accuracy of measurements carried out using global navigation satellite systems (GNSS) and with support from precise position determination system based on active geodesy network (ASG-EUPOS) may make an automatic landing significantly easier. Achieving appropriate accuracy of altitude and position measurements made with a GNSS receiver as well as obtaining precise information about the landing field surface (in the form of a digital terrain model) should enable determination of the aircraft altitude above ground level within an accuracy of single centimeters. Such an accuracy may significantly enhance the safety of maneuvers and reduce the weight and costs of an unmanned aircraft (once the additional ultrasonic distance sensors or a radio-altimeter have been removed) or enable extra redundancy of measurement results.

The typical structure of UAV control system is presented in Fig. 8. The system consists of two basic elements: Flight Control System (FCS) and Ground Flight Control Station (GFCS) [4] [9]. The task of the airborne part is to automatically stabilize the aerial vehicle's parameters as well as to automatically navigate it according to the programmed flight path. The task of the GFCS is to prepare the UAV for its mission (checking procedure before a flight, programming of waypoints, choosing procedures in case of data transmission link loss, etc.) and to monitor the mission during which there is a possibility of making systematic modifications of the flight plan as well as operating additional equipment mounted on the board of the aircraft (which quite often includes e.g. observation cameras) [12]. An example of GFCS graphic software interface worked out by the authors of this paper is presented in Fig. 9.

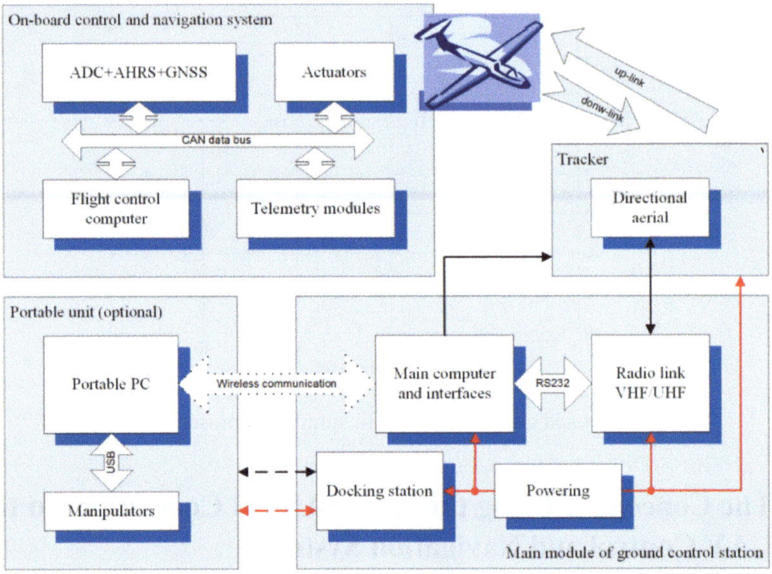

Fig. 8. Structure of the unmanned aerial vehicle control and navigation system

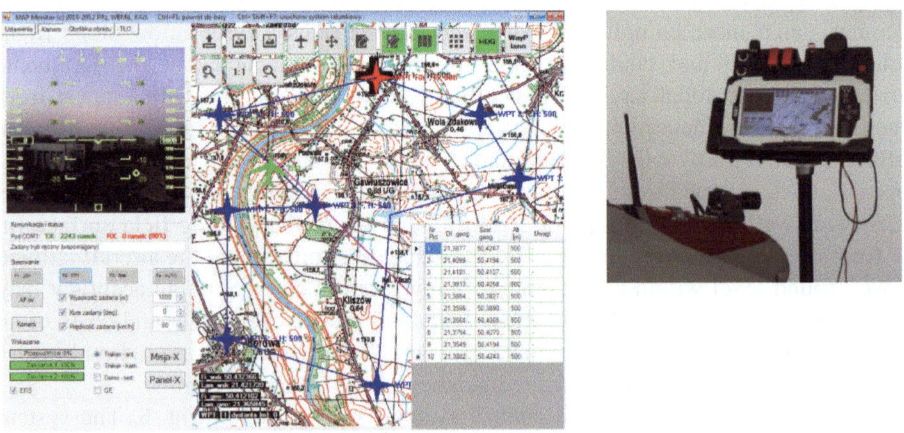

Fig. 9. GFCS graphic software interface (left) and UAV with portable unit (right)

Take-off and landing stages in most cases involve a pilot-operator. Light UAVs are started manually in general while larger objects require the use of a launcher to take off or are started in a classical way. What presents a much more serious problem is the landing process. Performing a touchdown in an uncontrolled way may result not only in the loss of the airframe but also of its highly costly specialized avionics and on-board equipment. Many manufacturers of unmanned systems use landing systems involving parachutes or special capturing nets in order to prevent crashing aerial vehicles into the ground. As far as larger objects are concerned, landing has to be performed in a conventional way and usually by an experienced pilot-operator.

Making an autopilot landing requires the use of suitable control algorithms and being equipped with a set of sensors supplying precise and continuous information about altitude referenced to ground level. One of the solutions which enable the information supply seems to be the use of satellite navigation system that offers accurate position data corrections and is linked to the digital model of the landing terrain.

Numeric simulations were carried out using a non-linear model of the unmanned aircraft automatic control and navigation system. The simulation environment was additionally equipped with elements modeling the GNSS receiver's work characterized by occurrence of short and long period errors. The simulations were carried out on the assumption that the values of position determination errors made by the GNSS receiver for 2σ (σ – standard deviation) in horizontal plane will come respectively to:

- 0 m for the perfect system,
- 22 m for the GNSS receiver without corrections,
- 0.2 m for the GNSS receiver receiving differential corrections from the ASG-EUPOS system.

The plan of the flight was drawn up in such a way (Fig. 10, left) to enable terrain scanning on the set sections of the route by means of apparatus mounted on the board of an unmanned aircraft (e.g. taking stereo-photogrammetric images, LIDAR scanning). The simulations were carried out under windless conditions (Fig. 10, right), at an invariable wind speed of 10 m/s with the direction of 315 and 270 degrees (Fig. 11) and at a variable wind speed of 5-15 m/s blowing from the same directions (Fig. 12). The simulation results, according to our expectations, indicate that the precision of controlling the unmanned aircraft on the determined route under windless conditions and with the invariable wind was significantly augmented when the navigation system cooperating with the GNSS system supported from ASG-EUPOS correction was applied (cross track from the set route amounts to single tenths of centimeters). Unfortunately, when the wind is gusty a supply of even highly accurate position information does not ensure precise navigation of the unmanned aircraft on the determined route. The cross track from the set route amounts to even 40-50 m when a wind speed changes rapidly in this case. Making precise measurement flights is dependent on stable meteorological conditions that determine practical benefits of applying the ASG-EUPOS system.

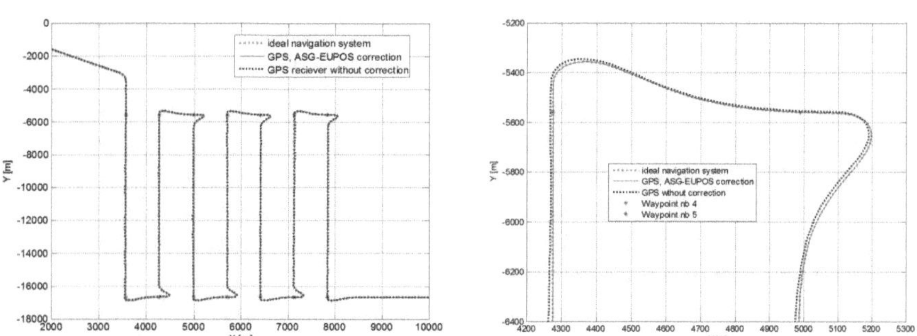

Fig. 10. Influence of the ASG-EUPOS correction system on the mini UAV control precision (windless conditions)

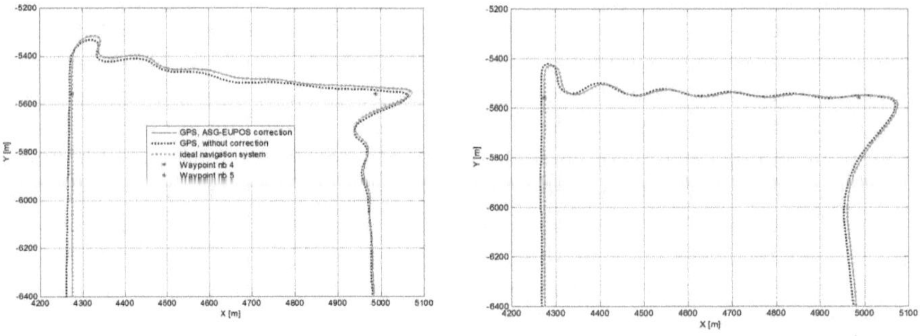

Fig. 11. Influence of the ASG-EUPOS correction system on the mini UAV control precision; invariable wind of 31, 10 m/s (left) and 270 degrees, 10 m/s (right)

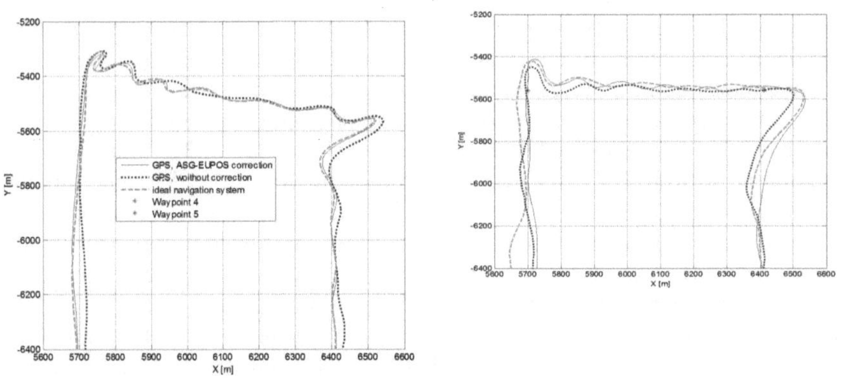

Fig. 12. Influence of the ASG-EUPOS correction system on the mini UAV control precision; variable wind of 315, 5-15 m/s (left) and 270 degrees, 5-15 m/s (right)

Achieving a high accuracy within which the position of an unmanned aerial vehicle is determined, will not only enhance the safety of flight operations, but will also make it easier to adjust the aerial vehicle to new tasks and applications. Aerial pictures for geodesic purposes are, for instance, taken at a high altitude with the use of high-accuracy and costly devices mounted on the board of an aircraft that is appropriately adjusted to that. Achieving a satisfying quality of the pictures can be ensured under suitable meteorological conditions related to, among other things, visibility, and cloud base and coverage. Such conditions do not occur every day at mid-latitudes. Moreover, there are periods of bad weather often lasting many weeks. Maintaining operational capability of both the crew and the equipment generates additional and often unpredictable costs.

Aerial pictures may be taken on the board of an unmanned aerial vehicle at a considerably lower altitude. The area covered by pictures taken during one flight will obviously be drastically reduced compared to flights made at a higher altitude. Nevertheless, an unmanned aerial vehicle will be able to carry out a laborious and monotonous process in an automatic and repeatable way. A detailed and long-lasting scanning of the terrain will be performed without fear for going over the working time limit of the pilot and deck operator or making them tired. It is worth mentioning that in a classical system the function of an operator may be performed only by a person of unusual endurance while working when the balance of visual and motor stimuli is upset.

6 The Automatic Landing

Landing is the most difficult stage of flight and requires the high precision control of the aircraft trajectory. Moreover, unfortunately light unmanned aircrafts and landing fields are not equipped with the special landing augmentation devices. UAV designers are conducting research on autonomous landing systems ensuring such operational features that will not require exceptional manual skills and specialized aviation training from the UAV operators. In last years, the autonomous flight control system for an attitude stabilization and maneuver tracking of the aircraft has been designed at Rzeszow University of Technology. The system is able to realize the defined mission from take-off to landing on the advisable landing field. This paper presents a proposal of specific, projected by research team from RUT control laws.

Landing consists of six stages: approach, flare-out, overshoot, touch-down point, landing run and braking (Fig. 13).

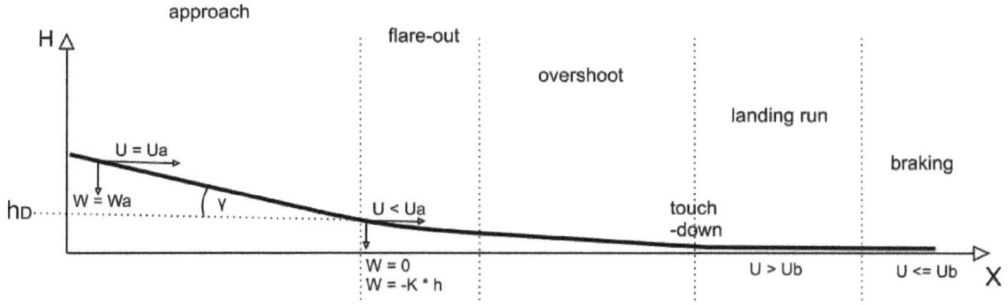

Fig. 13. Landing trajectory executed by autonomous control system

When the UAV achieves the final waypoint of the pre-planned airway the approach stage is executed. During the approach UAV control system maintains following flight parameters: airspeed U, vertical speed w and desired path angle γ (15).

$$W_a = [-U_a \sin\gamma + K_w * (d * \sin\gamma - h)]_{-3}^{3}, \tag{15}$$

where:

W_a - vertical speed during approach,
U_a - airspeed during approach,
γ - path angle during approach,
K_w - proportional coefficient determined experimentally,
d - distance from the theoretical point of touch-down,

h - height above the landing surface,

$[]^a_b$ - saturation range.

If the decision high (hD) is crossed over the landing surface, control system executes beginning of the flare which is a temporary stage before reaching new state of the flight. The flare is the most difficult part of landing. In this paper the method of automated control of the flare path was considered. It consists in the principle of proportional vertical speed of the aircraft in relation to height above the ground (16).

$$w = -K * h, \tag{16}$$

where:

w - desired vertical speed,

K - proportional coefficient determined experimentally,

This control requires precise measurement of flight altitude in relation to landing surface. In this method control system works like pilot during flare, that is, minimizes thrust and maintains minimum, negative value of vertical speed of an aircraft. Such procedure causes a reduction in the lift coefficient and decrease in true air speed to the level of stall speed. The stall of aircraft begins at low height about 0.3 m above the landing surface. The last stage of landing before the touch-down occurs in both methods and is called overshoot. It is executed with very small (close to zero) vertical speed, decreasing airspeed, and a pitch angle appropriate for contact of the aircraft with landing surface. After the touch-down, the last stage of landing - landing run - is performed. During this stage control system keeps the direction of aircraft along the central axis of the landing surface and reduces the airspeed to braking speed (Ub). Aircraft heading is corrected from the beginning of the approach to the end of the braking stage of landing so that its longitudinal axis is parallel to the runway central line.

The block diagram of the flight control system during the landing phase is presented in Fig. 14. Landing executed according to trajectory model presented in Fig. 13 requires stabilization of selected flight parameters during all stages. In this paper, attention is limited to longitudinal motion of aircraft so a small part of the control system will be considered only. The flight control system consists of such parts as: angular velocity, pitch, vertical speed, true airspeed and controller. These blocks represent a modules of the control system responsible for stabilization of pitch angular rate, pitch angle, airspeed and vertical speed. They use the control laws based on the PID and PI controller with saturations and modifications. The block named controller represents a module that is responsible for generating commands used to control of the aircraft trajectory.

During the approach stage vertical velocity module employs the typical PI controller. In this case the desired value of vertical speed changes and atmospheric disturbances are relatively small and fast. The input signal of PI controller is a step function. So that principle of control ensure high quality of vertical speed stabilization. However, during flare stage the desired value of vertical speed changes slowly. The input signal of PI controller is similar to a linear function. In this case the quality of control process is not satisfactory. The transitory process of regulation is delayed (Fig. 15). Therefore, the last section (overshoot) of the landing before the touch-down is too short and executed with too high vertical speed.

This paper suggest a modification of control law for the vertical speed module after beginning the flare. These changes include the addition of two elements: second integral part and proportionality factor in feedback loop. The modified controller signed PI2 is presented in Fig. 16. In this way the quality of vertical speed stabilization during flare is comparatively better and the delay in the transitory process of regulation disappears (Fig. 17).

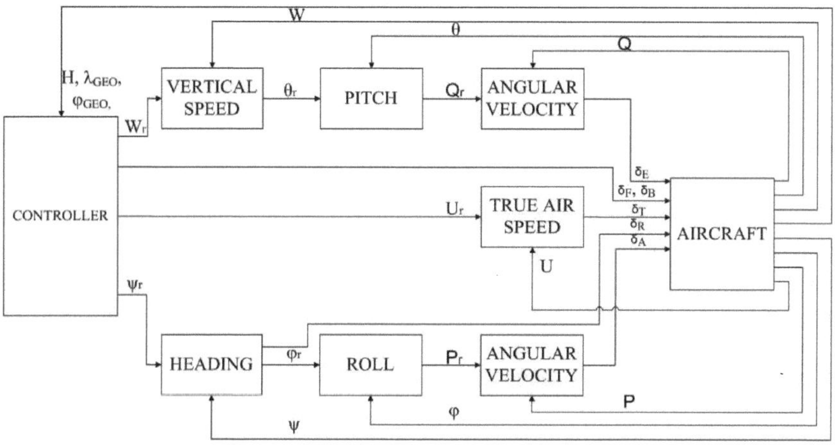

Fig. 14. Block diagram of the flight control system, where: H - altitude above the ground; λGEO, φGEO - longitude, latitude; Wr, W- required and actual vertical speed, θr , θ - required and actual pitch, Qr, Q- required and actual pitch velocity, Ur, U- required and actual airspeed, ψr, ψ - required and actual heading, φr, φ - required and actual roll, Pr, P- required and actual roll velocity, δE, δF, δB, δT, δR, δA - control signals of elevator, flaps, brakes, thrust, rudder and aileron

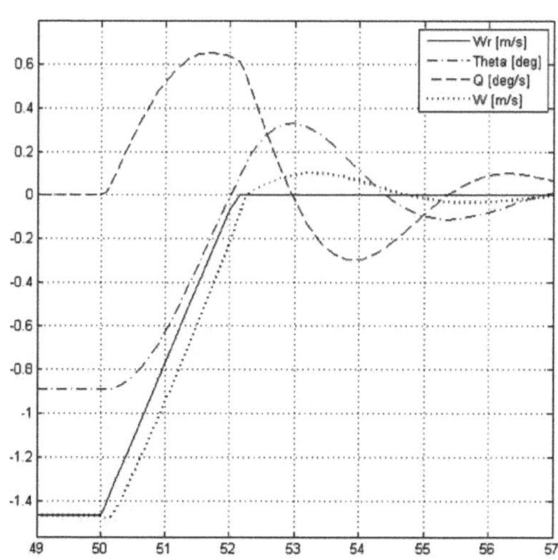

Fig. 15. The time history of flight parameters at beginning of flare maneuver with typical PI controller in the vertical speed channel

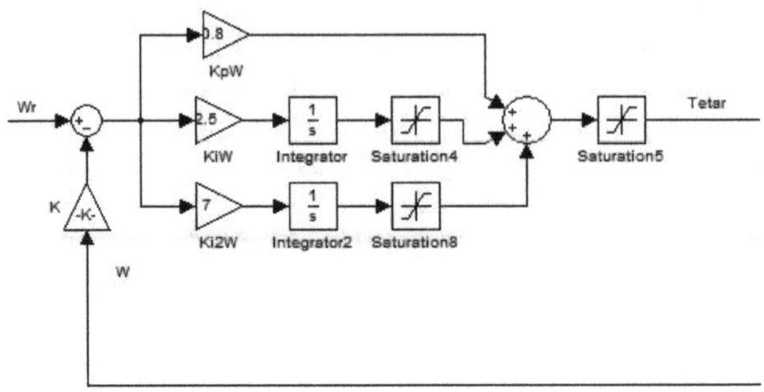

Fig. 16. The block diagram of the vertical velocity module during flare stage of landing

Fig. 18 presents the process of aircraft control during landing with vertical speed module employing the modified PI2 controller after beginning of flare maneuver. At the height of decision (hD) equal to 2 meters above ground the flight control system changes the value of vertical speed of an aircraft proportional in relation to ground and executes the flare stage. Then, in the overshoot stage the aircraft descends with small (close to zero) vertical speed and touching-down.

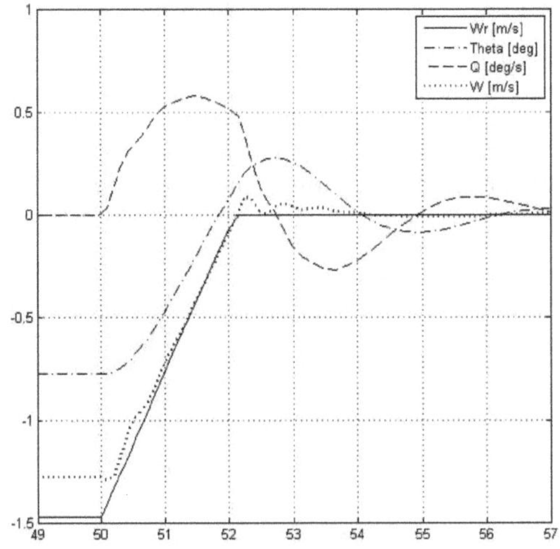

Fig. 17. Flight parameters at beginning of flare maneuver with modified PI2 controller in the vertical speed module

The simulations were performed using a special test stand consisting of the flight simulator X-Plane and onboard computer with xPC control system. For the purpose of analysis, synthesis, and initial computer simulations, linear models of aircraft, engine and actuators have been used.

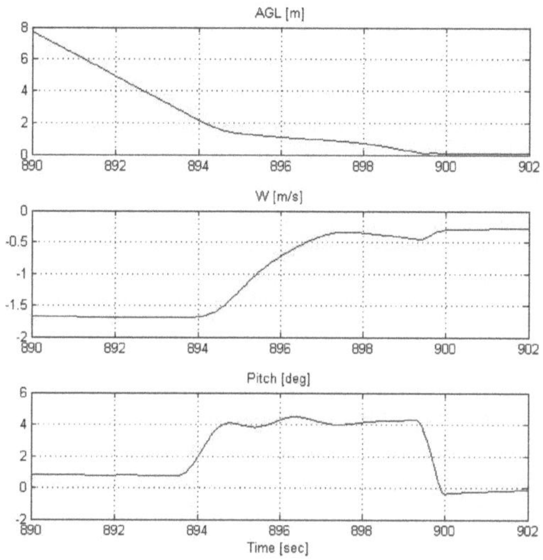

Fig. 18. The plot of the flare maneuver

7 Summary

The paper presents selected innovated solutions prepared by research team from Rzeszow University of Technology. All of them have been created during last years and are implemented into modern control systems for UAVs developed by RUT. Moreover they have been tested in real flight trials.

Presented solutions are unique and are being still developed.

References

[1] Åström, K.J., Murray, R.M.: Feedback Systems. Princeton University Press, Princeton (2008)

[2] Chudy, P., Tomczyk, A., Rzucidlo, P.: Safety enhanced digital flight control system. Aircraft Engineering and Aerospace Technology 81(5, 4), 416–423 (2008)

[3] Chudy, P., Rzucidlo, P.: Real-time simulations of environmentally friendly flight control system. In: 9th IEEE International Symposium on Applied Machine Intelligence and Informatics, Smolenice, Slovakia, vol. Article number 5738867, pp. 161–164 (2011)

[4] Chudý, P., Dittrich, P., Rzucidlo, P.: HIL Simulation of a Light Aircraft Flight Control System. In: Digital Avionics System Conference, AIAA/IEEE Digital Avionics Systems Conference - Proceedings, Williamsburg, VA, US, vol. art. no. 6382395, pp. 1–11, 6D11–6D113 (2012)

[5] Dołęga, B., Gruszecki, J., Rogalski, T., Rzucidło, P., Tomczyk, A.: The Possibilities of Chosen Applications of Control and Navigation Systems in General Aviation Aircraft and Unmanned Air Vehicles. Aviation 11(2), 23–30

[6] Dołega, B., Rogalski, T.: Control System For Medium-Sized Flying Target. Aviation 13(1), 11–16 (2009), doi:10.3846/1648-7788

[7] Kopecki, G.: Aircraft Trajectory Model Following Control With The Use of Linear Quadratic Regulator Control Laws. In: AIAA Guidance Navigation and Control Conference, Minneapolis, Minesota, August 13-16, AIAA Paper 2012-4828 (2012)

[8] Kopecki, G., Pieniazek, J., Rogalski, T., Rzucidło, P., Tomczyk, A.: Proposal for navigation and control system for small UAV. Aviation 14(3), 77–82 (2010)

[9] Kopecki, G., Pieniążek, J., Rogalski, T., Rzucidło, P.: Pokładowy system sterowania samolotem bezzałogowym bazujący na magistrali CAN ("Unmanned aircraft airborne control system based on CAN"), High-Speed Tracklaying Vehicles (23) no 1. Institute of Research and Development of Mechanical Devices OBRUM, Gliwice (2008) (in Polish)

[10] Kopecki, G.: Analysis of Control Quality of Aircraft Lateral Motion During Approach with the Use of Different Control Laws. Aviation X(3), 543–550 (2006)

[11] McRuer, D., Ashenas, I., Dunstan, G.: Aircraft Dynamics and Automatic Control. Princeton University Press, Princeton (1973)

[12] Mikulski, J.: Information flow in unmanned aircraft control system and communication with its environment. In: Ciecinski, P., Pieniążek, J., Rzucidło, P. (eds.) Advances in Transport Systems Telematics, pp. 65–72. WKŁ, Warszawa (2008)

[13] Nelson, R.C.: Flight Stability and Automatic Control. Mc Graw Hill, Boston (1998)

[14] Rogalski, T., Tomczyk, A.: Low-Cost Attitude Reference System for Unmanned Air Vehicle, AIAA paper No. AIAA-2005-6965, conference Infotech@Aerospace, Arlington, VA, USA (2005)

[15] Stevens, B.L., Lewis, F.L.: Aircraft Control And Simulation. John Willey& Sons, Inc., New York (1992)

[16] Turkoglu, K., Ozdemir, U., Nikbay, M., Jafarov, E.M.: PID Parameter Optimization of an UAV Longitudinal Flight Control System (2012), `http://www.waset.org/journals/waset/v45/v45-61.pdf` (cited January 14, 2012)

Design and Implementation of Mobile Ground Base Station for UGV

Paweł Iwaneczko, Karol Jędrasiak, Krzysztof Daniec, and Aleksander M. Nawrat

Institute of Automatic Control, Silesian University of Technology,
Akademicka 16, 44-100 Gliwice, Poland

Abstract. The purpose of this article is to present the designed and implemented Mobile Ground Base Station for UGVs. It was designed by the team from Silesian University. The MGBS (Mobile Ground Base Station) was tested using two different types of loopbacks: SIL (Software In the Loop) and HIL (Hardware In the Loop). After performing tests in virtual reality the MGBS was testes in the integration with the real UGV object. All tests were passed successfully.

Keywords: mobile ground base station, UGV, tele-operation, UGV control.

1 Introduction

Mobile ground base station (MGBS) is a system, that allows the operator to manage and remote control of the unmanned ground vehicle (UGV). In addition to manual control of the object, operator can watch the live on-board cameras and telemetry data view in the same time. Thanks to this person who control the object can faster and more accurately make a decision related to the specific task of the UGV. Unmanned ground vehicles are often used to recognize the area where there is a risk of loss of life or for recognition the enemy objects. UGV are also often used to search for and evacuate the wounded from the battlefield and for transporting ammunition or weapon systems to the soldiers. For this purpose, the base station must have a functionalities such as: monitoring and management of the UGV and UAV, display of the video stream from the object on-board cameras in real time, emergency remote manual control, it should have a touch display screen and dedicated key panel to operate the system functions. It should be noted, that MGBS should be highly mobile, should have a reinforced casing structure, should provide a secure encrypted communication with the environment and be easy to use with gloves.

2 Design

Most of mobile ground base stations solutions is based on the case structure [1-6]. In the article [7] and [8] author is connecting virtual reality to the ground base station to perform software simulation.

A.M. Nawrat (ed.), *Innovative Control Systems for Tracked Vehicle Platforms*,
Studies in Systems, Decision and Control 2,
DOI: 10.1007/978-3-319-04624-2_4, © Springer International Publishing Switzerland 2015

Construction of the mobile ground base station is based on the case structure. Casing is made of reinforced material. The case is equipped with a touch-screen monitor with high brightness, which is built-into the lit of the case. The key panel consists of 17 function keys, 3 function switches and emergency stop key. Case also includes 2 manual controllers (joysticks). Power switch of the case is protected with the key lock (ignition). Case is presented on the fig. 1 in two options: a shows the opened view, and b shows the closed case.

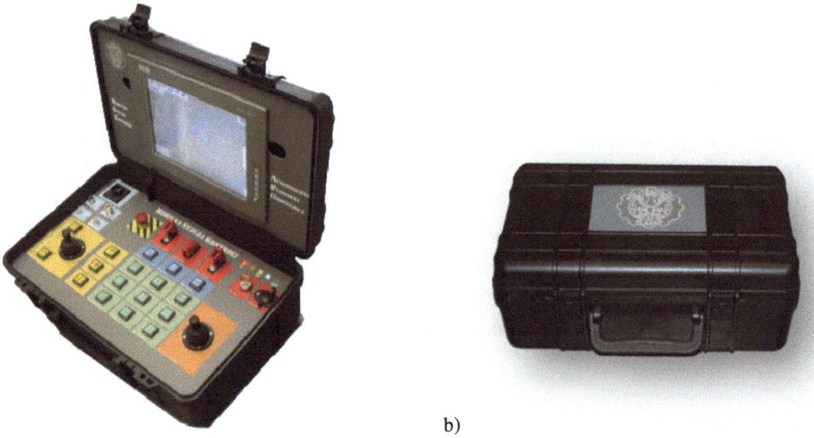

a) b)

Fig. 1. Mobile Ground Base Station view: a) opened, b) closed

Under the keyboard panel there is placed dedicated electronic which is responsible for a stable and reliable work of all system components. Electronic system supports up to 40 digital input keys (including built-in LEDs) and up to 16 input analog joysticks. It is connected to a separate electronic module, with GPS sensor, and a separate radio link module (RF 869 MHz), which are used to communicate with and navigate the APG or a group of objects [13]. On the top left of the keyboard panel are installed four expansion ports for connecting the external antennas or mobile devices. Basic microcontroller controls the input ignition, which turns on power supply for all components, and holds it until all other parts is not turned off. Internal gel rechargeable battery (12V) have a system protection against too deep discharge and it can be charged from external AC power socket. Above the power socket is installed 5 LEDs that indicate the percentage of battery level, which is fully checked by the dedicated voltage measuring system. The battery must be charged with a constant current and constant voltage. Inside of the case there is also installed Wi-Fi router, an industrial PC computer and high brightness LCD monitor. It can be separately switched on/off to manage energy savings. All these components are shown in fig. 2.

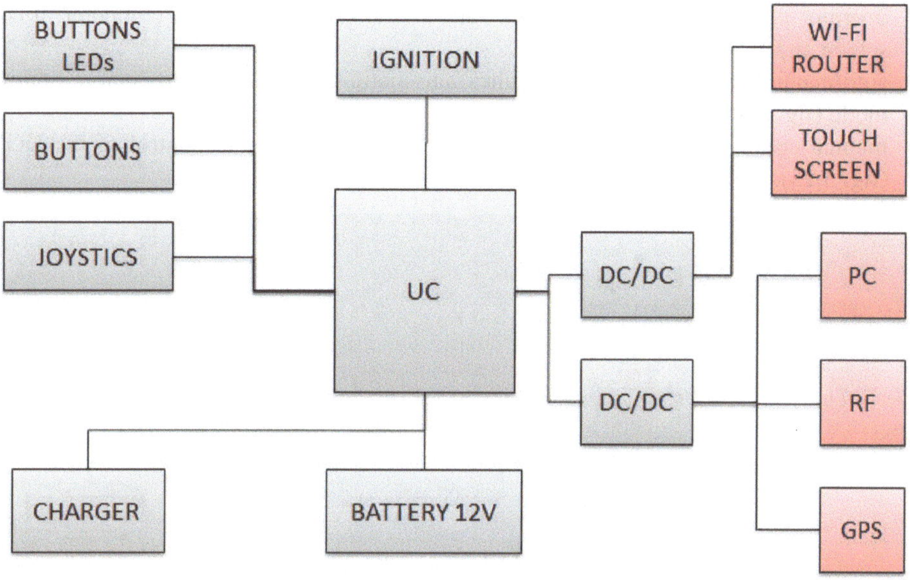

Fig. 2. Internal components connection schema

3 Functionality and Operation Instruction

The main functionality of the MGBS is the remote control of APG platform. For this reason case can be turned on in two modes. The first mode is without operator panel, which implicate the power saving. This is because the fact, that the PC, touch screen and Wi-Fi router, which consume the most energy, are switched off. The secondary functionality of the MGBS is the management of the APG platform. In this mode operator can use the control application to:

- show the location and azimuth orientation of the UGV,
- visualize the cameras,
- visualize the telemetry data on charts or dedicated sensor gauges,
- manage the unmanned objects,

3.1 UGV Control without Operator Panel

The UGV control mode without operator panel was designed for intuitive remote control of unmanned APG object and was highly simplified. After turning the ignition switch on (fig. 3.5), dedicated electronics that manages the power of all system components - starts the radio link module, GPS module and input/output module, which handles keys and the joysticks. Active key panel components is as follow:

- emergency stop switch – that stops all APG control systems, and its default value is opened (fig. 3.1),

- operator panel enable switch – that runs and stops industrial PC computer, touch screen and Wi-Fi router (fig. 3.2),
- auto/manual switch – which switches automatic control mode to remote control mode (fig. 3.4),
- manual control joystick – that allows the operator to remote control of the UGV (fig. 3.25).

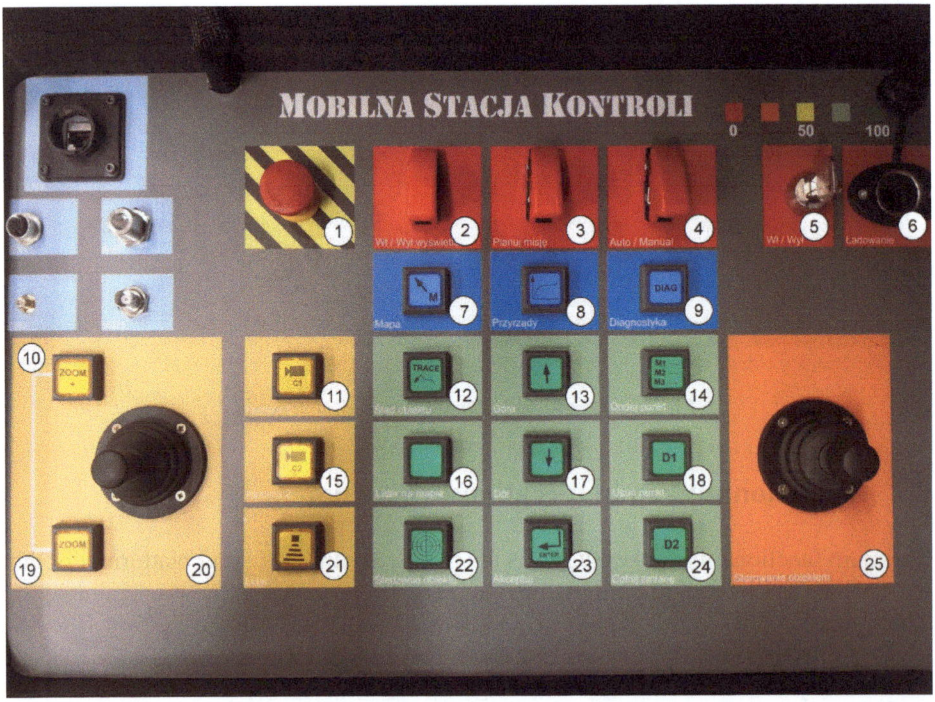

Fig. 3. Keyboard panel with footnotes

3.2 UGV Management with Operator Panel

The UGV control mode with operator panel was designed for management the unmanned APG object. When operator switches the operator panel switch on (fig. 3.2), there will be launched all most energy consuming components of the case system: industrial PC computer, Wi-Fi router, and LCD touch screen. When operator switches the ignition switch off (fig. 3.5), when PC computer is turned on, the power LEDs are blinking for 10 seconds to inform the operator, that system on the PC is not closed properly. In this mode all keys and joysticks are enabled and are used to handle the PC operator application:

- zoom in button – this button is responsible for map control zooming in, alternatively with a camera zooming in (fig. 3.10),

- zoom out button – this button is responsible for map control zooming out, alternatively with a camera zooming out (fig. 3.19),
- left joystick – this joystick is used for move horizontally and vertically the map control, alternatively with rotating camera view (fig. 3.20),
- first camera button – toggles on/off the video stream window from the front APG camera (fig. 3.11),
- second camera button – toggles on/off the video stream window from the PTZ observation camera (fig. 3.15),
- LIDAR workspace button – changes the active workspace to LIDAR sensor view (fig. 3.21),
- map workspace button – changes the active workspace to navigation map control view (fig. 3.7),
- gauges workspace button – changes the active workspace to gauges control view (fig. 3.8),
- diagnostics workspace button – changes the active workspace to diagnostics chart control view (fig. 3.9),
- object trace button – toggles on/off the buffered trace of the unmanned object on the navigation map control (fig. 3.12),
- map LIDAR button – toggles on/off the LIDAR sensor view of the UGV on the navigation map control (fig. 3.16),
- track object button – toggles on/off tracking the object on the map, what means that the map automatically changes its position, so that the object is still visible in the middle of the screen (fig. 3.22),
- add waypoints button – changes the tap on the screen mode to adding a new waypoint (GPS location) on the navigation map (fig. 3.14),
- remove waypoints button – changes the tap on the screen mode to removing the waypoint from the navigation map (fig. 3.18),
- undo button – undoes the last 3 edits on the map waypoint mission (fig. 3.24),
- mission switch – toggles on/off the right mission panel, where operator can choose and show on the map one of the saved missions (fig. 3.3),
- up button – up navigation button, which is used to select previous mission from the opened mission panel (fig. 3.13),
- down button – down navigation button, which is used to select next mission from the opened mission panel (fig. 3.17),
- select button – accept button, which is used to send the selected mission from the mission panel to the unmanned object (fig. 3.23).

In order to maintain the integrity of the graphical user interface, last four of the above described buttons have alternatively functions in the diagnostic workspace. This allows the operator to intuitively and quickly, using the keyboard, even with thick gloves, to select multiple of telemetry data on real time chart.

Most of functions are binded both to keys described above and to the icons, which are added to the visualization of the PC operator application (PCOA). Industrial computer has installed Windows 7 Embedded system, and the operator application is implemented in C#.NET language using the WPF and DirectX components.

Visualization of the PCOA is divided into 4 separated workspaces with different components and 2 independent windows. Those four workspaces are:

- navigation map workspace – that visualizes the UGV object on the map in its GPS coordinates, and allows operator to manage the unmanned object autonomous route (fig. 5),
- LIDAR gauge workspace – that visualizes the *Light Detection and Ranging* sensor, which is installed in the front of the APG platform (fig. 4.a),
- telemetry gauges workspace – that visualizes: the left crawler speedometer, right crawler speedometer, main speedometer, horizon gauge, compass gauge, oil and engine temperature gauges, accumulator voltage gauge and fuel level gauge (fig. 4.b),
- diagnostic chart workspace – that allows to visualize all telemetry data on the one real-time chart (fig. 4.c).

a) b) c)

Fig. 4. Telemetry workspaces: a) LIDAR gauge, b) telemetry gauges, c) diagnostic charts

PC operator application includes two management icons layer. Main layer is located in the top-left corner of the screen, it is visible on each of the workspaces in front of the all controls and displayed windows (fig. 4, 5), and it is used to management the active workspace and open camera windows. The second icons layer is closely linked to the navigation map. Icons located in the main management layer are as follows:

- first camera icon – toggles on/off the video stream window from the front APG camera (fig. 5.1),
- second camera icon – toggles on/off the video stream window from the PTZ observation camera (fig. 5.2),
- LIDAR workspace icon – changes the active workspace to LIDAR sensor view (fig. 5.3),
- map workspace icon – changes the active workspace to navigation map control view (fig. 5.4),
- gauges workspace icon – changes the active workspace to gauges control view (fig. 5.5),
- diagnostics workspace icon – changes the active workspace to diagnostics chart control view (fig. 5.6),
- battery control – shows the battery percentage level of the case electronic systems (fig. 5.8).

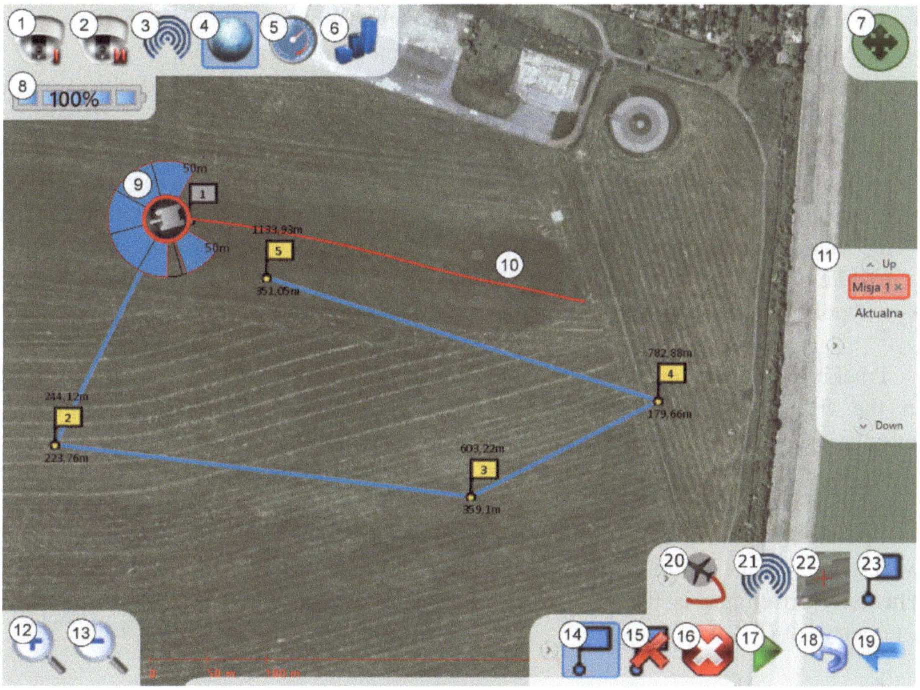

Fig. 5. Navigation map control view with footnotes

Icons located in the map navigation management layer are as follows:

- map movement icon – toggles on/off the possibility of moving the map view (fig. 5.7). Operator can move the map view by grabbing and dragging the touch screen in the clear area of map images
- mission menu – allows the operator to select and show one of the previously saved missions on the map (fig. 5.11). Mission menu contains: horizontal arrow - which is used to toggle the visibility of the menu, missions list, and up and down arrows - which are used to scrolling the missions list. The last mission from the list is an actually edited mission, which operator can create and save on the computer by sending it to the UGV object,
- zoom in icon – this icon is responsible for map control zooming in (fig. 5.12),
- zoom out icon – this icon is responsible for map control zooming out (fig. 5.13),
- add waypoints icon – changes the tap on the screen mode to adding a new waypoint (GPS location) on the navigation map (fig. 5.14),
- remove waypoints icon – changes the tap on the screen mode to removing the waypoint from the navigation map (fig. 5.15) ,
- abort mission icon – cancel and delete the currently executed mission from the UGV (fig. 5.16),
- send mission icon – sends actual selected mission from the mission menu to the UGV (fig. 5.17)

- undo icon – undoes the last 3 edits on the map waypoint mission (fig. 5.18),
- UGV menu icon – shows the UGV management icons (fig. 5.19),
- object trace icon – toggles on/off the buffered trace of the unmanned object on the navigation map control (fig. 5. 20). Buffered trace is shown in the fig. 5.10,
- map LIDAR icon – toggles on/off the LIDAR sensor view of the UGV on the navigation map control (fig. 5.21). Map LIDAR sensor is shown in the fig. 5.9,
- track object icon – toggles on/off tracking the object on the map, what means that the map automatically changes its position, so that the object is still visible in the middle of the screen (fig. 5.22).
- waypoints menu icon – shows the waypoints management icons (fig. 5.23).

Tracking the UGV on the map, while the object changes its coordinates, is shown in the fig. 6. When tracking is disabled the map don't changes its position automatically (fig. 6.a-b). When tracking is enabled the map automatically changes its position, so that the object is still visible in the middle of the screen (fig. 6.c-d). It is possible to select and track objects according to the algorithms presented in [9,10].

Operator application also includes management layer of the diagnostic chart workspace, which is almost identical as mission menu in navigation map workspace. The only difference is the possibility of select multiple chart data series from the telemetry list (fig. 4.c).

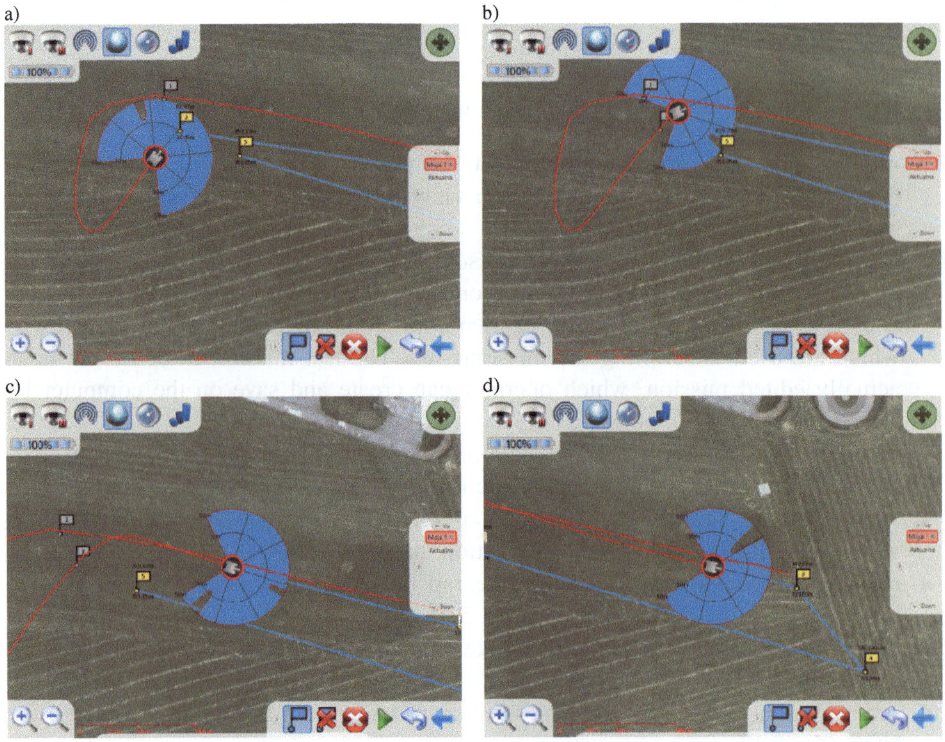

Fig. 6. Tracking object views

In addition to the four separate workspaces, the application allows the operator to run two independent windows. These windows are visible on top of other controls, can be minimized, maximized, resized and moved by the operator to a convenient position on the screen (fig. 7). These windows displays live video stream from two cameras placed on the APG platform. The stream can be also archived as an evidence [11,12].

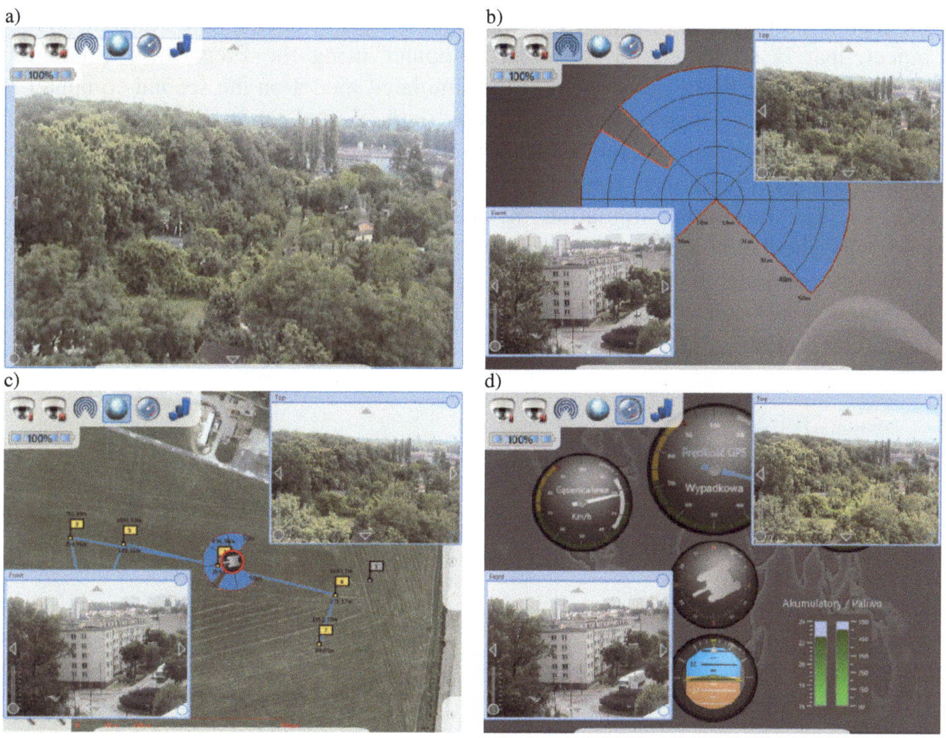

Fig. 7. Cameras view on different workspaces

4 System Integration with Environment

MGBS case system is integrated with the real UGV object and with the Prepar3D® Lockheed Martin simulation software. This is because the control application software have to be well-proven and reliable. To provide a good verification of all functions of the control application, it was implemented in three ways, to integrate with: *Software In the Loop* (SIL) simulation, *Hardware In the Loop* (HIL) simulation and with the real UGV object.

4.1 HIL and SIL Integration with the Prepar3D® Lockheed Martin Simulation Software

Software In the Loop (SIL) simulation controls a physical process model (the simulator) using the software, which is a model of the regulator, and it's running on the same hardware platform. In the *Hardware In the Loop* (HIL) simulation process involves a real regulator, which communicates with the simulator through the appropriate interfaces. In both SIL and HIL loops – control application is executed on the mobile ground base station (MGBS), and it is used to manage the unmanned APG object, that is simulated on the other computer using the Prepar3D® simulation software. Control and management of the simulated model on the second computer is provided by control application, which was implemented so as to simulate the real object control systems. Communication interface between the MGBS operator application and Prepar3D® control software is based on 2.4 GHz wireless technology (fig. 8). The hardware simulation includes all electronic devices in the case, in particular the keyboard panel and microcontroller, which are connected with PC industrial computer by serial port. The difference in the SIL simulation is that the keyboard is not connected to the PC and it is displayed on the touch screen in the extending bottom menu.

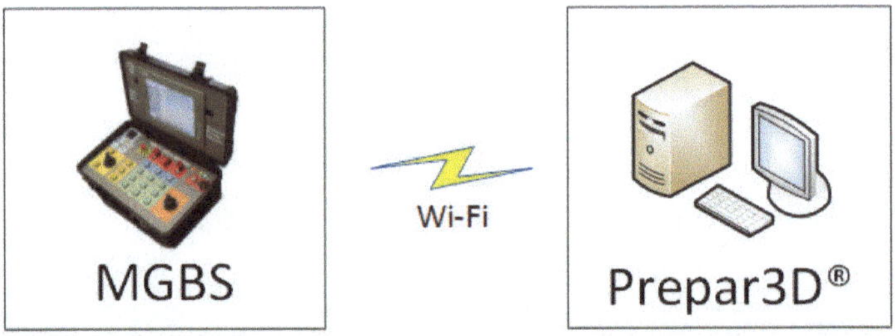

Fig. 8. HIL and SIL simulation

4.2 Integration with the Real Object

Communication between the mobile ground base station and the real unmanned ground vehicle object is also based on 2.4 GHz wireless technology. In the real object integration, case system parts are used for manual control of APG as well as for management of the autonomous control mode of the vehicle. All control systems of the UGV are implemented on a dedicated electronics systems. The control system of the APG consists of four electronic modules, where on each microcontroller is working an μCOS operating system. All modules communicate with each other via CAN bus (fig. 9). One of the modules is responsible for Ethernet connection with environment and is used as management unit and as link between the base station and APG control systems. Implementation of this module is very similar to SIL and HIL control application, with the difference, that first one is implemented in C language, and the second one in C# .NET language.

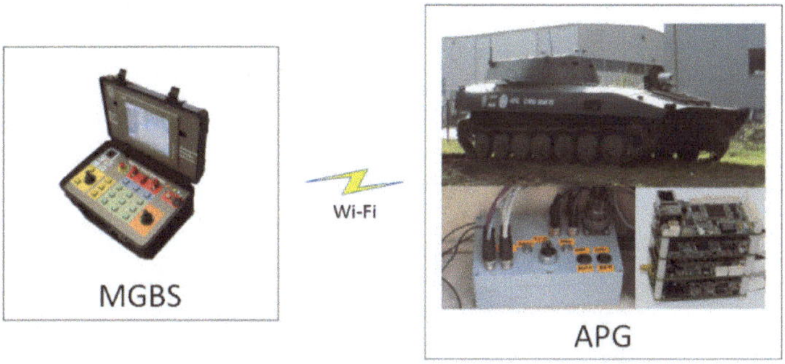

Fig. 9. Real object and MGBS integration

5 Tests

The work of all MGBS components and the integration with environment, as it was mentioned in the chapter 5, have to be well-proven and reliable. In order of that, tests were performed using the real object as well as simulated object in Prepar3D® Lockheed Martin simulation software. Unfortunately, not all of the MGBS functionalities can be identically mapped by the simulator and the real object, so all the tests performed in the simulation were also repeated on the outside training area with real UGV.

5.1 Control of the UGV Object Simulated in the Prepar3D® Lockheed Martin Software

Control and management tests of the simulated object, using the Prepar3D® Lockheed Martin simulator, were performed in the laboratory room in combination of the MGBS with the desktop PC computer (fig. 10). Tests were performed both in the SIL simulation loop – without using the operator key panel and microcontroller in the case, and in the HIL simulation loop – with all electronic components, which are located in the case. Most of performed tests in this mode was related to the visualization of the MGBS software and the correct interpretation of incoming telemetry data. Performed tests:

- visualization of the LIDAR control on the map and separate workspace,
- telemetry data displayed on the chart control,
- navigation of the map control (moving and zooming),
- waypoints management,
- mission creation, edition and sending it to the object,
- animation of the gauges needles,
- tracking object on the map,
- show trace of the object.

Fig. 10. Control and management the UGV object in simulation software

5.2 Control of the Real Object

Tests, that was performed in the integration with the real object was also performed in simulation mode, but only the real object test can correctly verify the functionalities associated with the management and control of autonomous systems. All driving test using the APG platform were performed on the outside training area (fig. 11). It was very important to keep all necessary precautions, and perform all tests away from the people. Performed tests:

- manual control,
- autonomous control,
- obstacle avoidance,
- camera video stream,

Fig. 11. Control and management the real UGV object

6 Conclusions

The article presents design, implementation and operation instruction of mobile ground base station for unmanned ground vehicles. The MGBS (*Mobile Ground Base Station*) was tested using two different types of loopbacks: SIL (*Software In the Loop*) and HIL (*Hardware In the Loop*). After performing tests in virtual reality the MGBS was testes in the integration with the real UGV object. All tests were passed successfully.

References

[1] Internet source, `http://zala.aero/en/electronic/gcs.htm`

[2] Internet source, `http://www.oilabsorbspill.com/oil-spill-monitoring-system/1035088.html`

[3] Internet source, `http://www.unmannedgroup.com/index.php?page=2&subpage=7&data=15`

[4] Internet source, `http://www.bayraktartech.com/Mobile-GCS`

[5] Internet source, `http://www.armedforces-int.com/article/mini-unmanned-aerial-vehicle.html`

[6] Internet source, `http://www.aaicorp.com/products/unmanned/portable_gcs`

[7] Walter, B.E., Knutzon, J.S., Sannier, A.V., Oliver, J.H.: Vr Aided Control of Unmanned Vehicles. Virtual Reality Applications Center. Iowa State University, Ames, IA 50011-2274

[8] Jovanovic, M., Starcevic, D.: Software Architecture for Ground Control Station for Unmanned Aerial Vehicle. In: Tenth International Conference on Computer Modeling and Simulation, Laboratory for Multimedia Communications, Faculty of Organizational Sciences. University of Belgrade,

[9] Kuś, Z., Nawrat, A.: Object Tracking for Rapid Camera Movements in 3D Space. In: Nawrat, A., Kuś, Z. (eds.) Vision Based Systems for UAV Applications. SCI, vol. 481, pp. 57–76. Springer, Heidelberg (2013)

[10] Kuś, Z., Nawrat, A.: Object Tracking for Rapid Camera Movements in 3D Space. In: Nawrat, A., Kuś, Z. (eds.) Vision Based Systems for UAV Applications. SCI, vol. 481, pp. 57–76. Springer, Heidelberg (2013)

[11] Werner, A., Bach, M., Duszeńko, A.: Vision Object Access and Management with Usage of the In-Memory Databases. In: Nawrat, A., Simek, K., Świerniak, A. (eds.) Advanced Technologies for Intelligent Systems. SCI, vol. 440, pp. 231–246. Springer, Heidelberg (2013)

[12] Bach, M., Duszenko, A., Werner, A.: Column-Oriented Metadata Organization of Vision Objects. In: Kapczyński, A., Tkacz, E., Rostanski, M. (eds.) Internet - Technical Developments and Applications 2. AISC, vol. 118, pp. 49–55. Springer, Heidelberg (2012)

[13] Babiarz, A., Bieda, R., Jaskot, K.: Vision System for Group of Mobile Robots. In: Nawrat, A., Kuś, Z. (eds.) Vision Based Systems for UAV Applications. SCI, vol. 481, pp. 139–156. Springer, Heidelberg (2013)

[14] Mezyk, A., Switonski, E., Kciuk, S., Klein, W.: Modelling and investigation of dynamic parameters of tracked vehicles. Mechanics and Mechanical Engineering 15(4), 115–130 (2011)
[15] Martynowicz, P., Kciuk, S., Mezyk, A.: Rotary-shock-absorber with magnetorheological valves. Advanced Materials Research 628, 505–511 (2013)
[16] Barnat, W., Panowicz, R., Niezgoda, T.: Numerical and Experimental Comparison of Combined Multilayer Protective Panels. Acta Mechanica et Automatica 6(1), 13–16 (2012)
[17] Barnat, W., Panowicz, R., Niezgoda, T., Dybcio, P.: Numerical analysis of IED detonation effect on steel plate. Acta Mechanica et Automatica 6(1), 10–12 (2012)
[18] Panowicz, R., Barnat, W., Niezgoda, T., Szymanczyk, L., Grzymkowski, J.: Acta Mechanica et Automatica 6(1), 49–52 (2012)
[19] Barnat, W., Niezgoda, T., Panowicz, R., Sybilski, K.: The Influence of Conical Composite Filling on Energy Absorbtion During The Progressive Fracture Process. WIT Transactions on Modelling and Simulation 51, 625–633 (2011)
[20] Niezgoda, T., Panowicz, R., Sybilski, K., Barnat, W.: Numerical Analysis of Missile Impact Being Shot By Rocket Propelled Grenades with Rod Armour. WIT Transactions on Modelling and Simulation 51, 625–633 (2011)
[21] Barnat, W., Dziewulski, P., Niezgoda, T., Panowicz, R.: Application of Composites to Impact Energy Absorbtion. Computational Materials Science 50(4), 1233–1237 (2011)
[22] Niezgoda, T., Barnat, W.: Analysis of Protective Structures Made of Various Composite Materials Subjucted to Imapct. Materials Science and Engineering A 483(1-2 C), 705–707 (2008)
[23] Daniec, K., Jedrasiak, K., Koteras, R., Nawrat, A.: Embedded Micro Inertial Navigation System. Applied Mechanics and Materials 249, 1234–1246 (2013)
[24] Jędrasiak, K., Bereska, D., Nawrat, A.: The Prototype of Gyro-Stabilized UAV Gimbal for Day-Night Surveillance. In: Nawrat, A., Simek, K., Świerniak, A. (eds.) Advanced Technologies for Intelligent Systems. SCI, vol. 440, pp. 107–116. Springer, Heidelberg (2013)
[25] Jędrasiak, K., Nawrat, A., Wydmańska, K.: SETh-link The Distributed Management System for Unmanned Mobile Vehicles. In: Nawrat, A., Simek, K., Świerniak, A. (eds.) Advanced Technologies for Intelligent Systems. SCI, vol. 440, pp. 247–256. Springer, Heidelberg (2013)
[26] Jędrasiak, K., Nawrat, A.: Image Recognition Technique for Unmanned Aerial Vehicles. In: Bolc, L., Kulikowski, J.L., Wojciechowski, K. (eds.) ICCVG 2008. LNCS, vol. 5337, pp. 391–399. Springer, Heidelberg (2009)
[27] Nawrat, A., Jedrasiak, K.: SETh System Spatio-Temporal Object Tracking Using Combined Color and Motion Features. In: WSEAS International Conference, Proceedings, Mathematics and Computers in Science and Engineering, vol. 9 (2009)
[28] Jedrasiak, K., Daniec, K., Nawrat, A.: The Low Cost Micro Inertial Measurement Unit. In: Industrial Electronics and Applications (ICIEA), pp. 403–408 (2013)
[29] Bereska, D., Jędrasiak, K., Daniec, K., Nawrat, A.: Gyro-Stabilized Platform for Multispectral Image Acquisition. In: Nawrat, A., Kuś, Z. (eds.) Vision Based Systems for UAV Applications. SCI, vol. 481, pp. 115–121. Springer, Heidelberg (2013)
[30] Daniec, K., Iwaneczko, P., Jędrasiak, K., Nawrat, A.: Prototyping the Autonomous Flight Algorithms Using the Prepar3D Simulator. In: Nawrat, A., Kuś, Z. (eds.) Vision Based Systems for UAV Applications. SCI, vol. 481, pp. 219–232. Springer, Heidelberg (2013)

[31] Babiarz, A., Bieda, R., Jędrasiak, K., Nawrat, A.: Machine Vision in Autonomous Systems of Detection and Location of Objects in Digital Images. In: Nawrat, A., Kuś, Z. (eds.) Vision Based Systems for UAV Applications. SCI, vol. 481, pp. 3–25. Springer, Heidelberg (2013)

[32] Bereska, D., Daniec, K., Fraś, S., Jędrasiak, K., Malinowski, M., Nawrat, A.: System for Multi-Axial Mechanical Stabilization of Digital Camera. In: Nawrat, A., Kuś, Z. (eds.) Vision Based Systems for UAV Applications. SCI, vol. 481, pp. 177–189. Springer, Heidelberg (2013)

The Prototype of an Unmanned Underwater Vehicle – Stability and Maintain a Specified Course

Krzysztof Jaskot, Piotr Ściegienka, and Aleksander M. Nawrat

Silesian University of Technology, Institute of Automatic Control,
Akademicka 16, 44-100 Gliwice, Poland
{krzysztof.jaskot,aleksander.nawrat}@polsl.pl

Abstract. The paper presents the modules of the control system remotely operated unmanned underwater vehicle used to stabilize the exchange rate and maintain the direction of motion. The authors present a simple control algorithms, and presented the verification of the assumed conditions of a real object. The control algorithms used information from the accelerometers, pressure sensors and an electronic compass.

Keywords: unmanned underwater vehicle, a pressure sensor, electronic compass, control algorithm.

1 Introduction

In modern times many qualified specialist divers perform underwater work. Their tasks include the maintenance of canals and collector sewers, inspection of hydraulic structures, perform bathymetric map of water bodies, technical evaluation of underwater structures and much more. With modern hydrographic equipment can even detect leakages early petroleum materials from tanks located on the seafloor [1]. However, sometimes harsh conditions of life-threatening, long time under water, or even just the hydrostatic pressure is difficult to overcome barrier to humans. Above a depth of 50 m there are further problems, the diver to breathe air can not use the gas only with a different volume content of oxygen. At greater depths, work by divers is very difficult. In situations where the divers are very hard to work because of difficult conditions or threats to life, the use of unmanned underwater vehicles are ROV [2] (Remotely Operated Vehicle called). ROV is a remotely operated vehicle operator, used primarily for inspection, observation of the underwater environment, but not only. The possibility of such a vehicle mounted on a variety of equipment (sonar, cameras, probes, manipulators, etc.) makes it almost unlimited possibilities.

In contrast to the diver, ROV can operate under water for many hours. Complex vehicle stability systems allow the operator to conveniently control of the vehicle even in very difficult conditions. In vehicles of this type of communication between the operator and the vehicle is done by a special cable. Use it to operator commands are sent to the vehicle and all the information from the vehicle to the operator, for example, the vision of the camera or other data from on-board equipment. Underwater vehicles are

A.M. Nawrat (ed.), *Innovative Control Systems for Tracked Vehicle Platforms*,
Studies in Systems, Decision and Control 2,
DOI: 10.1007/978-3-319-04624-2_5,

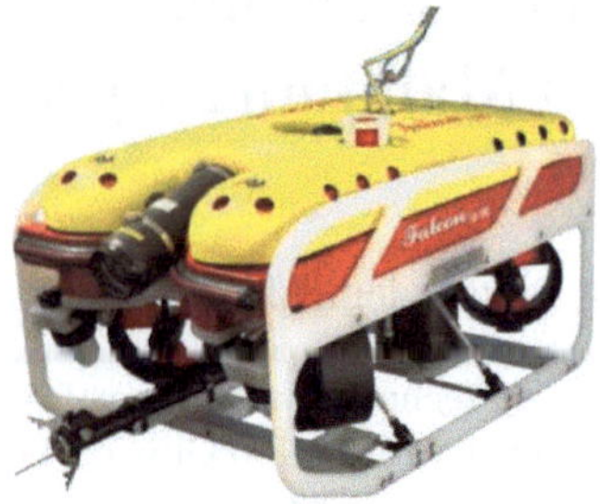

Fig. 1. ROV - SeaEye Falcon [24]

often supplied with oil rigs, ships and rescue ships. In the case of the Navy are often used not only ROV, but also other varieties of underwater vehicles, typically adapted for military purposes, such as UUV (Unmanned Underwater Vehicle called) and AUV (Autonomous Underwater Vehicle called), which can be fired directly from the gaps torpedo submarines. Vehicles of this type to be able to work in difficult conditions, are complex systems stabilize their spatial position, controlling the rate, depth and drift of the vehicle. Counteract external factors by work propulsors to maintain the vehicle in a fixed position. An example is used by the Polish company Petrobaltic, Falcon portable ROV SeaEye, shown in Figure 1. Its systems allow you to maintain the stability of a fixed rate with an accuracy of +/-0.5°, and maintain a stable depth of immersion in the 300m [24], even in difficult sea conditions. Drive motors provide the power to him within 50kgf (500N), which with its 50kg gives it a very large dynamic motion and speed of more than 3 knots. Very good visibility thanks to its powerful 75W illumination and a camera with high resolution and sensitivity. This is a very mobile vehicle used primarily for the inspection and observation. Vehicles designed to operate at much greater depths provide a much more powerful. 6000 Victor French is able to perform work at a depth of 6000m. It is a 4 ton ROV, shown in Figure 2. It has two robotic arms, each capable of lifting 100kg. It has eight cameras and eight lamps with total power of 5kW. Such a vehicle may be equipped with a variety of on-board equipment, sensors and sonar. Thus, these types of vehicles allow us to study the flora and fauna at depths inaccessible to man. In 1995, Japanese ROV Kaiko descended to a depth of almost 11.000m at the bottom of the Mariana Trench and took bacteria samples from the bottom, leading to the discovery of previously unknown enzymes, microbes and genes. Kaiko is not the only vehicle that has made such a step. American Nereus, shown in Figure 3 was another. It has an internal power supply with 4000 lithium batteries, and communication takes place via an optical fiber with a thickness of a human hair [26]. These are very expensive, professional vehicles, but there are also amateur projects which represent a much cheaper solution. Amateur vehicles often are small structures made of PVC, having remote control, simple thrusters and cameras to allow for observation. Often, however, do not have electronic systems to ensure a stable position of the vehicle in difficult conditions. The purpose of this paper is to present the prototype of an unmanned underwater vehicle utilizing a ROV stabilization system, built

Fig. 2. ROV - Victor 6000 [25]

Fig. 3. ROV - Nereus [26]

from the position and acceleration sensors, static and dynamic, and the depth and pressure sensors. The system of sensors will be connected to the microprocessor system and will work with the system remote control and visualization (PC application). This vehicle is inherently low-budget project, whose building blocks are possibly cheap solution, but effective. The vehicle will be able to work in fresh water at depths to 25 m. It will have four ballast tanks, increasing the buoyancy of the vehicle which will allow for the lifting of the vehicle weight to 30 kg above his own weight. Mounted on a moving camera will allow underwater observation, and 6 propellers mounted in different planes of the vehicle will ensure him sufficient mobility. Electronic modules ensure control of the vehicle, and control algorithms implemented sufficient stability of the vehicle under water. In this work attention is focused mainly on system stability. Immersion depth on specific systems require stabilization of buoyancy control of the vehicle, its speed to maintain proper vertical and acceleration. The waves and water currents acting on the vehicle and cause it to drift bolting systems, vehicle stability by controlling propulsors must maintain it in spite of the target position. Uneven loading of the vehicle causing it to tilt, making it difficult to maneuver and observations. In this case, stability systems controlling the buoyancy trim tank (ballast) must maintain the vehicle at an angle given by the operator. The question of stability of the vehicle ROV is therefore very complex, requiring multiple complex control algorithms [3]. In this paper we will address only the grounds for its stability, given the problems that are encountered with concrete solutions and methods for their solution, and further plans for development and improvement of the vehicle in terms of its stability under water. Prototype ROV system stability, discussed in this work, the whole system is basically stable, which includes several algorithms currently regulated. These regulators controlling the basic functions of car set it in a specific spatial position by the operator with sufficient accuracy [4]. By the spatial position of the vehicle is understood the depth of immersion, trim and the current rate [5], as shown in Figure 4. To stabilize the vehicle

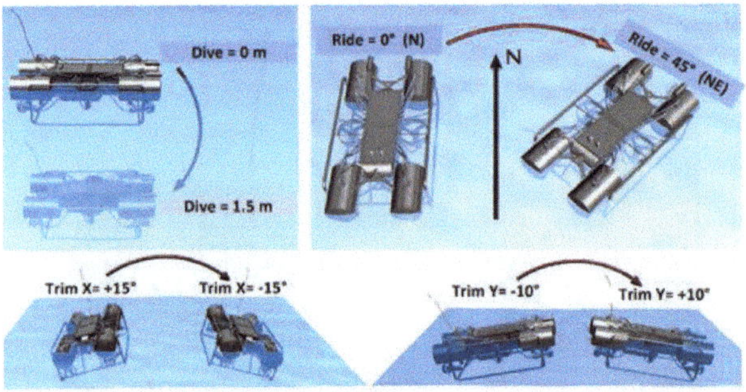

Fig. 4. The spatial position of the vehicle - examples

is also classified as speed descent / ascent, because it is part of the algorithm maintain the required depth of immersion [6]. In the remainder of the work are references to the central unit and motor driver. The central body will be called the main driver of the vehicle based on the ATmega128 chip. This driver is inside the rear of the hull of the vehicle with a bundle of cables and wires at the place marked on Figure 5. His task is to take any action automatic control of all vehicle peripherals, communication with all modules and sensors, and the operator panel. The main processor or microcontroller unit will be called the ATmega128 microcontroller in the central unit. Motor controller will be called Direct Control module propulsors based on ATmega168 microprocessor that accepts commands from the CPU. It is located in the rear of the hull vertically in the slot next to the main battery, its location is marked in Figure 6. Described later in work stability modules control propulsors indirectly through the motor controller [7]. Before describing the various modules of stabilization operator panel will be described, by which these modules are tuned. An operator from the operator panel, in addition to basic vehicle control functions, also including the ability to set the parameters of the vehicle stability systems, setting controls and manual control of the vehicle without the

Fig. 5. The location of the central unit inside the vehicle

Fig. 6. The location of the engine driver inside the vehicle

use of automatic control. In addition to the opportunity to observe the individual parameters of the vehicle by means of appropriate indicators. Figure 7 shows a sample view of this panel. This panel is a PC application written in VC + +, which is the center of the underwater vehicle control. Some graphical user interface consists of two parts: the visualization and control. The first one is used for observation of the second vehicle to control it. Both of the panel are described in turn below. Some visualization is highlighted in Figure 8. It includes indicators of the current state of the vehicle. They are grouped with the so-called. the frames. These include:

- Engines
- Air Pressure
- Depth
- The water level in ZB (in the ballast tanks).
- The Compass
- Roll and pitch
- Frame graphics.

In the frame of "engines" visible indicators showing a 6, with which the power of the individual thrusters currently operate. The values are shown on a scale of 0-100%. Figure 9 shows a sample view of the frame. This value is in fact determined to fill the rectangular PWM signal (called Pulse-width modulation), which is powered by a propeller. At any rate sagittal whether the information is currently working propeller. These

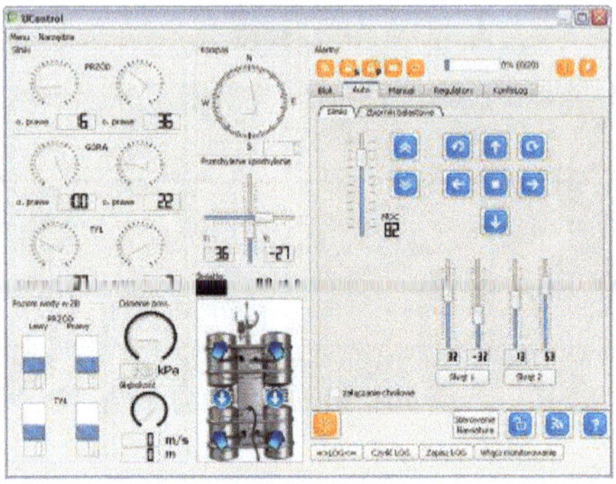

Fig. 7. The operator panel - general view

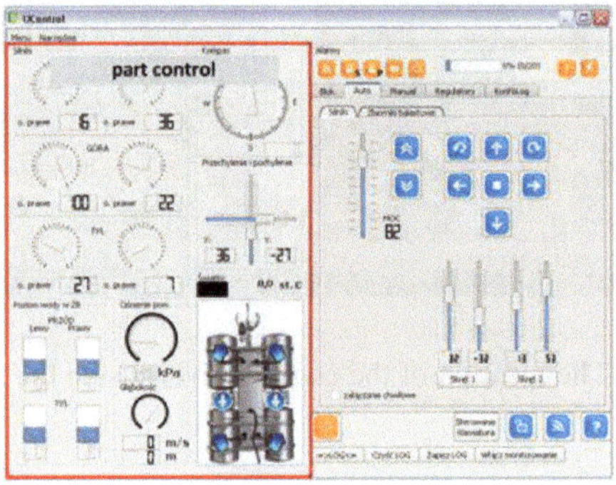

Fig. 8. The operator panel - part of the visualization

indicators are divided into 3 sections: "FRONT", "UP", "REAR". This division is appropriate propulsors front, top (for vertical movement) and the rear. The position of these propulsors in the vehicle structure is shown in Figure 10.

2 Indicator of Air Pressure in the Manifold

Below, in the "air pressure", is the indicator showing the current air pressure in the manifold. Figure 11 presents a view of this indicator. The collector, in which pressure is measured, is behind the supply air pressure regulator from the cylinder with compressed air. Pressure indicator on the operator panel allows the operator to monitor changes to this pressure, if he followed any leaks of air masses due to a leak or if he dropped the pressure in the cylinder supply. The air pressure in the collector is measured by means of an analog sensor MPX5700GP, whose output signal is transmitted to the A/D converter in the central unit of the vehicle. The maximum measuring range of the sensor relative to the ambient pressure is 7 at, so manifold pressure regulator set cannot be greater. Because of the assumptions, the vehicle is to submerge to a depth of 25 m, the pressure in the manifold must be 3 at above atmospheric (hydrostatic pressure at that depth is 2.5

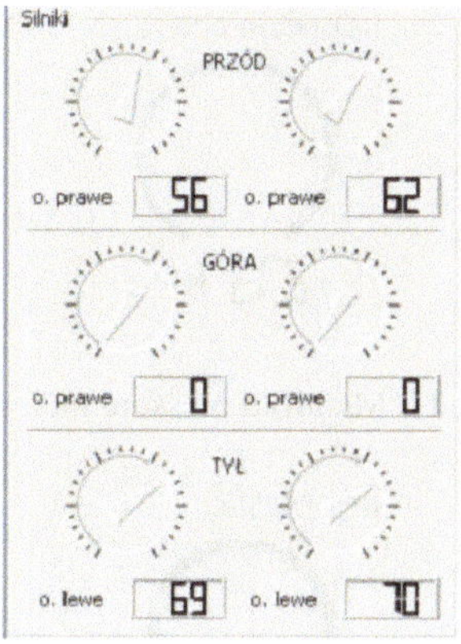

Fig. 9. Indicators of the propellers

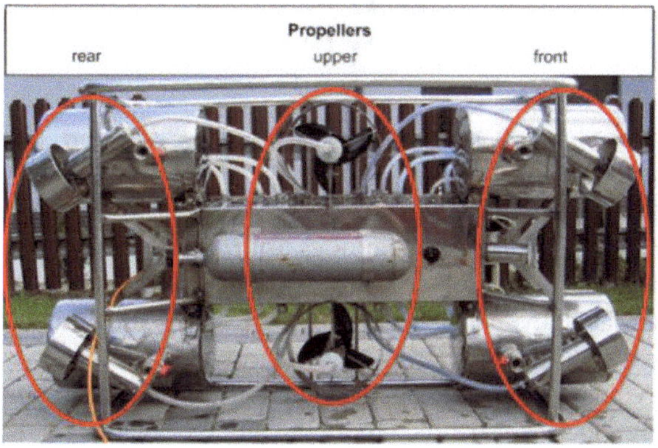

Fig. 10. The location of propulsors

at), so, if necessary, you could push water from ballast tanks being in total immersion. Frame "depth" contains a pointer of the current depth of immersion of the vehicle and the vertical velocity (descent / ascent). It is shown in Figure 12 The values of immersion are positive, because the depth is the vertical distance from surface water to the facility which is under water [8]. This value is always positive. This is shown in Figure 13 Speed of vertical movement of this index is an absolute value, therefore, is positive for both the ascent and descent of the vehicle.

3 Water Level Indicators in the Ballast Tanks

In the frame of "The water level in ZB," there are four indicators of the current level of water in each ballast tank. They are shown in Figure 14. The value of the water level in the reservoir is needed to determine when the tank is empty and when full of water. This is important when filling the tank with water or emptying. During filling, it is

Fig. 11. Manifold pressure indicator

Fig. 12. Submersion indicator and vertical speed

important when the tank is already full, so that water does not get into the tubes inside the hull, while the emptying, so that does not consume unnecessarily push air into the water when the tank is already empty. During the construction of the vehicle, created two types of water level measuring device in the ballast tanks. The first device was designed based on the sensor GP2D120 Sharp. This sensor measures the distance by triangulation. Sends an infrared signal, which after reflection from the object being measured returns to the sensor is read where the angle of incidence. Based on the angle of the return signal is determined by the distance, which is represented on the output voltage [9]. In this case the object from which the infrared beam bounces off a moving float in a tube placed in the tank. This is illustrated in Figure 15. Ultimately, this approach was abandoned because, during testing it was found that water vapor condensing from the tank into the glass sensor strongly distort the measurement. The second version of the measurement of water level in the ballast tanks, is based on the resistive ladder with reed relays. This ladder is part of the voltage divider. Shown in Figure 16 is attached float Neodymium magnet. Moving in the tube, reed switches closest to him, thus bringing to the circuit located in series with the resistor. As a result, the float at different heights in the circuit includes resistors of different resistance, which is associated with the change in the output voltage level of the system. On this basis, is recognized in the reservoir water level. This method is cheap and simple in design. It is currently used in the vehicle.

4 Current Exchange Rate Index

Frame "compass" contains an analog current exchange rate of the vehicle. Below it is a digital display showing the current rate levels. This is illustrated in Figure 17. Course,

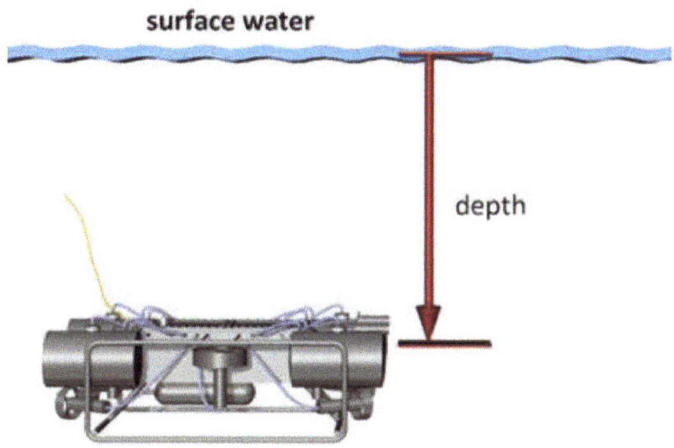

Fig. 13. The value of the depth of immersion

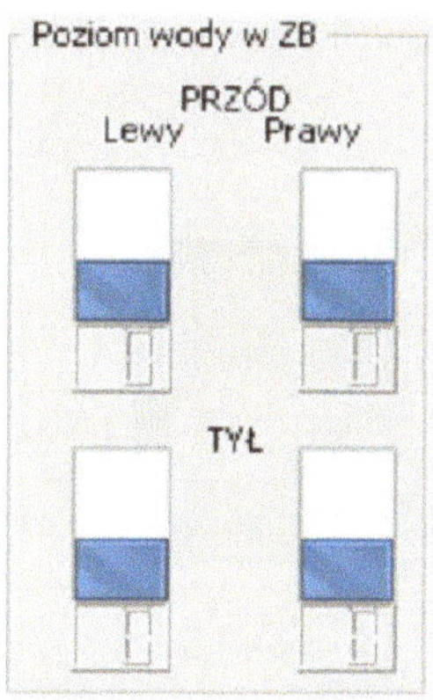

Fig. 14. Water level in the ballast tanks

this is a direction determined by the angle contained between the northern part of the NS line, and the bow (front) part of the diametrically hull [10]. The value of the current exchange rate, in addition to the presentation of the operator, is also used in the module to maintain a specified rate for automatic operation of propellers, so as to maintain the course set by the vehicle operator. Trim is to determine the inclination of the vehicle forward or backward, and its tilt to the right or left. Now trim the vehicle is presented in the "roll and pitch," the operator panel. The Y-axis indicates the slope trim, where positive values mean the inclination of the vehicle in front, while the negative - tilt back. The X-axis indicates the tilt of the vehicle to the side, where positive valuesindicate tilt of the vehicle to the right, while negative to the left. The configuration of these axes is shown in Figure 18. However a sample of the index of the operator panel trim is shown in Figure 19. The value of the current trim, in addition to the presentation of the operator, is used as a stabilizer trim unit, to maintain a constant preset pitch and roll of the vehicle.

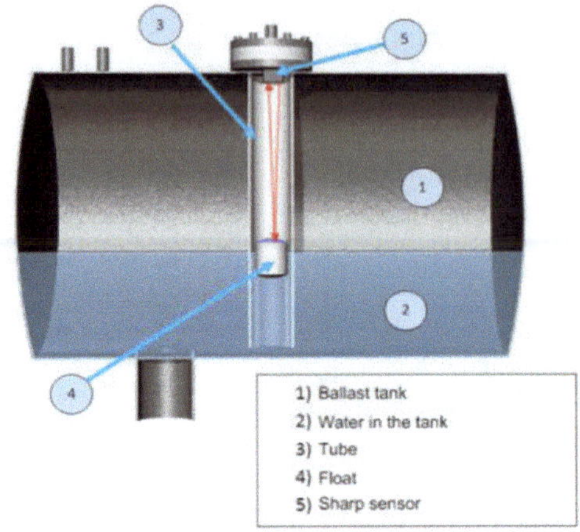

Fig. 15. Measurement of water level sensor Sharp

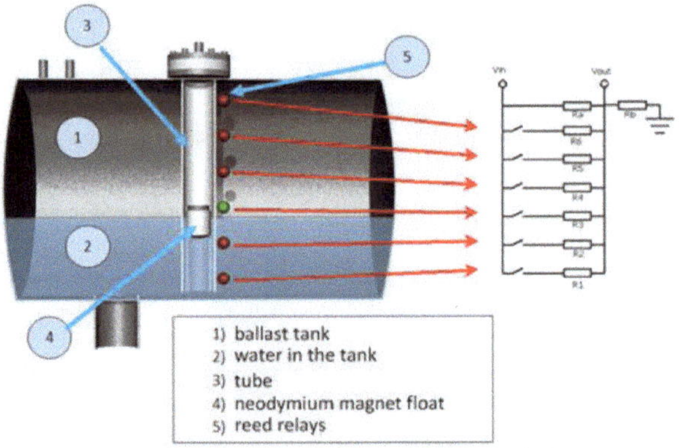

Fig. 16. Measurement of water level reed relay sensor

5 Part of the Operator Control Panel

Some of the graphical control panel allows the operator to manually control the particular vehicle, setting the control parameters and calibration of sensors. Some control is highlighted in Figure 20. It consists of an alarm panel, the panel themed bookmarks, separating the functions of individual departments control panel buttons and pop.

5.1 Auto Control Panel

In the "Auto" is the panel automatically control the propulsors and automatic control panel, solenoid water ballast tanks. They are presented in Figure 21. The automatic control panel propulsors, with buttons placed there, or from a computer keyboard, you can issue orders to the central unit, such as:

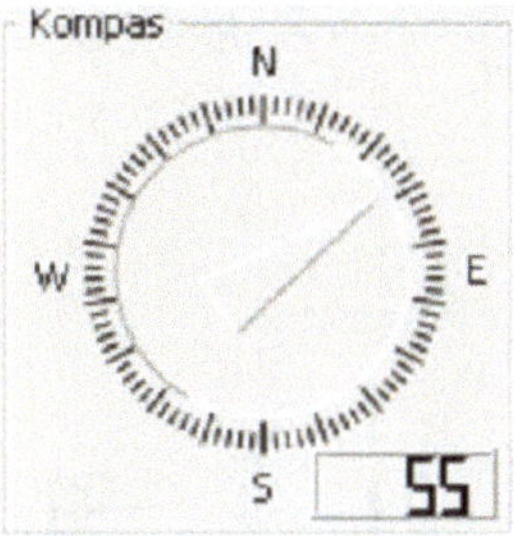

Fig. 17. Current exchange rate index

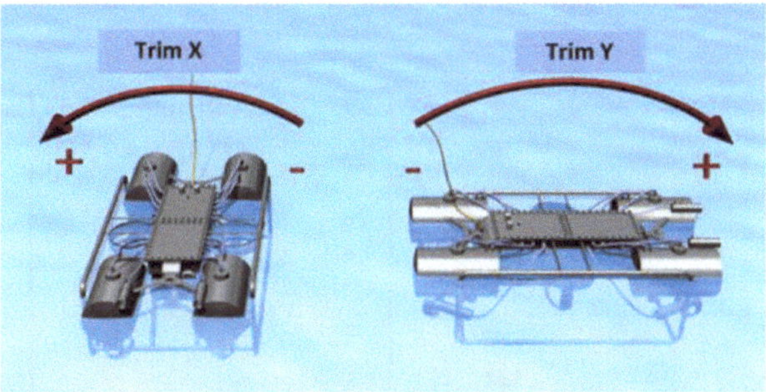

Fig. 18. Trim axle configuration relative to vehicle

- Move forward - backward,
- Move right - left,
- Turn in place at the right - left,
- Turn in a curve to the right - left,
- Work propellers for ascent - descent,
- Rretention of all propellers.

The automatic control panel, ballast solenoid operator is able to fill the ballast tanks of the vehicle with active control of the vehicle leveling and adjustable hysteresis parameters of the controller. The panel automatically control solenoid is also an option emptying ballast tanks by the time 100 ms.

5.2 Manual Control Panel

In the "Manual", shown in Figure 22, there are three panels: "engines", "valves" and "camera". In the panel "Engines" operator can manually set, with which power is to operate each individual propeller, and whether the right or left swing. These options are useful primarily for testing purposes propulsion module [7]. The panel of "taps" is a number of options which allow the operator to turn any of the solenoids located in the vehicle and determining the time of their activation, after which the stop. In this panel, there are also buttons for launching buoys start solenoid, causing the launch of buoys.

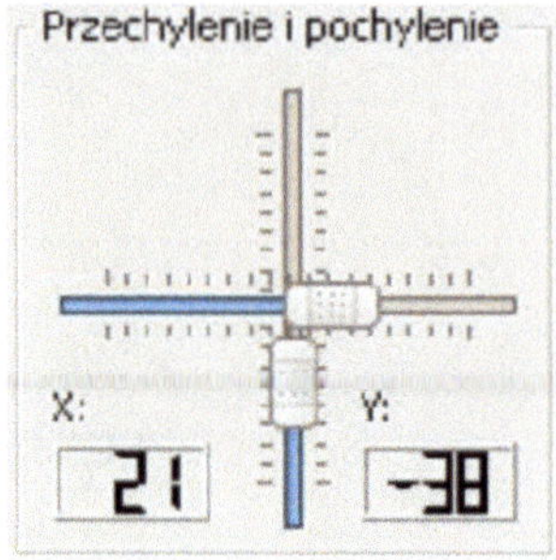

Fig. 19. Trim indicators

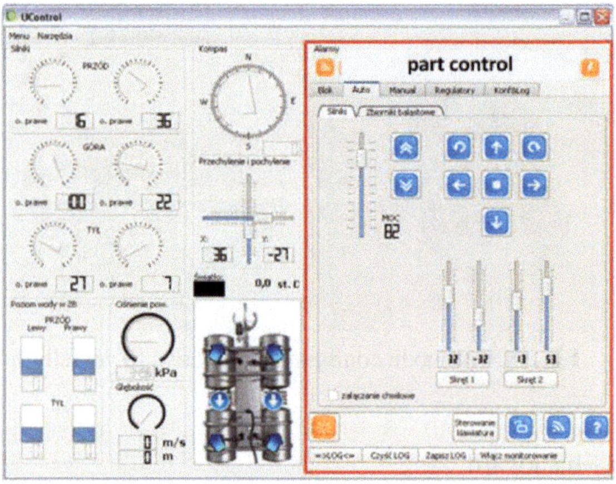

Fig. 20. The operator panel - part of the control

In firing launchers buoys is not given on time solenoids, because time has already been experimentally determined in advance on such that even at low pressure manifold (about 1 at) pistons had time to eject the launcher completely throwing out buoy. Figure 23 is a wiring diagram shown launchers buoys and their cross-section. In the panel "camera" operator has the options to rotate the drive mechanism of light/camera. These options to move "up", "stop" and move "down". In addition, there is no possibility of exceeding the range of rotation, because switches have been applied to prevent the possibility of turning him out of allowed range.

5.3 The Panel Set the Major Regulators

In the "regulators", the operator can start and stop controls to maintain specified descent and baud rate and to define the control parameters for them. View this panel is shown in Figure 24. In the vehicle there are 3 main systems stabilization:

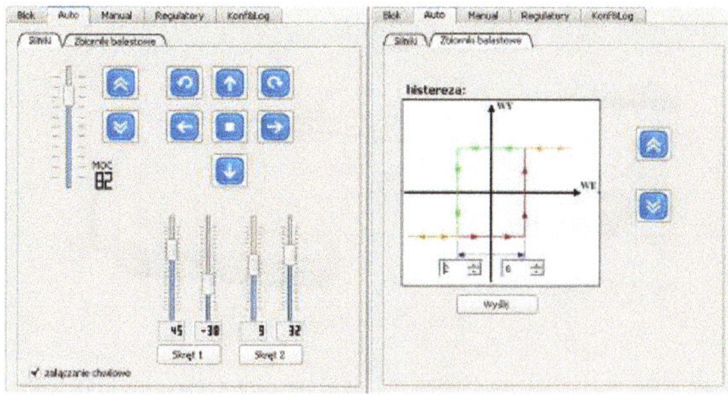

Fig. 21. The operator panel - the panel "Auto"

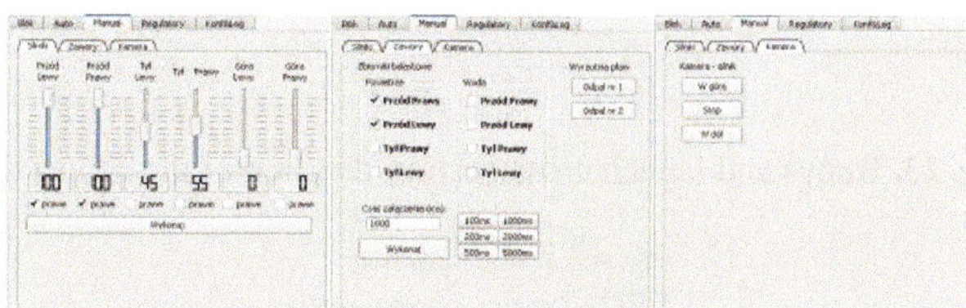

Fig. 22. The operator panel - the panel "Manual"

1. First trim,
2. Second depth of immersion (2 versions),
3. Third the course.

In the "regulators" are available options for setting the control parameters for the other two stabilization systems, the depth of immersion and the course. In the case of stabilization of the depth of immersion, the operator can determine the target depth, along with the allowable deviation. In addition, for the first version of the controller, the operator sets the percentage of engine power, which are used during adjustment, and for the second version, the range of vertical speed. In the case of the regulator to maintain the course, the operator can define the target rate, which is to move the vehicle and the allowable deviation from this size.

6 Module Stability of the Vehicle Trim

This module allows you to trim any setting (tilt and inclination) of the vehicle by the operator and allows for automatic leveling it without operator intervention [11]. Trim is measured in OX and OY axes of the vehicle, where the values of X mean tilt trim (positive to the right, negative to the left), and the slope of the Y value of the vehicle (positive forwards, negative backwards). Vehicle axle configuration is presented in Figure 25, the tilt and inclination example in Figure 26 and 27. This module allows you to include presentations on the trim in the operator panel. Through the vehicle trim module stability is understood:

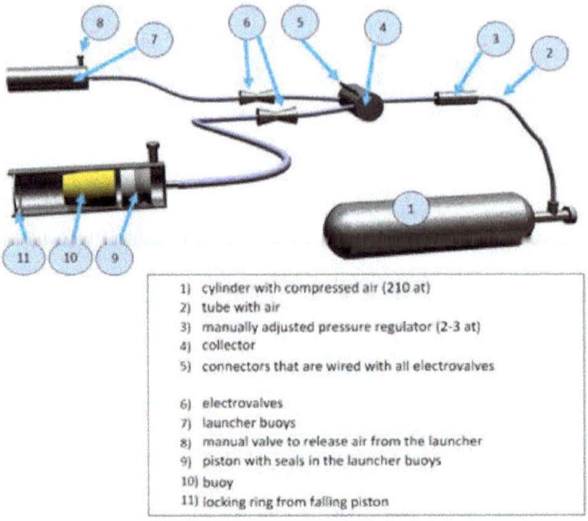

Fig. 23. Buoys and launch cross-section diagram of the connection

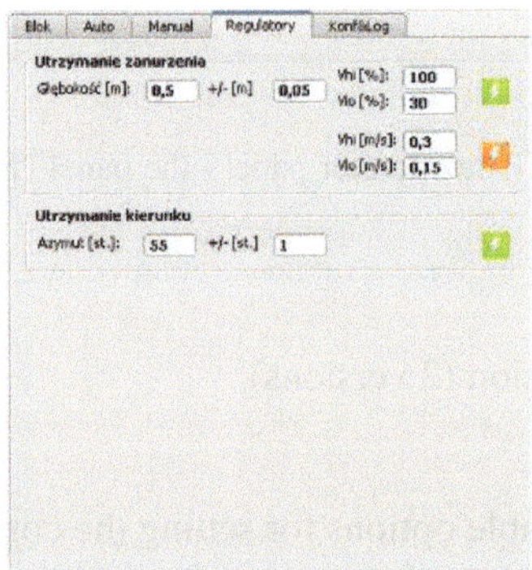

Fig. 24. The operator panel - panel "regulators"

- an electronic system that allows activation of solenoid water ballast tanks,
- implemented the algorithm in this system allows manual control valves with electro-level PC applications,
- implemented the controller automatically controls the vehicle trim solenoid,
- electronic circuit for measuring current in the trim of the vehicle axes OX and OY used as feedback during the control and regulation,
- and the communication interface with the CPU module

Sample image of the vehicle prepared for immersion and persistent trim set is shown in Figure 28. In case of improper balance of the vehicle or the vehicle by raising his arm heavy objects in the water, the vehicle changes its trim. Trim can also be changed when the vehicle will be flooded ballast tanks of water during heavy wave action. Move the vehicle tilting is often undesirable because it impedes the observation and maneuverability. In this case, the regulation of ballast water to compensate for trim is necessary. This module is used to stabilize the trim, which in addition to manual control, currently

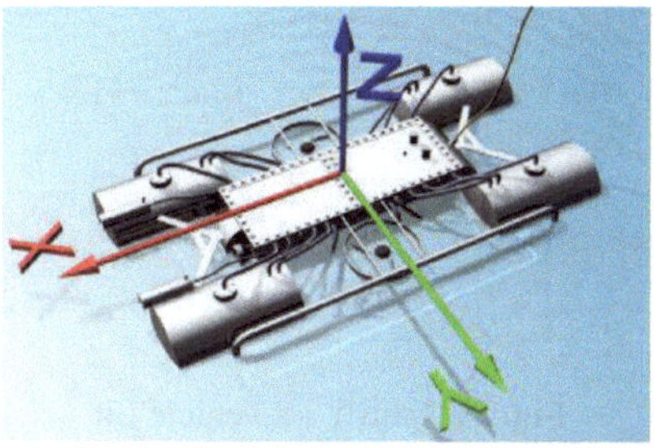

Fig. 25. Configuring the axis of the vehicle

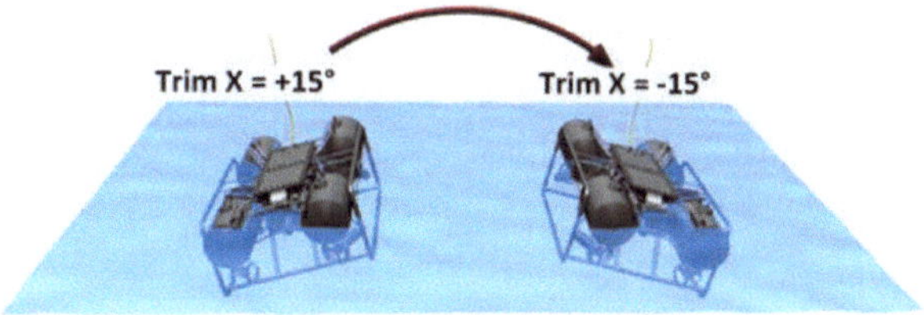

Fig. 26. Trim X - a vehicle tilt

allows automatic adjustment of trim while reducing the buoyancy of the vehicle (for subsequent immersion) [12,13].

6.1 Specification of the Algorithm

The algorithm is implemented in the central unit of the vehicle and is part of the overall system stability. Currently trim stabilization module has two functions: measurement of actual trim of the vehicle with the availability of these data to the operator panel, and automatic adjustment of trim during the flooding ballast tanks in order to soak the vehicle. This algorithm works in the OX and OY axes of the vehicle. The operator defines the position (trim) the target for the vehicle in which it is to sink (it is mostly horizontal position), while the controller based on data from the measuring system controls valves vent trim ballast tanks in order to obtain the target trim during the descent. If ballast tanks are flooded in order to obtain a zero buoyancy of the vehicle to the draft, then vent tanks are open. For various reasons (eg, water waves, uniform load) tanks may fill unevenly. The result is that the vehicle will not be in a horizontal position. However, if the operator turns on the regulator trim, set the target balanced trim control, it is during flooding reservoirs in order to obtain a zero buoyancy of the vehicle, the controller will shut vent these tanks, in which the vehicle is tilted side. So they just flooded the other tanks until the vehicle is improperly skewed. Thanks to trim the vehicle will be approaching the target. In addition to automatic control, the module also allows manual control valves on the tanks. The operator of the panel is free to open and close valves and supply vent at any time, in order to manually set the trim and buoyancy of the vehicle. The module provides the current value of the operator panel trim tilt in the form

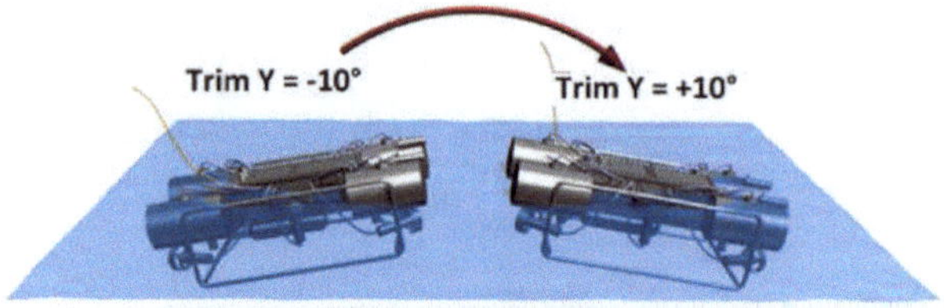

Fig. 27. Trim Y - a vehicle tilt

Fig. 28. ROV - a stable position with flooded reservoirs

of indicators, which facilitates the orientation of the operator in the current position of the vehicle.

6.2 The Measuring System

Now trim the vehicle is measured by the measuring system and passed to trim stabilizer trim module. This system is a DE-ACCM3D, 3-axis accelerometer module, which is based on the ADXL330 sensor is used to measure static and dynamic acceleration. Here is an example of measurement (DE-ACCM3D module) and static acceleration data for the vehicle rotated by an angle of -30°in the y-axis to the horizontal, as shown in Figure 29. Assumptions:

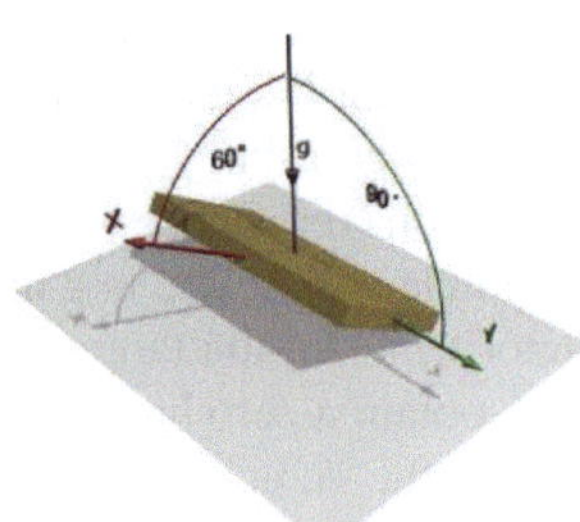

Fig. 29. DE-ACCM3D - an example of measurement

- Accelerometer module power supply voltage of 3.3V,
- Analog to digital sampling the analogue signal has a resolution of 10 bits and the reference voltage of 2.56V.

In order to determine the value of the slope facility, accelerometer module vectors X and Y are projected onto the vector acceleration of gravity g. Let the projected vectors are called the X_R and Y_R, respectively. Using the formula for the projection of the vectors 1:

$$a = g\cos(\alpha) \tag{1}$$

vectors are calculated:

$$X_R = \cos(120°)g = -\frac{1}{2}g \tag{2}$$

$$Y_R = \cos(90°)g = 0g \tag{3}$$

In the case of the power module $V_S = 3.3$V, according to the formula 4

$$V_a = 1.66V + a\frac{0.33V}{1g} \tag{4}$$

at the output of the module X and Y appear appropriate analog signals

$$V_X = 1.66 + X_R\frac{0.33V}{1g} = 1.5V \tag{5}$$

$$V_Y = 1.66 + Y_R\frac{0.33V}{1g} = 1.66V \tag{6}$$

Then both signals are transmitted to the entry of the analogue-digital microcontroller unit. This transducer has a resolution $R_{es} = 210$ and the reference voltage $V_{REF} = 2.56$V. Let the voltage analog signals converted to digital data call, respectively, X_{AC} and Y_{AC}. After sampling analog signals V_X and V_Y, according to the formula 7:

$$d = \frac{a}{V_{REF}}R_{es} \tag{7}$$

digital data output from the transmitter are as follows:

$$X_{AC} = \frac{V_X}{V_{REF}}2^{10} = 600 \tag{8}$$

$$Y_{AC} = \frac{V_Y}{V_{REF}}2^{10} = 663 \tag{9}$$

The values in this form are passed to the stabilization of Trim in the central unit and the controller operates the trim values. However, in the operator panel, the data are converted to degrees in order to improve the presentation of their operator. Let the trim values presented in degrees called X_D and Y_D, respectively. Then using the formula 10:

$$\alpha = \arccos \frac{V_{REF} \frac{d}{R_{es}} - 1.66V}{0.33V} \tag{10}$$

data are presented in the operator panel are calculated as follows:

$$X_D = \arccos(0.4848) = 119^\circ \tag{11}$$

$$Y_D = \arccos(-0.0076) = 90.4^\circ \tag{12}$$

The results obtained are not accurate in relation to the actual values. The inaccuracy is due to the finite resolution A/D converter that samples the analog signals from the outputs of the module DE-ACCM3D. The resulting difference, however, is negligibly small for the application in this case.

6.3 Practical Tests Module

For testing purposes, a vehicle located on the surface of the water, was flooded ballast tanks in order to obtain a zero buoyancy. Flooding took place in tanks with trim control incorporated, whose task was to control flooding of reservoirs in such a way as to keep the vehicle level. Was defined as the level of trim with a value of X=640 Y=676 (unit of these values in the scale corresponds to the values of the output of the transmitter AC). The approximately linear for small roll change this value by 1 in relation to the value adopted as the level corresponds to a change in trim of 0.45° relative to 90°. Hysteresis switch on the controller has a width d = 3 (a = 0, b = 3). The vehicle was artificially tilted in different directions, which forced the response regulator in the form of closing / opening the individual solenoid valves to compensate for the level. The results from the accelerometer and the electro-valves were saved to a file, under which the generated waveforms. Figure 30 and 31 show the 15-second portion of the test. We can see that the response regulator, and thus drive the valve is almost immediate and satisfactory for the purpose of stabilization.

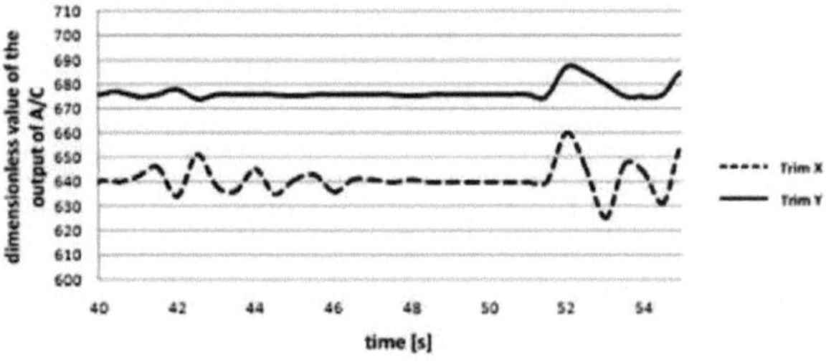

Fig. 30. Stability Test Trim - output A/D converter

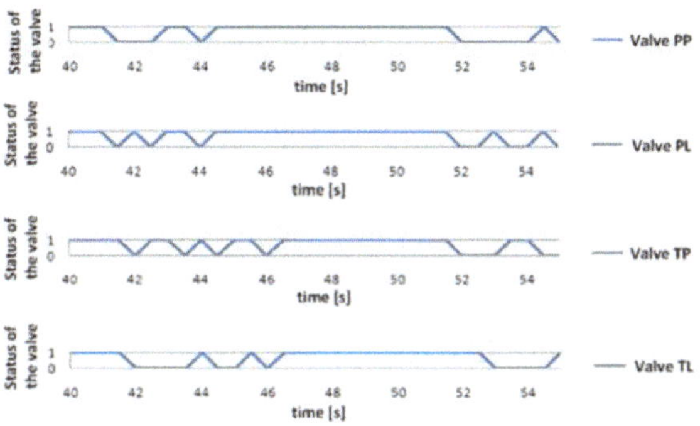

Fig. 31. Trim stability test - switching valves

7 Stabilization Module Immersion Depth

Stabilization of the depth of immersion is one of the most important issues underwater vehicle control [14,15]. The vehicle must be able to maintain the desired depth by the operator with high precision [16]. Otherwise, perform underwater work or even the observation of the underwater environment would be very difficult. Discussed in this paper prototype was equipped with a module to maintain desired depth. It is the most powerful module in the vehicle stability. Below is a general classification of its individual components:

- Control algorithm implemented in the vertical motors in microcontroller unit,
- an algorithm for calculating the depth of immersion on the basis of hydrostatic pressure implemented in microcontroller unit,
- calibration algorithm implemented in the microcontroller unit,
- depth of immersion controller implemented in microcontroller unit,
- the second controller with a depth adjustable speed descent / ascent implemented in the microcontroller unit,
- hydrostatic pressure measurement module.

Sample pictures during the descent of the vehicle are shown in Figure 32. The depth of immersion underwater vehicle can be measured in several ways. If the vehicle is selected method, which consists in measuring the total pressure of water at depth

Fig. 32. ROV - immersion of the vehicle

(hydrostatic pressure + atmospheric), and deducting it on previously measured atmospheric pressure at the surface. Knowing the formula for the hydrostatic pressure of the liquid [17]:

$$\rho_{hydr} = \rho g h \tag{13}$$

where: ρ - density of the liquid $[kg/m^2]$, g - acceleration due to gravity $[m/s^2]$, h - depth of immersion in the liquid $[m]$ can calculate the pressure increase when changing the water column depth of immersion of 10m:

$$\rho_{10m} = 1000\frac{kg}{m^2}9.806\frac{m}{s^2}10m = 98066.5Pa = 1at \tag{14}$$

The calculations show that for every 10 m depth the hydrostatic pressure increases by one atmosphere of support (1 at). Thus, by measuring the atmospheric pressure at the surface and the total pressure at the depth, depth can be calculated using the formula:

$$H_{immersion} = \frac{\rho_{total} - \rho_{atm}}{98066.5Pa}[m] \tag{15}$$

In the case of this vehicle in this way is calculated the depth of immersion. When the vehicle is emerged, once read current external pressure as reference pressure (pitch = 0 m). Then, in the submerged vehicle external pressure is measured periodically and comparing them with the reference pressure determines the depth of immersion.

7.1 Specification of Control Algorithm 1

The module is equipped with two controllers, the first, which maintains the desired depth and direction of the controlling power of engine, the second, which maintains the desired depth controlling the speed of descent/ascent. This algorithm implements a simple five-position controller whose schema is shown in Figure 33. In order to minimize unnecessary calculations at the level of the microcontroller, it tries to maintain a constant external pressure without target depth calculation. The operator defines the depth of immersion, which is automatically converted to the total external pressure, then the corresponding unit in the scale values of the output of A/D converter and the controller works of this size. The controller receives the feedback data directly from the output of A/D converter of the measuring system and the set value is defined in this scale. However, in the operator panel pressure value is converted to depth in meters and in this form is presented. The operator at the control mode also defines the values of engine

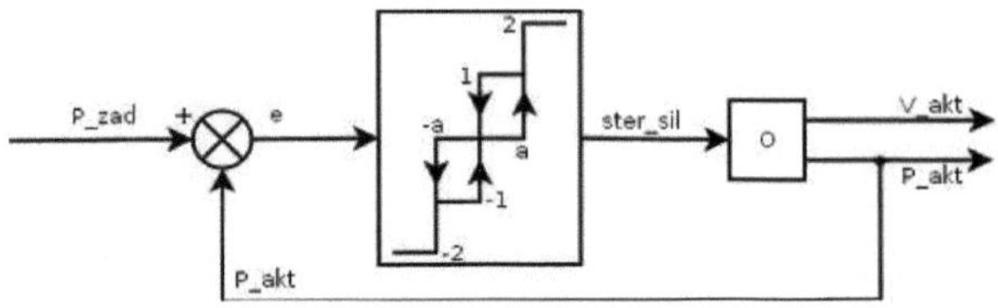

Fig. 33. Diagram of the controller to maintain a desired depth

power for descent/ascent, if the vehicle is far from the desired depth (beyond the scope of hysteresis) and less powerful engines for descent/ascent, when the vehicle is close to the desired depth.

7.2 Specification of Control Algorithm 2

This control, as previous described, operates directly on the values of the output from the A/C, corresponding to the values of external pressure. The main role of the regulator is, like its predecessor, maintaining desired depth of immersion. However, its function is the extra speed descent/ascent. Furthermore, in contrast to the previous controller, after reaching the target depth of the vertical speed of the vehicle decelerates to zero by controlling motors, respectively, and not simply disable the motors. Diagram of this controller is shown in Figure 34.

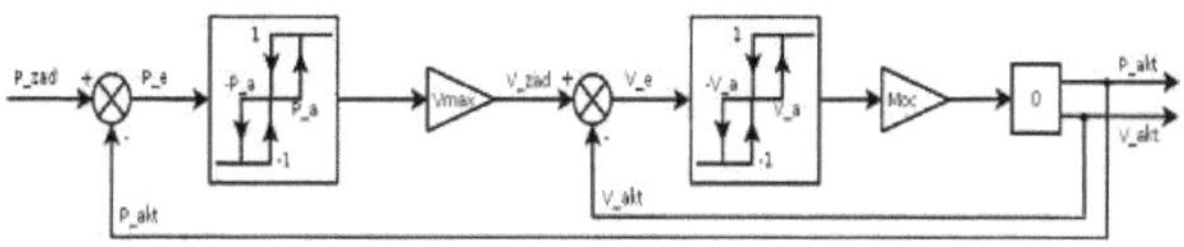

Fig. 34. Scheme controller 2 to maintain desired depth

7.3 Maintaining the Desired Depth – The Practical Test Control Algorithm

Subsequent tests were conducted on the basis of maintaining desired depth using a control algorithm first To measure the depth of immersion was used, mounted in the vehicle, the module-based pressure sensor MPX4250GP. Two measurements were carried out with the following control settings: Test 1:

- setpoint H_{set} depth = 1.5m,
- permissible deviation from the desired depth of immersion ,
- the power of the first propellers power1 = 20%, —item second power propulsorspower2 = 40%.

The vehicle was submerged by the regulator to maintain a desired depth. Depth reference is H_{set} = 1.5m accuracy . Engine power set by the regulator power2 = 40%, power1 = 20% Figure 35 shows the mileage of the vehicle at the time of immersion, while Figure 36 how an individual values needed for calculations. The measurement data obtained were calculated the average velocity of descent (to a depth of 1.4 m):

$$V_{des} = \frac{dH_Z}{dT_Z} = \frac{1.4m}{9.3s} = 0.15\frac{m}{s} \tag{16}$$

average speed of ascent of the vehicle (from a depth of 1.4 m to 0.9 m):

$$V_{asc} = \frac{dH_W}{dT_W} = \frac{0.5m}{4.8s} = 0.1\frac{m}{s} \tag{17}$$

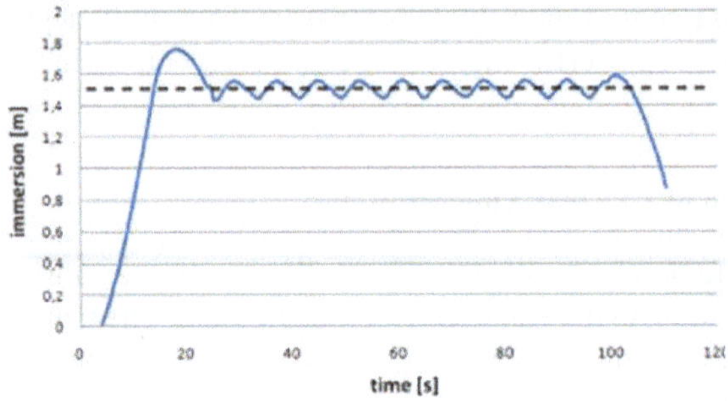

Fig. 35. Mileage Test No. 1 to maintain the depth of immersion

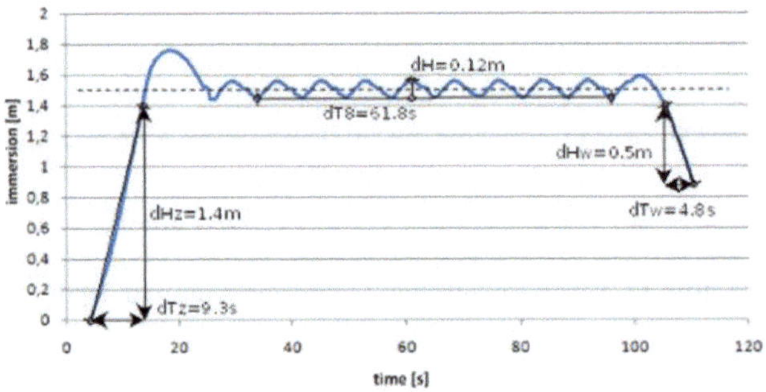

Fig. 36. Mileage Test No. 1 to maintain the depth of immersion - the calculations

and the duration and frequency of oscillation of the depth of immersion around the setpoint:

$$T_{osc} = \frac{dT8}{8} = \frac{61.8s}{8} = 7.72s \tag{18}$$

$$f_{osc} = \frac{1}{T_{osc}} = \frac{1}{7.72s} = 0.129s^{-1} \tag{19}$$

and amplitude (peak to) these oscillations:

$$A_{pp} = dH = 0.12m \tag{20}$$

Test 2: The vehicle was submerged by the regulator to maintain a desired depth. Depth reference is H_{set} = 1.5m. The other settings were changed during the test, initially (0 s) is set:

- permissible deviation from the desired depth of immersion $\Delta H1 = 0.05m$,
- the power of the first propellers power11 = 20%,
- second power propulsors power21 = 30%,

then in the 63s test set:

- permissible deviation from the desired depth of immersion $\Delta H2 = 0.02m$,
- the power of the first propellers power12 = 20%,
- second power propulsors power22 = 30%,

then in the 93s test set:

- permissible deviation from the desired depth of immersion $\Delta H3 = 0.01m$,
- the power of the first propellers power13 = 10%,
- second power propulsors power23 = power13,

Figure 37 shows the mileage of the vehicle at the time of immersion. Figure 38 shows how the values were determined for various stages of the calculations for this test.

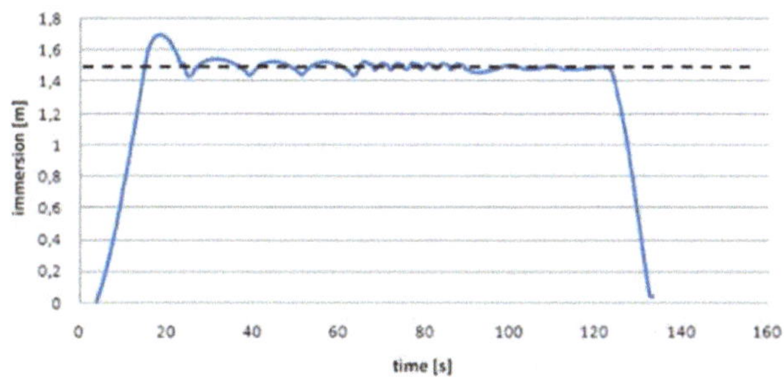

Fig. 37. Mileage Test No. 2 to maintain the depth of immersion

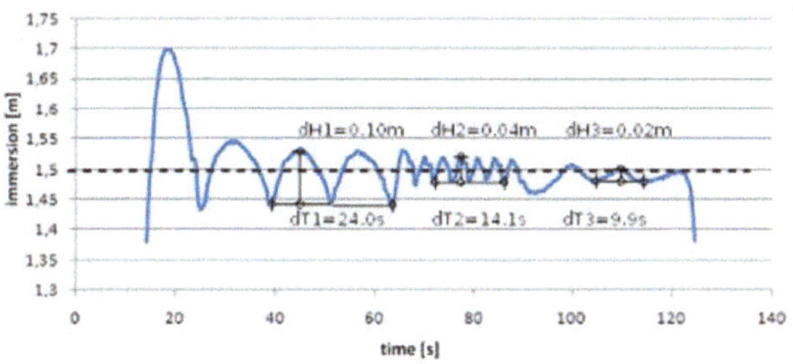

Fig. 38. Mileage Test No. 2 to maintain the depth of immersion - the calculations

From these measured data and calculated period of oscillation frequency of the depth of immersion around the setpoint:

$$T_{osc1} = \frac{dT1}{2} = \frac{24s}{2} = 12s \tag{21}$$

$$f_{osc1} = \frac{1}{T_{osc1}} = \frac{1}{12s} = 0.083s^{-1} \tag{22}$$

$$T_{osc2} = \frac{dT2}{4} = \frac{14.1s}{4} = 3.52s \tag{23}$$

$$f_{osc2} = \frac{1}{T_{osc2}} = \frac{1}{3.52s} = 0.284s^{-1} \tag{24}$$

$$T_{osc3} = dT3 = 9.9s \tag{25}$$

$$f_{osc3} = \frac{1}{T_{osc3}} = \frac{1}{9.9s} = 0.101s^{-1} \tag{26}$$

and amplitude (peak to) these oscillations:

$$A_{pp1} = dH_1 = 0.10m \tag{27}$$

$$A_{pp2} = dH_2 = 0.04m \tag{28}$$

$$A_{pp3} = dH_3 = 0.02m \tag{29}$$

Throughout the duration of the test, set the oscillations do not exceed the established range. Adopted a target range of the depth of immersion is very small, thus minimizing the engine power to 10% allowed for the maintenance of oscillations in this regard. In addition, the period of these oscillations is close to 10 seconds, so the traffic is very smooth and barely noticeable in the whole range of oscillations (up to 2 cm). At this stage, the test clearly shows a non-zero buoyancy of the vehicle. Regulator vehicle plunged into the depth of the target, then turn off the engines. Since the speed of descent was very small (10% engine power), the vehicle stopped almost perfectly on the set depth. However, after excluding the positive buoyancy of the vehicle engine gave it a speed which resulted in the ascent of the vehicle until it exceeded the scope of hysteresis. Then re-enable the propellers caused the re-immersion to a depth reference. In all tests, the depth of immersion undying oscillations around the setpoint value is the result of simple adjustment five-position switch. Of course the ideal solution would also be an ideal vehicle to achieve zero static buoyancy through mutual balance of these forces, but in practice, achieving the ideal of zero static buoyancy is impracticable. This module still needs improvements, but at the present moment allows you to maintain a high degree of accuracy at any immersion depth not exceeding 5m. Figure 39 is shown a sample picture of the vehicle sustained permanent immersion 0.3 m.

8 Module to Maintain a Specified Rate

When the vehicle is under water, the operator is not possible to discern the direction in which the vehicle is directed or visually monitor where the fluid. The ideal solution would be to install in a vehicle GPS unit, but radio waves under water is quickly suppressed, so the GPS antenna would have to be on the water surface. Water does not suppress the magnetic field, so in order to facilitate orientation of the vehicle control,

Fig. 39. ROV - to maintain a depth of 0.3 m

was installed in the electronic compass module, used to measure the course. Module to maintain a specified rate, is based on the communication module is the compass. Module to maintain a specified rate to the operator panel provides the current value of the course given in degrees, and enables the controller to maintain the course. The operator defines the target rate and accuracy, and the controller adjusts the propellers work to maintain a specified rate. Figure 40 shows a schematic diagram of the controller. Controller algorithm is implemented to maintain the course of the microcontroller unit

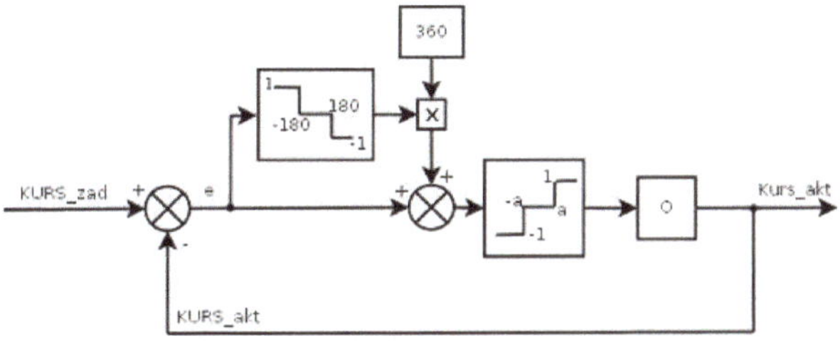

Fig. 40. Diagram of the controller to maintain a specified rate

and is activated and set in the operator panel. This is a simple controller based primarily on three-position relay without hysteresis. Because most courses are counted in the system of 0-360°(to the right, according to a clockwise direction) [25], in the later work, the course will be expressed as a measure of the angle 0 - 360°, where 0°is north direction, and angle values are rising in a clockwise direction. Azimuth, as in the case of course, will be called the angle expressed as a measure of 0 - 360°, where values are rising in a clockwise direction, with the difference that will apply to a given (arbitrary) horizontal direction, and not only the direction indicated by the drastic hull. As a result of this algorithm, the vehicle moves straight ahead, when the slope of the course, one of the front propellers (adequate to the situation) has reduced power, resulting in the accident vehicle torsional torque. The vehicle will tend to the destination course. After receiving the target rate of horizontal movement of all thrusters are set to the same labor force and the vehicle moves straight ahead at the designated course. Current measuring system is a module of the course CMPS03.

9 Summary

Ensuring a satisfactory stability of the vehicle is a complicated issue. First of all requires accurate and rapid measurement devices, control algorithms which ensure appropriate feedback in the toughest conditions. Stabilization systems typically need to be adapted to the vehicle structure. Currently used very precise electronic compass turned out to be inadequate when applied to the vehicle due to measurement error in his roll. These errors involve a deterioration of stability to maintain the course, because measured values for the feedback control algorithm. So the next step will be testing other solutions to measure the rate of the vehicle, the Earth's magnetic field sensors less susceptible to tipping, or even such complex and high-tech devices like Inertial Measurement Unit [18]. This will involve the construction of a new module to maintain a specified rate and implementation of new algorithms based on the data sensors. In the case of stabilization of the vehicle's position is very important to selecting the appropriate control algorithm. Current algorithms used to maintain the required depth shall bring the vehicle in the undying, small oscillations in a vertical position. This is due to too simple adjustment. Used to stabilize the vehicle roll while filling the ballast tanks work very well, but the whole system stability trim requires expansion to stabilize the trim work at any time. During tests in the water tank in a strong wind observed the vehicle drift and its torsion. This is due to water waves interacting with the vehicle. ROV must be able to work under such conditions, it will be necessary to create a lasting stabilization of the vehicle unit in a fixed position, even in strong waving of water in the tank water. This will require the use of very sensitive sensors for the detection of even small, unexpected changes in the position of the vehicle and the use of algorithms that will control the vehicle's propulsors in such a way that he returned immediately to the previous position of maintaining a given trim, course and immersion. With time, the construction may be imperfect and to improve vehicle stability will need to rebuild it. In the future, after fine tuning the system for stabilizing the vehicle and provide him sufficient stability in the shallow waters, you need to dive deeper into it. You will be a series of new problems associated with a higher water pressure and leaking structures. Upgrading and extension of the stabilization system of the vehicle requires, therefore, still many months of work and testing.

References

1. Dyrcz, C., Grabiec, D., Olejnik, A.: Identyfikacja wraku "Engels" - przykladem wykorzystania mozliwosci wspolczesnego hydrograficznego sprzetu pomiarowego do badania wrakow i zanieczyszczen na morzu, Konferencja Morska "Aspekty bezpieczenstwa nawodnego i podwodnego oraz lotow nad morzem", Gdynia, pp. 51–56 (2004)
2. Christ, R.: ROV Manual. Butterworth Heinemann, Oxford (2007)
3. Refsnes, J.E., Sorensen, A.J.: Design of control system of torpedo shaped ROV with experimental results. In: MTTS/IEEE TECHNO-OCEAN, OCEANS 2004, vol. 1, pp. 264–270 (2004)

4. Ming-Chung, F., Yu-Lun, H.: The simulation of the ROV motion with anti-pitch control in uniform current. In: Symposium on Underwater Technology and Workshop on Scientific Use of Submarine Cables and Related Technologies, pp. 120–125 (2007)
5. Kinsey, J.C., Whitcomb, L.L.: In Situ Alignment Calibration of Attitude and Doppler Sensors for Precision Underwater Vehicle Navigation: Theory and Experiment. IEEE Journal of Oceanic Engineering 32(2), 286–299 (2007)
6. Dai, J., Zhao, X., Tan, M.: Fuzzy logic control in autonomous ROV navigation. In: IEEE Region 10 Conference on Computers, Communications, Control and Power Engineering, TENCON 2002, vol. 3, pp. 1566–1569 (2002)
7. Sciegienka, P.: Ukad sterowania bezzaogowego pojazdu podwodnego, praca magisterska, Gliwice (2009)
8. Sakagami, N., Kanayama, T., Ueda, T., Hashizume, H., Shibata, M., Onishi, H., Murakami, S., Kawamura, S.: Design and development of an attitude control system for a human-sized ROV. In: 11th International Conference on Control Automation Robotics and Vision (ICARCV), pp. 2141–2146 (2010)
9. Jodkowski, B., Sydor, K.: Wstep do uzytkowania modeli GP2D12 iGP2Y0A02, Kolo Naukowe Robotykow, Wrocaw (2006)
10. Czajewski, J.: Nawigacja zeglarska, wydanie 2; wyd. WK, Warszawa (1986)
11. Heping, L., Zhenbang, G., Min, L.: Sliding mode control of ROV based on RBF neural networks adaptive learning. In: 3rd International Conference on Intelligent System and Knowledge Engineering, pp. 590–594 (2008)
12. Savaresi, S.M., Previdi, F., Dester, A., Bittanti, S., Ruggeri, A.: Modeling, identification, and analysis of limit-cycling pitch and heave dynamics in an ROV. IEEE Journal of Oceanic Engineering 29(2), 407–417 (2004)
13. Inoue, T., Suzuki, H., Shimamura, T., Nakajima, K., Shioji, G.: Experimental research on horizontal rotation of ROV induced by external forces near sea surface. OCEANS, 1–6 (2008)
14. Heping, L., Zhenbang, G.: Disturbance Fuzzy Approach-based Sliding Mode Control on the Working Attitude Adjusting Device of ROV. In: International Asia Conference on Informatics in Control, Automation and Robotics, pp. 235–239 (2009)
15. HarsamizadehTehrani, N., Heidari, M., Zakeri, Y., Ghaisari, J.: Development, depth control and stability analysis of an underwater Remotely Operated Vehicle (ROV). In: 8th IEEE International Conference on Control and Automation, pp. 814–819 (2010)
16. Massimo, C.: Laser-Triangulation Optical-Correlation Sensor for ROV Slow Motion Estimation. IEEE Journal of Oceanic Engineering 31(3), 711–727 (2006)
17. Massalski, J., Massalska, M.: Fizyka dla inzynierów cz 1. WNT, Warszawa (2006)
18. Jaskot, K., Babiarz, A.: Uklad inercyjny do pomiaru orientacji obiektow. Przeglad Elektrotechniczny 86(11a), 323–333 (2010)
19. Jedrasiak, K., Daniec, K., Nawrat, A.: The Low Cost Micro Inertial Measurement Unit. In: Industrial Electronics and Applications (ICIEA), pp. 403–408 (2013)
20. Bereska, D., Jędrasiak, K., Daniec, K., Nawrat, A.: Gyro-Stabilized Platform for Multispectral Image Acquisition. In: Nawrat, A., Kuś, Z. (eds.) Vision Based Systems for UAV Applications. SCI, vol. 481, pp. 115–121. Springer, Heidelberg (2013)
21. Daniec, K., Iwaneczko, P., Jędrasiak, K., Nawrat, A.: Prototyping the Autonomous Flight Algorithms Using the Prepar3D Simulator. In: Nawrat, A., Kuś, Z. (eds.) Vision Based Systems for UAV Applications. SCI, vol. 481, pp. 219–232. Springer, Heidelberg (2013)
22. Babiarz, A., Bieda, R., Jędrasiak, K., Nawrat, A.: Machine Vision in Autonomous Systems of Detection and Location of Objects in Digital Images. In: Nawrat, A., Kuś, Z. (eds.) Vision Based Systems for UAV Applications. SCI, vol. 481, pp. 3–25. Springer, Heidelberg (2013)

23. Bereska, D., Daniec, K., Fraś, S., Jędrasiak, K., Malinowski, M., Nawrat, A.: System for Multi-Axial Mechanical Stabilization of Digital Camera, Vision Based Systems for UAV Applications. In: Nawrat, A., Kuś, Z. (eds.) Vision Based Systems for UAV Applications. SCI, vol. 481, pp. 177–189. Springer, Heidelberg (2013)
24. http://www.seaeye.com/products.html
25. http://www.ifremer.fr/fleet/systemessm/engins/victor.html
26. http://www.whoi.edu

Part II
Sensoric and Vision Based Information for UGVs

Many research groups works on design and implementation of control systems for unmanned vehicles. One of the most challenging tasks is to design robust control system that allows human operator stay away from operation zone which can be very dangerous. There are few approaches to this problem. One common concept of those approaches is structure of unmanned system. In all described in details in this chapter approaches there are two main elements: ground control station and vehicle control system. Those two element communicate to each other via link. In most cases it is wireless link. In underwater systems communication is performed by wired connection.

First proposition of control system is remote control. Human operator is away from vehicle and has full control on the vehicle. Operator controls signals which are send directly to vehicle actuators. Control signals are transferred from ground station to vehicle by wireless link. System on board of unmanned vehicles set control values on actuators. This approach is to replace the human factor in danger zone with unmanned robotized vehicles One of the most important tasks is to design control system is presented, which allows remote manual control of a number of robots by means of chosen input devices. Main focus is placed on the problem of user friendly control, reducing operator workload and computer controlled intervention increasing safety of subject vehicle.

In teleportation approach human operator has visual feedback from cameras mounted on vehicle. Operator don't need direct visual contact with vehicle and he is limited only by communication link power. This approach guarantees security for human operator.

Final approach is to implement autonomy functions on vehicle which is able to do tasks without human intervention. That option requires full sensor system with control processor unit on the vehicle. It's also required advanced data processing and control algorithms. Role of human operator is limited to sets mission goals and observe how vehicle achieving them.

One of the most important component of the unmanned vehicles are control base stations. They are necessary for all of mentioned approaches. Most of them uses rugged computers but some are designed directly for selected type of unmanned vehicle. Hardware of base control station should have two parts. First allows for basic control of vehicle and should be safe, redundant and robust. Second part can have advanced futures. In most cases its computer program with modules such as map module, sensor visualization module.

This chapter includes a number of important challenges in the fields mentioned above. At the same time valuable suggestions and conclusions from authors are presented and discussed in detail.

Modern unmanned vehicles are equipped in more and more advanced vision systems. Vision systems has day light cameras, thermo vision cameras, night vision cameras and other spectral cameras. To achieve autonomy for unmanned objects many research groups works on control algorithm based on vision sensors. From cameras we can get huge amount of data. From those data it is possible to get very complex information about surrounding environment of the unmanned vehicle. This is the reason for continue research on vision based control. Properly designed control algorithm based on vision data will be the fastest way to achieve autonomy on unmanned vehicles.

One of the approaches described in this chapter is control of camera head system applied to object tracking. The study assumed that in order to track objects moving on the ground, the UAV with the mounted camera would be used herein. During the flight of the UAV, there may appear various disturbances such as rapid rotations and shifts of the UAV. Furthermore, we assumed that the object was moving on the flat ground.

Another important problem is significance of the velocity limitations in the camera head control system. A prerequisite for the control system was that it should guarantee the tracking of an object moving on the ground, even in the case of disturbances such as the rapid rotations and shifts of the UAV. Equipping the UAV with the flight trajectory control system allowed to model the rotations and shifts of the UAV in the particular axes as independent control channels. What is more, the study examined behavior of the control system for different values of the disturbances.

Another problem we can met in unmanned vehicle camera systems is camera calibration. This is very important issue because even best algorithm will not be working when cameras would be performed improper. Because of that all camera systems should be calibrated to get a proper video data.

Also important issue is to store gathered video data. The activity of video systems is concerned with the need of large amounts of LOB data types storage. In order to satisfy efficient access to such data it's necessary to develop adequate mechanisms of data storage and processing. Because of the fact that traditional relational systems may be inefficient, the possibilities of using NoSQL database that may speed up the video content management are described in the paper.

Another important problem in transmission of video stream. one of the popular wireless technologies, use to transfer different types of data, is the IEEE 802.11 standard. This popular "WiFi" technology provides low-cost wireless Internet capability for end users and also is used in modern measurement and control distributed systems. Wireless medium is a shared medium, so as more and more devices demand the bandwidth, concentrating on performance becomes critical.

In this chapter authors showed research performed in area of using vision information for control of unmanned vehicles.

Camera Head Control System with a Changeable Gain in a Proportional Regulator for Object Tracking

Zygmunt Kuś and Aleksander M. Nawrat

Silesian University of Technology, Institute of Automatic Control,
Akademicka 16 St., Gliwice, Poland
{zygmunt.kus,aleksander.nawrat}@polsl.pl

Abstract. The goal of the paper was to examine the changeable gain in the camera head control system which is applied to object tracking. The study assumed that in order to track objects moving on the ground, the UAV with the mounted camera would be used herein. During the flight of the UAV, there may appear various disturbances such as rapid rotations and shifts of the UAV. Furthermore, we assumed that the object was moving on the flat ground. The UAV's velocity was far greater than the maximum velocity of the tracked object. Due to equipping the UAV with flight trajectory control system, it allowed to model the rotations and shifts of the UAV in the particular axes as independent control channels. The behaviour of the control system was also examined for the different values and types of disturbances. Developing the function which defined the relationship between the error signal and gain in the control system let us track the object appropriately. Conclusively, the results indicated on the fact that using changable gains in the camera head control system, allowed to track the object for both, the small and major disturbances. Owing to the changeable gain, it was possible to eliminate rapid reactions of the control system during major disturbances. As a consequence, there was not any overshoot - the object was being tracked even during high values of the disturbances. Simultaneously, it was noticed that the disturbances which shift the UAV in the (x, y) plane, had less impact on the appriopiate object tracking than the disturbances which rotated the UAV in the horizontal or vertical planes.

Keywords: Unmanned Aerial Vehicles, Object Tracking, Control System, Image Processing Methods, Changeable Gain.

1 Introduction

Tracking objects which move on the ground, can be done by means of a UAV with a mounted camera. There are two main tasks for the UAV's control system and camera head control system. The first one has to guarantee short distance between the UAV and tracked object whereas the second one has to guarantee proper turn of the camera lens towards the object. Therefore the pattern recognition and image processing methods [1,2,3,4,5,6,7] can be used only if the object is within the camera field of view.

The aforementioned control systems are utilised to make the object visible in the camera field of view. However, this task can be hindered by the disturbances. We assume

A.M. Nawrat (ed.), *Innovative Control Systems for Tracked Vehicle Platforms*,
Studies in Systems, Decision and Control 2,
DOI: 10.1007/978-3-319-04624-2_6, © Springer International Publishing Switzerland 2015

that the UAV's velocity is greater than the maximum velocity of the object. Owing to this fact, the disturbances may be the cause of losing the object from the picture.

Consequently, the image processing methods [14] - [23] do not help us to follow the object.

The role of the control system is to counteract rapid disturbances. These disturbances are the changes of the location and orientation of the UAV caused by the rapid gust of wind. Owing to the fact that the dynamics of the camera head is faster than the dynamics of the whole UAV, counteracting disturbances lies in compensating the change of the UAV's location and orientation by means of turning the camera head. However, it is not necessary for the UAV to restore UAV's orientation and location to the previous values. For the stationary or slow-moving object, such an approach guarantees tracking without losing the object from the camera field of view. In this way, we obtain the control system resistant to disturbances.

According to [25] and [26], it was assumed that the P regulator, alternatively with the PD compensator, could compensate rapid and considerable disturbances. The results proved that the necessary angular velocities of the camera head are lower than the feasible to achieve maximum angular velocities [27], [28]. However, one may ask whether it is possible to use the margin of the angular velocity to compensate far greater or more rapid disturbances.

2 Modelling of the Dynamics of the UAV Equipped with the Flight Trajectory Control System

We can distinguish various types of UAVs. One of the examples of a UAV is a helicopter. Certain models of UAVs (helicopters) were presented in [8,9,10,11,12,13]. Each of these UAV's types requires a different flight control system which guarantees following the set flight trajectory. For various types of UAVs, it is possible to define the state of a UAV by means of providing the location in the coordinate system which is connected with the ground (x_h, y_h, z_h), the azimuth angle β_h and the angle α_h which describes the deviation in the vertical plane. The aforementioned angles and location are presented in the figures 7, 8 and 9.

This study assumes that a UAV is equipped with the control sytem of the helicopter trajectory which is adjusted to a given device. Thus, in the UAV's model we take into account only the influence of the disturbances and the influence of the set value on the UAV. The rapid gusts of wind can be treated as disturbances. Therefore this type of disturbances causes the rapid changes of UAV's location and orientation. The basic property of this model is a quick response of the UAV to the disturbance and slow return of the UAV's state to the proper value. The model which is discussed in the paper has these dynamic properties.

A UAV is a multivariable controlled plant; however, due to the implemented control system, we may considerably simplify the model. For the examined model of the plant, the output signals are location (x_h, y_h, z_h) and the UAV's rotation angles (α_h, β_h). During the control system modelling we use the decoupling property. It means that the input signal for a given channel does not influenc the other outputs . This results in simplifying the model. Therefore we will present the modelling for a single control channel. We will use the same type of the model for each control channel.

As a model which complies with the above-mentioned requirements, we propose a model which consists of first order lag, intergrator and feedback loop and with the disturbance in the plant's input. The bock scheme of this model is presented in the figure 1. The methods presented in [8,9,10,11,12,13], taking into account the trajectory control system, let us conclude that such a model guarantees proper representation of the UAV's dynamic properties.

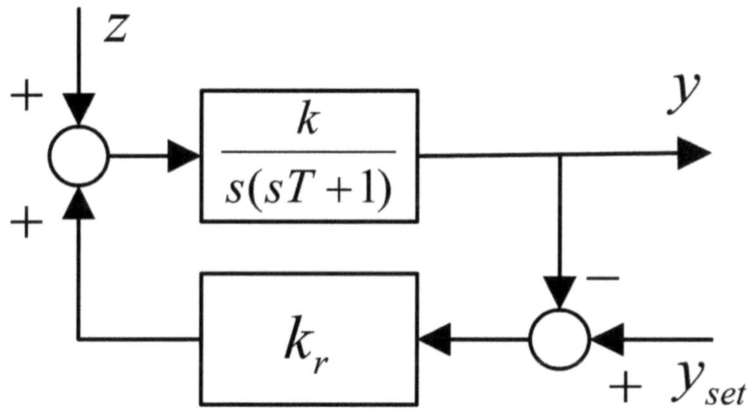

Fig. 1. The model of the effects of disturbance and set value signal on a controlled plant equipped with a control system

The signals marked in the figure 1, which presents a block diagram for a one-dimentional case, have the following meaning:

y_{set} - the set point of signal y,
$z(t)$ - disturbance,
$y(t)$ - a parameter of a UAV's flight (e.g. UAV's orientation or location).

The last stage of the model synthesis is a right selection of the values k_r, T and k. We choose these values in order to obtain the simulation transients equal to the UAV's transients obtained during the flights. Conclusively, we obtain the model of the UAV sensitive to disturbance and properly following the signal w(t).

3 Modelling of the Dynamics of the Camera Head with a Mounted Camera

We have assumed herein as simple as possible model of the camera head which depicts its dynamic properties. What is more, the same model is assumed for both rotation axes. The parameters of the examnined camera head model were based on the physical properties of the particular camera head [27], [28].

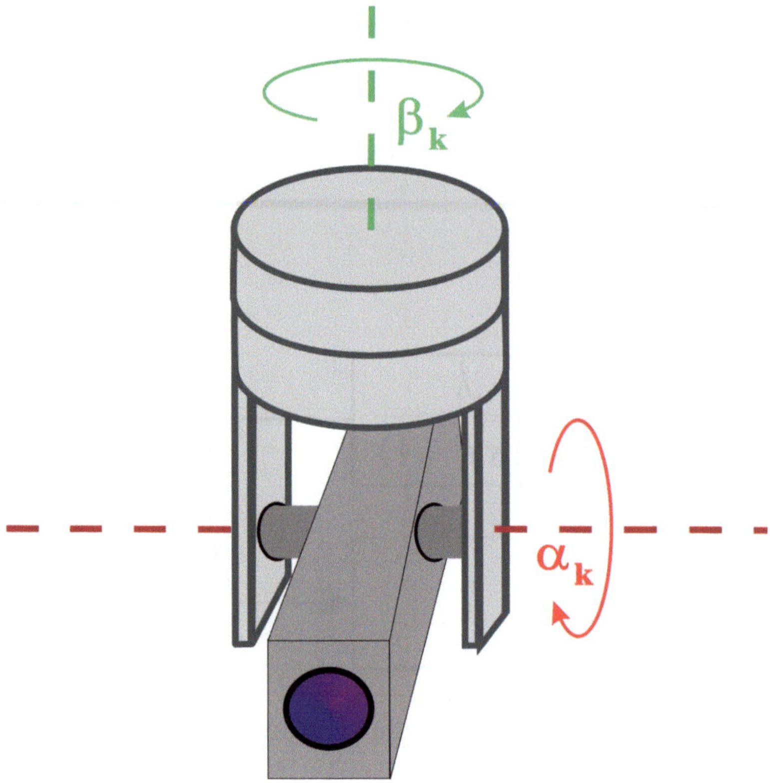

Fig. 2. The pictorial diagram of the camera head with two possible rotation axes

The figure 2 presents the pictorial diagram of the camera head with two possible rotation axes.

The figure 3 presents the rotation angle of the camera head in the a vertical plane.

The figure 4 presents the rotation angle of the camera head in the horizontal plane (x, y).

The figure 5 presents the control system of the camera head for the angle β.

The figure 6 presents the control system of the camera head for the angle α.

According to the figures 5 and 7, behaviour of the camera head both in horizontal and vertical axes was modelled by means of first order lag with integrator. The parameters which denote the kalfa and kbeta gain allow to model the rotation velocity of the camera head. In the case of engines of greater power, the angular velocities achieve bigger values. The time constants Talfa and Tbeta on the hand allow to model the load of the engines because of the weight of the camera and camera head elements. The greater time constant corresponds to the greater weight of these elements. However, in the case of examinig the real camera head with a mounted camera, it is indispensable to identify the parameters for a given device.

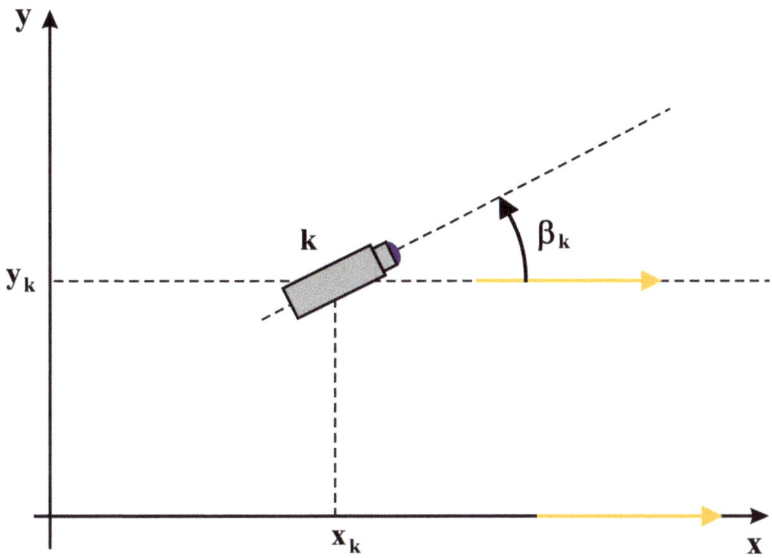

Fig. 3. The rotation angle of the camera head in the a vertical plane

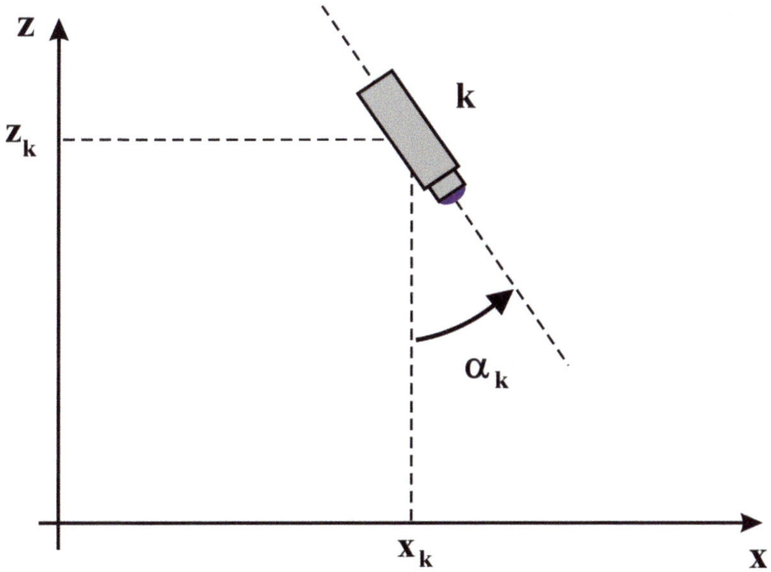

Fig. 4. The rotation angle of the camera head in the horizontal plane (x, y)

4 Location and Orientation of the UAV, Camera and Tracked Object in 3D Space

The following section will elaborate on the geometrical relationships between the camera and tracked object location and orientation. In this study a helicopter is used as an

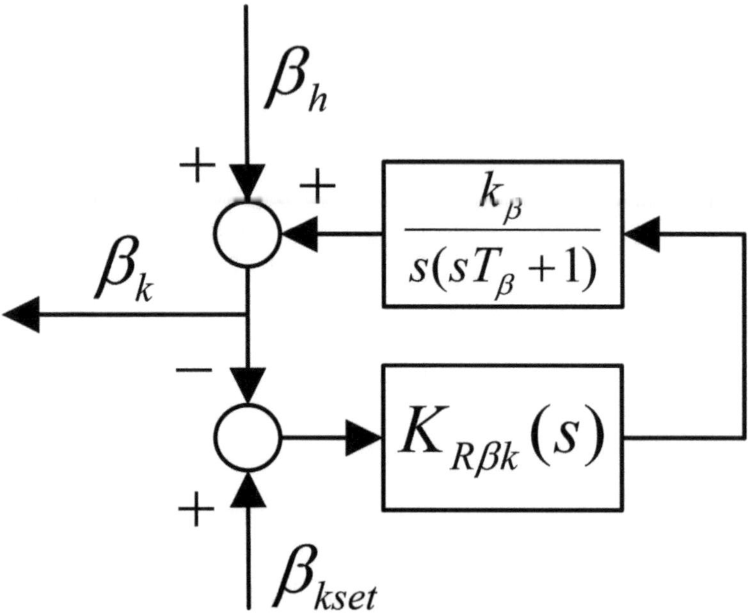

Fig. 5. The control system of the camera head for the angle β

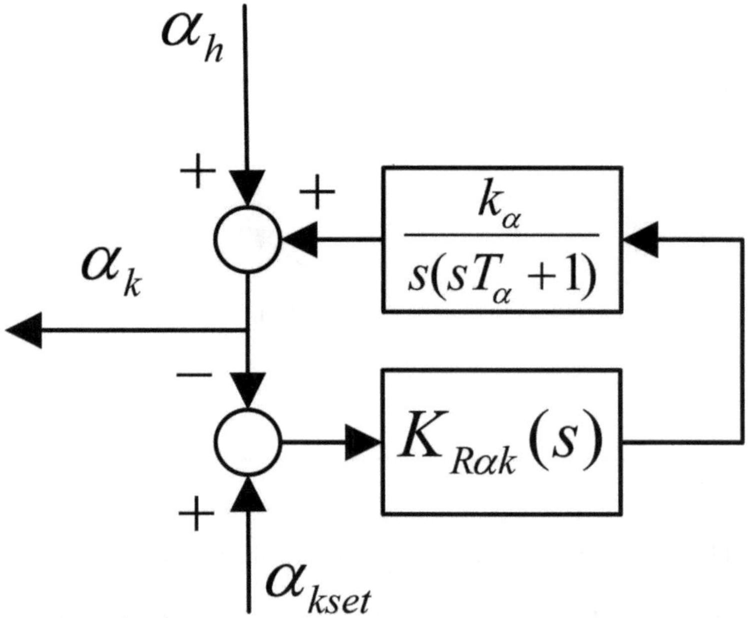

Fig. 6. The control system of the camera head for the angle α

example of a UAV. Basic assumptions concerning geometrical relations for the examined problem were presented in [26]. The main assumptions state that the helicopter position equals camera position and the tracked object moves on a flat ground ($z_o = 0$).

The camera head control algorithm is based on geometrical relationships between camera and tracked object locations. Similarly, as it was presented in [26]. the current object position is calculated only in the moment when pattern recognition algorithms give information that the object is in the picture.

The figure 7 presents the scheme of the object, camera and UAV's location in 3D space in the case when the camera is turned towards the tracked object.

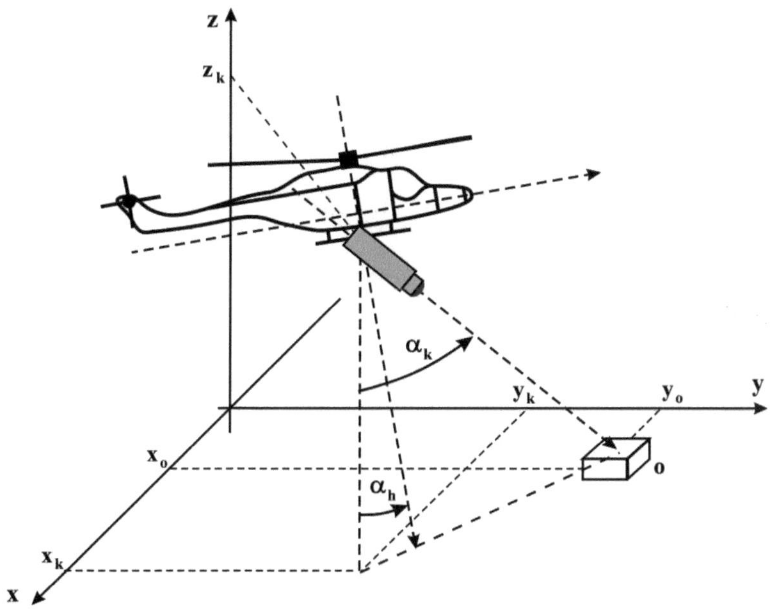

Fig. 7. General view of the 3D scene for a helicopter as an example of a UAV

The rotation angles α_k and α_h in the vertical plane are presented in figure 7.

The figure 8 presents the UAV, camera and tracked object as a projection on (x, y) plane.

The figure 9 presents the camera and tracked object's location and orientation in the vertical plane. The distance d_{xy} is the projection of the distance d on the plane (x, y).

The above presented geometrical relationships are used during calculations in the control system.

5 The Modelling of the Control Systems for the UAV and the Camera Head

Section 3 elaborated on the camera head control system with the constant gain - figures 5 and 6. In this section we will suggest a modification of the control system which

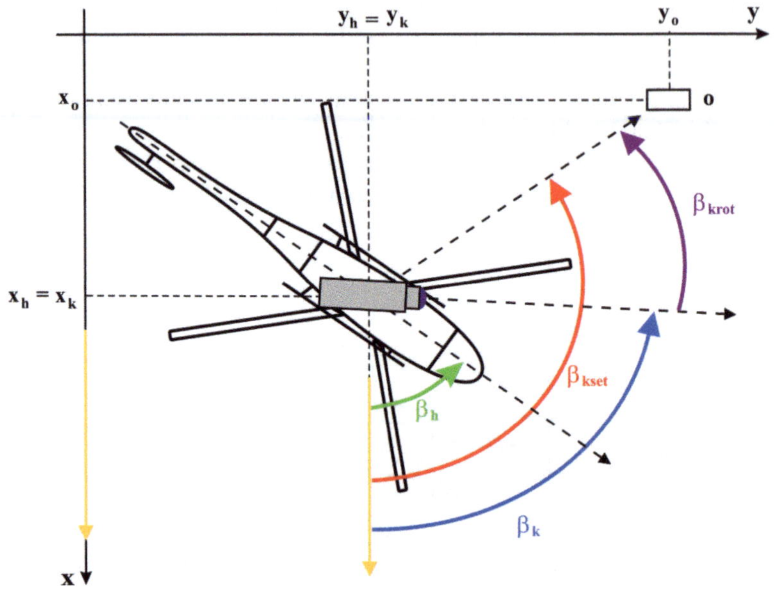

Fig. 8. The UAV, camera and tracked object as a projection on the (x, y) plane

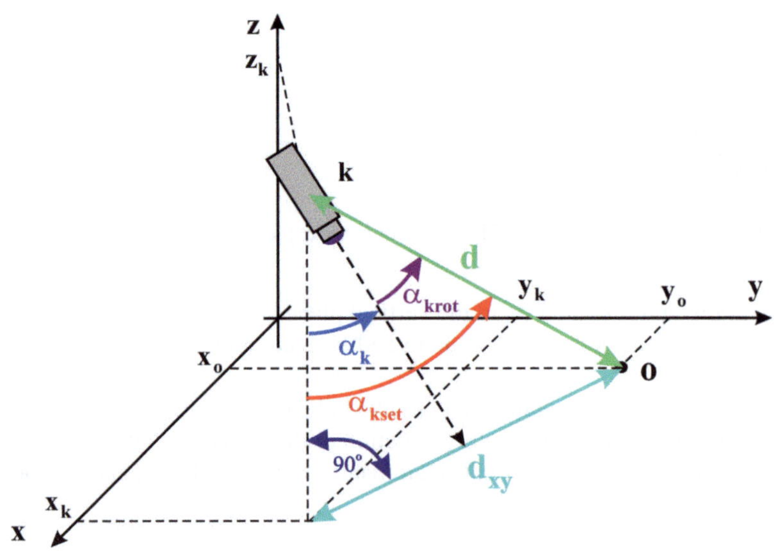

Fig. 9. Camera and tracked object's variables defining location and orientation in the vertical plane

consists of applying the changeable gain in the proportional regulator. The value of the gain depends on the value of the angle α_k or β_k error signal. Therefore this function has to meet the following requirements:

a) for e_{alphak} approaching zero, $k_{r\alpha k}$ approaches the minimum value;
b) for e_{alphak} approaching the infinity, $k_{r\alpha k}$ approaches the maximum value;
c) when both, negative and positive e_{alphak} have the same absolute value then $k_{r\alpha k}$ assumes the same value.

The same requirements have to be met for the control system of the angle β_k.

After the experiments we can state the function $k_{r\alpha k} = 10^{|e_{\alpha k}|}$ meets the aforementioned requirements. Moreover, we have used this function further in this study. Additionally, the upper limit and lower limit for the value of $k_{r\alpha k}$ and $k_{r\beta k}$ were utilised herein. The figure 10 presents the control system of the UAV and camera head control system with a changeable gain in the controller $K_{r\beta}(s) = k_{r\beta k}$.

The same control system is applied to controlling the angle α_k.

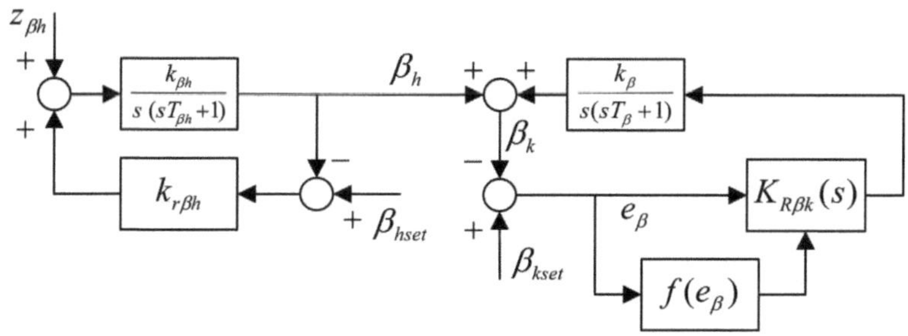

Fig. 10. The UAV and camera head's control system with the changeable gain in the proportional regulator

Generally, the basic steps of the control algorithm can be described as presented below.

The first step of the control algorithm is to correct rotation angle in a horizontal plane as we could see in the figure 8. The next step comprises of correcting rotation angles in a vertical plane - cf. fig. 9. The figure 9 presents relationships between the camera rotation angles in a vertical plane after correcting the rotation angles in the horizontal plane. According to the figure 9, the angle α_{krot} at which the camera has to be turned, can be calculated on the basis of the current camera rotation angle and the angle towards the object.

At the beginning of the control algorithm, we compute the object location when the control algorithm works with the object visible in the picture.

The position of the object (x_o, y_o) is calculated on the basis of rangefinder data, GPS and IMU data as it was presented in (1).

$$\left\{\begin{array}{l} x_o = x_k + d_{xy} \cdot \sin(90^o - \beta_k) \\ y_o = y_k + d_{xy} \cdot \cos(90^o - \beta_k) \end{array}\right\} \quad \text{for} \quad \beta_k \in \langle 0^o, 180^o \rangle$$

$$\left\{\begin{array}{l} x_o = x_k + d_{xy} \cdot \sin(90^o + \beta_k) \\ y_o = y_k - d_{xy} \cdot \cos(90^o + \beta_k) \end{array}\right\} \quad \text{for} \quad \beta_k \in \langle -180^o, 0^o) \tag{1}$$

$$\text{where} \quad d_{xy} = z_k \cdot \tan(\alpha_k)$$

We observe the distortion when the object is lost from the picture and the camera position or orientation changes considerably. Then the UAV's control system turns the helicopter to the position and orientation calculated by the UAV's control system. The UAV's orientation should be almost horizontal and the distance between the UAV and the tracked object should not be too long. Afterwards $x_k, y_k, z_k, \alpha_k, \beta_k$ are measured. The required values of the camera's rotation angles α_{kset} and β_{kset} are calculated on the basis of the object and camera's positions. These angles are calculated by the equations (2) stated below:

$$\begin{array}{ll} \alpha_{kset} = \arctan(\frac{d_{xy}}{z_k}) & \\ \text{where} \quad d_{xy} = \sqrt{(x_k - x_o)^2 + (y_k - y_o)^2} & \\ \beta_{kset} = -90^o - \arctan(\frac{x_o - x_k}{y_o - y_k}) & \text{for} \quad y_k \geq y_o \\ \beta_{kset} = 90^o - \arctan(\frac{x_o - x_k}{y_o - y_k}) & \text{for} \quad y_k < y_o \end{array} \tag{2}$$

In the next step, we compute the correction angles α_{krot} and β_{krot}. The camera has to be turned at these angles. These angles are calculated as shown in the equation (3).

$$\begin{array}{l} \alpha_{krot} = \alpha_{kset} - \alpha_k \\ \beta_{krot} = \beta_{kset} - \beta_k \end{array} \tag{3}$$

6 Examples of Control System Operation for the Constant and Changeable Gain in the Proportional Regulators

The following examples illustrate that the changeable gain in the camera head control system guarantee the proper tracking of an object.

Experiment 1

The experiment 1 will be conducted for minor disturbances and constant gains $k_{r\alpha k} = 2{,}5 * 3$ and $k_{r\beta k} = 0{,}25 * 3$.

The figure 11 presents the UAV's and tracked object's location. In the figure, we can clearly observe the moments of disturbances.

The figure 12 presents the transients of $\alpha_k, \alpha_{kset}, e_{\alpha k}, k_{r\alpha k}, u_1$, and u_2.

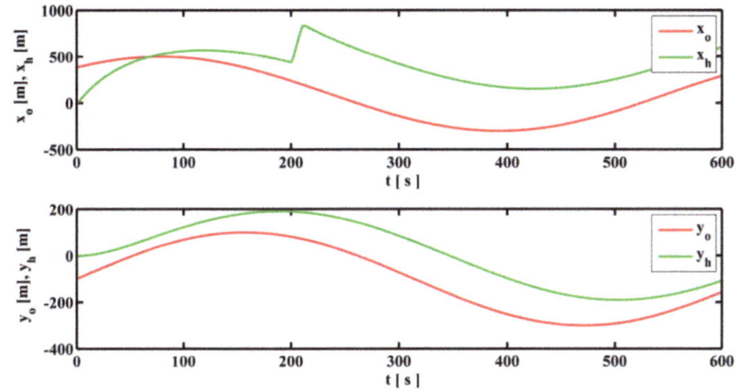

Fig. 11. The UAV's and tracked object's location in the experiment 1

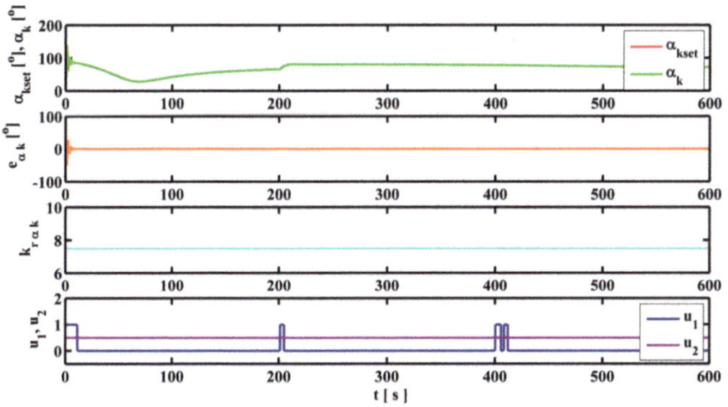

Fig. 12. The transients of $\alpha_k, \alpha_{kset}, e_{\alpha k}, k_{r\alpha k}, u_1$, and u_2 in the experiment 1

In this particular case, $k_{r\alpha k}$ is constant - independent of $e_{\alpha k}$. The value of the signal u_1 equal 1 means that the object is out of the camera field of view. The value of the signal u_2 equal 1 means that the object is lost by the control system therefore it is necessary to track the object with a help of the human operator. This situation will be referred to as the "manual operation" further in this paper. u_2 equal 0 during all simulation denotes the correct operation of the control system. The signal u_2 in all figures is increased in such a way to make the figure clear. The value of the singal u_2 equal 0 corresponds the value equal $0,5$ in a figure and the value of the singal u_2 equal 1 corresponds the value equal $1,5$ in a figure.

The figure 13 presents the transients of $\beta_k, \beta_{kset}, e_{\beta k}, k_{r\beta k}, u_1$, and u_2.

In this particular case, $k_{r\beta k}$ is constant - independent of $e_{\beta k}$. We can observe the proper operation of the camera head control system for constant $k_{r\alpha k}$ and $k_{r\beta k}$.

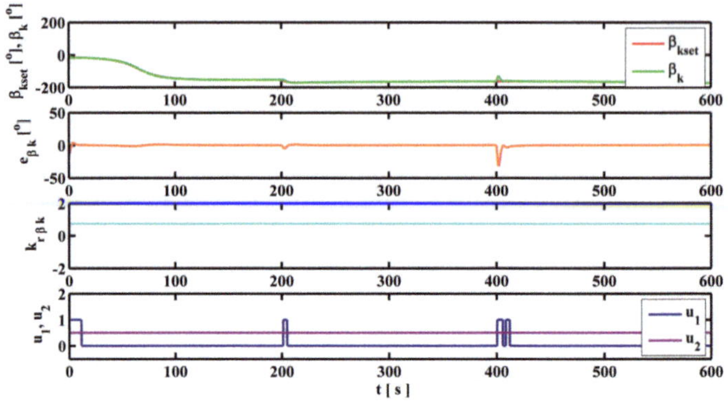

Fig. 13. The transients of $\beta_k, \beta_{kset}, e_{\beta k}, k_{r\beta k}, u_1$, and u_2 in the experiment 1

Experiment 2

The experiment 2 will be conducted for disturbances greater than in the experiment 1 and constant gains gains $k_{r\alpha k}$ and $k_{r\beta k}$.

Experiment 2a

At the beginning of the experiment 2, we assumed the following constant values of the gains: $k_{r\alpha k} = 2,5 * 3$ and $k_{r\beta k} = 0,25 * 3$.

The figure 14 presents the UAV's and tracked object's location. In the figure, we can clearly observe the moments of disturbances.

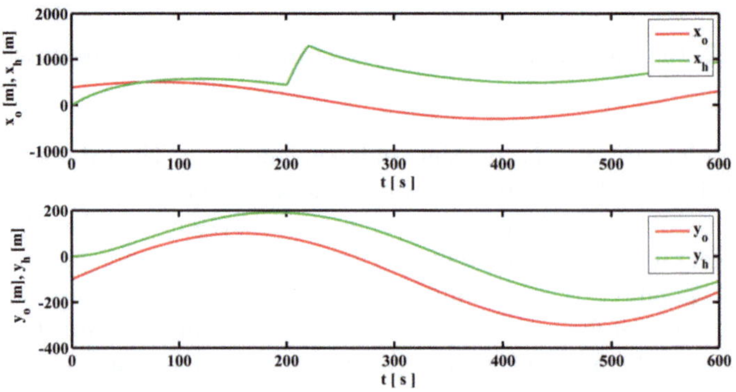

Fig. 14. The UAV's and tracked object's location in the experiment 2a

The figure 15 presents the transients of $\alpha_k, \alpha_{kset}, e_{\alpha k}, k_{r\alpha k}, u_1$, and u_2.

We can observe that u_1 equals 1 when the disturbances occur, and the value of the signal u_2 equals 1 for some time during the simulation. It means that the object is lost

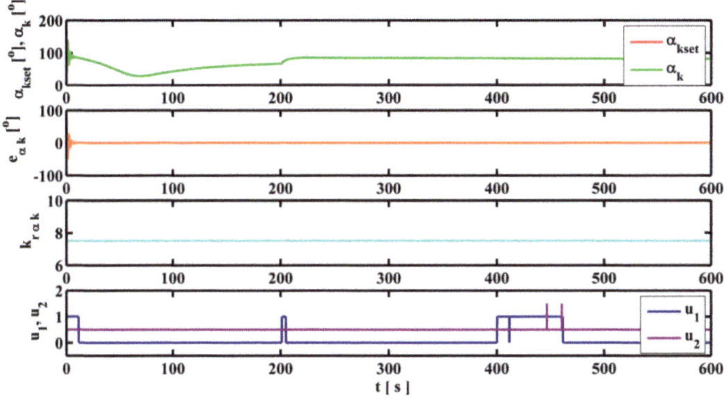

Fig. 15. The transients of $\alpha_k, \alpha_{kset}, e_{\alpha k}, k_{r\alpha k}, u_1$, and u_2 in the experiment 2a

from the field of view and the camera control system cannot turn the camera towards the object.

The figure 16 presents the transients of $\beta_k, \beta_{kset}, e_{\beta k}, k_{r\beta k}, u_1$, and u_2.

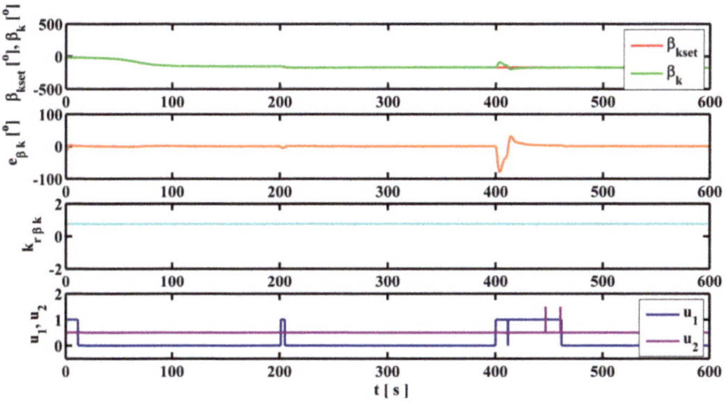

Fig. 16. The transients of $\beta_k, \beta_{kset}, e_{\beta k}, k_{r\beta k}, u_1$, and u_2 in the experiment 2a

We can observe that u_1 equals 1 when the disturbances occur, and the value of the signal u_2 equals 1 for some time during the simulation. It means that the object is lost from the field of view and the camera control system cannot turn the camera towards the object.

Experiment 2b

In order to improve the operation of the control system, we increase the values of $k_{r\alpha k}$ and $k_{r\beta k}$. We assumed the values of the gains: $k_{r\alpha k} = 2,5 * 6$ and $k_{r\beta k} = 0,25 * 12$.

The figure 17 presents the UAV's and tracked object's location. In the figure, we can clearly observe the moments of disturbances.

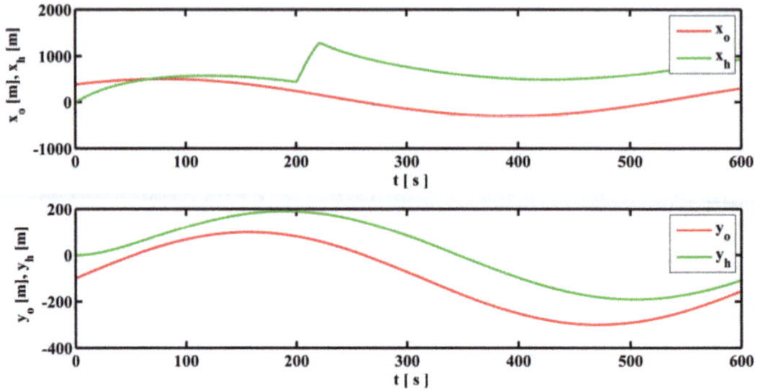

Fig. 17. The UAV's and tracked object's location in the experiment 2b

The figure 18 presents the transients of $\alpha_k, \alpha_{kset}, e_{\alpha k}, k_{r\alpha k}, u_1$, and u_2.

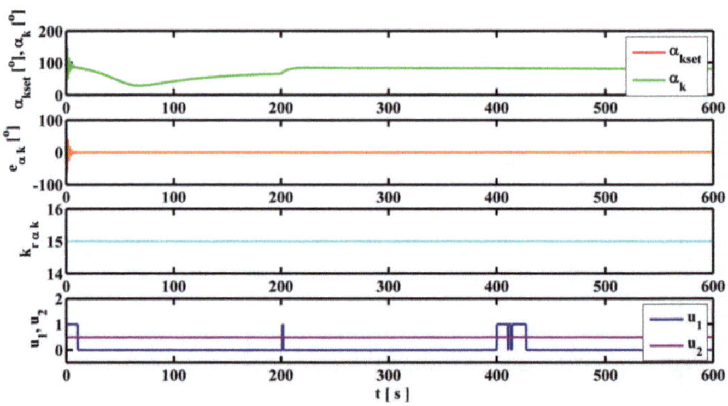

Fig. 18. The transients of $\alpha_k, \alpha_{kset}, e_{\alpha k}, k_{r\alpha k}, u_1$, and u_2 in the experiment 2b

We can observe that u_1 equals 1 when the disturbances occur; however, u_2 equals 0 during all simulation. It means that the object is lost from the field of view, but the camera control system operates correctly and the object appears in the camera field of view again.

The figure 19 presents the transients of $\beta_k, \beta_{kset}, e_{\beta k}, k_{r\beta k}, u_1$, and u_2.

We can observe that u_1 equals 1 when the disturbances occur; however, u_2 equals 0 during all simulation. It means that the object is lost from the field of view, but the camera control system operates correctly and the object appears in the camera field of view again.

The transients alfak and betak demonstrate the proper operation of the control system.

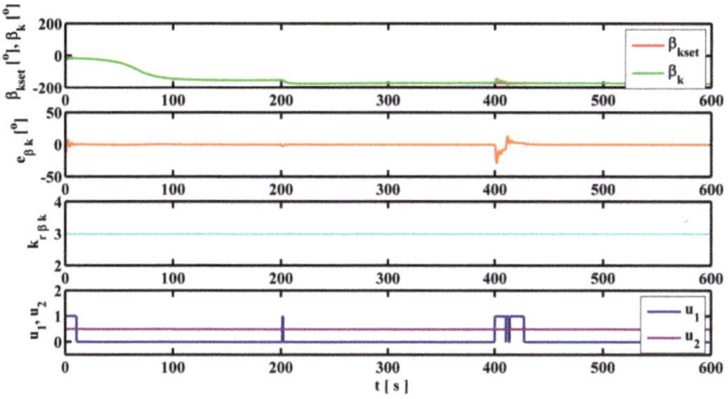

Fig. 19. The transients of $\beta_k, \beta_{kset}, e_{\beta k}, k_{r\beta k}, u_1$, and u_2 in the experiment 2b

Experiment 3

The experiment 3 will be conducted for disturbances greater than in the experiment 2.

Experiment 3a

At the beginning of the experiment 2, we assumed the following constant values of the gains: $k_{r\alpha k} = 2,5 * 6$ and $k_{r\beta k} = 0,25 * 12$.

The figure 20 presents the UAV's and tracked object's location. In the figure, we can clearly observe the moments of disturbances.

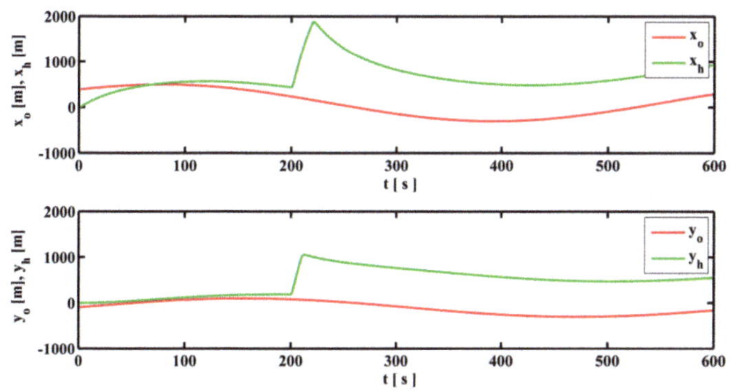

Fig. 20. The UAV's and tracked object's location in the experiment 3a

The figure 21 presents the transients of $\alpha_k, \alpha_{kset}, e_{\alpha k}, k_{r\alpha k}, u_1$, and u_2.

The figure 22 presents the transients of $\beta_k, \beta_{kset}, e_{\beta k}, k_{r\beta k}, u_1$, and u_2.

According to figures 21 and 22, u_1 equals 1 when the disturbances occur, and the value of the signal u_2 equals 1 for some time during the simulation. It means that

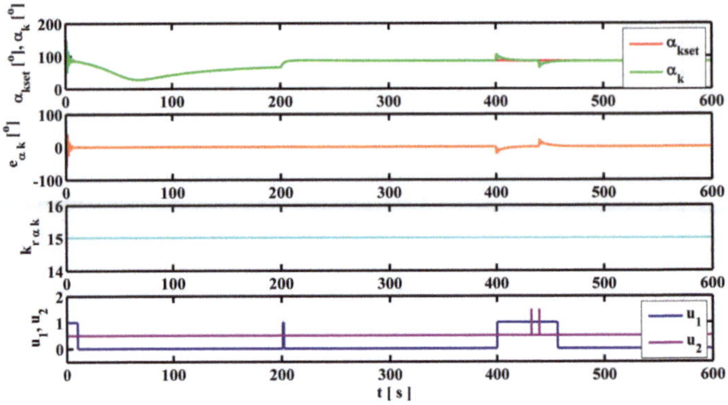

Fig. 21. The transients of $\alpha_k, \alpha_{kset}, e_{\alpha k}, k_{r\alpha k}, u_1$, and u_2 in the experiment 3a

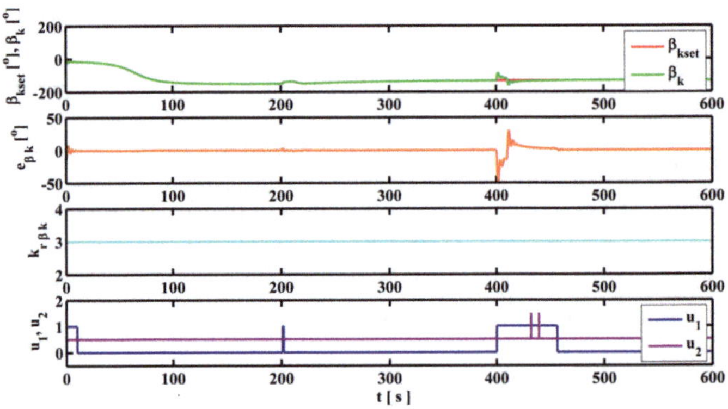

Fig. 22. The transients of $\beta_k, \beta_{kset}, e_{\beta k}, k_{r\beta k}, u_1$, and u_2 in the experiment 3a

the object is lost from the field of view and the camera control system cannot turn the camera towards the object.

Experiment 3b

We try to improve the camera control system by means of increasing the values of the gains, new values are as follows: $k_{r\alpha k} = 2,5 * 12$ and $k_{r\beta k} = 0,25 * 45$.

The figure 23 presents the UAV's and tracked object's location. In the figure, we can clearly observe the moments of disturbances.

The figure 24 presents the transients of $\alpha_k, \alpha_{kset}, e_{\alpha k}, k_{r\alpha k}, u_1$, and u_2.

The figure 25 presents the transients of $\beta_k, \beta_{kset}, e_{\beta k}, k_{r\beta k}, u_1$, and u_2.

According to figures 24 and 25, the object is lost from the camera field of view. The camera control system reacts to the disturbances and camera turns towards the object. The object appears in the camera field of view, but there is overshooting and the camera turns too much which results in losing the object. We lose the object beacause

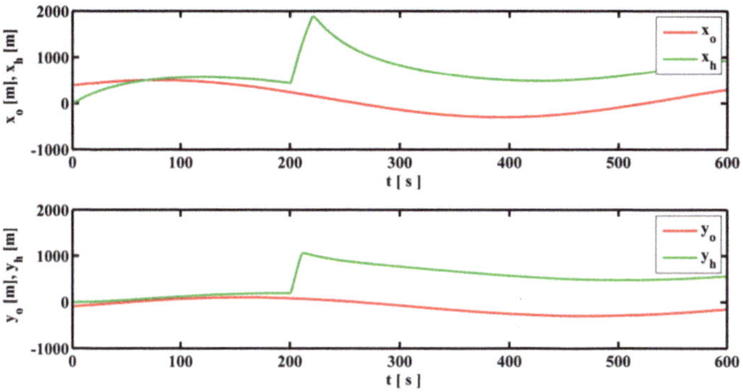

Fig. 23. The UAV's and tracked object's location in the experiment 3b

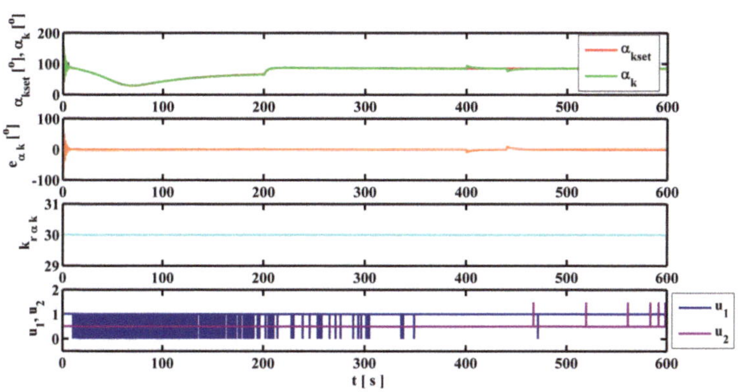

Fig. 24. The transients of $\alpha_k, \alpha_{kset}, e_{\alpha k}, k_{r\alpha k}, u_1$, and u_2 in the experiment 3b

the camera cannot stop turning at a right moment, the moment when the object is in the centre of the camera field of view.

Experiment 3c

The paper puts forward the solution of the abovementioned problem by using the changeable gains in the camera control systems. The function defining these gains was described in the section 5. The following figures present the transients for the control system with changeable gains.

The figure 26 presents the UAV's and tracked object's location. In the figure, we can clearly observe the moments of disturbances.

The figure 27 presents the transients of $\alpha_k, \alpha_{kset}, e_{\alpha k}, k_{r\alpha k}, u_1$, and u_2.

The figure 28 presents the transients of $\beta_k, \beta_{kset}, e_{\beta k}, k_{r\beta k}, u_1$, and u_2.

According to figures 27 and 28, the object is lost from the camera field of view. The camera control system reacts to the disturbances and camera turns towards the object. Owing to the gain, which is greater for a greater error and lower for a lower error, the

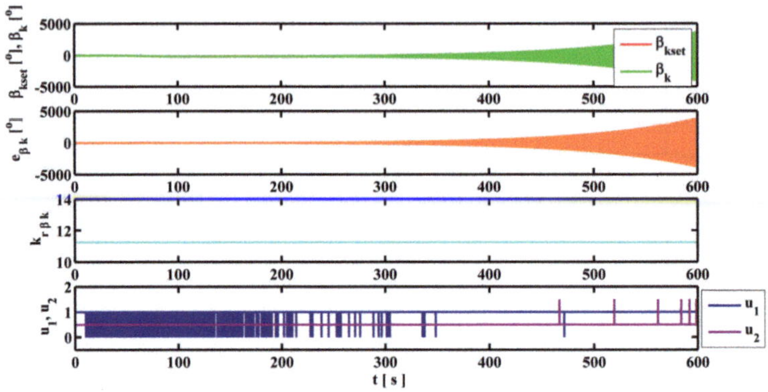

Fig. 25. The transients of $\beta_k, \beta_{kset}, e_{\beta k}, k_{r\beta k}, u_1$, and u_2 in the experiment 3b

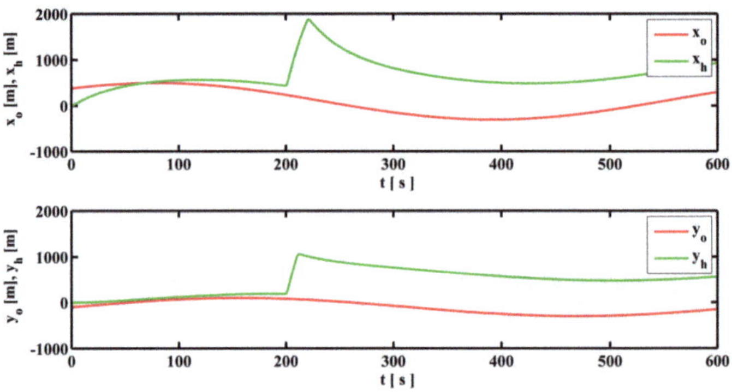

Fig. 26. The UAV's and tracked object's location in the experiment 3c

camera turns fast towards the object when the error is great. Contrary, the camera turns slowly towards the object when the object is near the center of the camera field of view.

The function $f(e)$, which defines the P controller gain, should be developed taking into account dynamical and statical properties of the camera head. Owing to the presented examples, we can state that the minor disturbances can be corrected by means of small gains in the camera head control system. On the other hand, in case of large disturbances the proper operation of the control system requires that the high values of the gains are applied. As the consequence, for the high values of the gain, the overshoots may appear during the control system operation. It means that the camer control system turns the camera towards the object with the great velocity. At last, the object apperars in the camera field of view; however, the camera head does not stop turning. The object moves across the picture and disappears. The overshoot may result in losing the object.

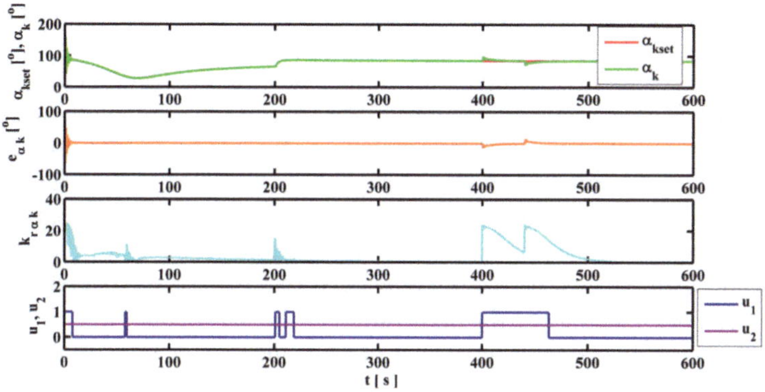

Fig. 27. The transients of $\alpha_k, \alpha_{kset}, e_{\alpha k}, k_{r\alpha k}, u_1$, and u_2 in the experiment 3c

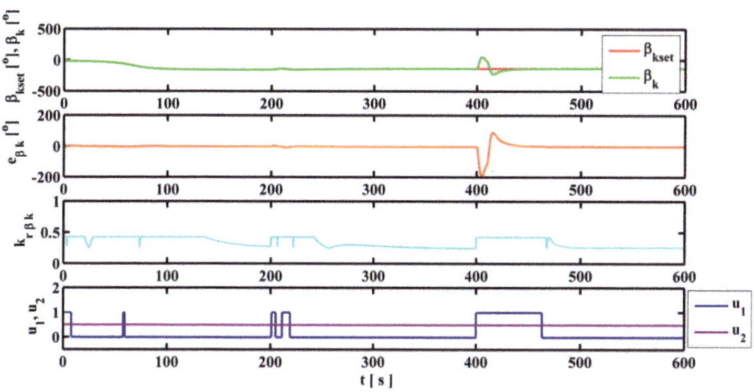

Fig. 28. The transients of $\beta_k, \beta_{kset}, e_{\beta k}, k_{r\beta k}, u_1$, and u_2 in the experiment 3c

7 Conclusions

This paper examined the changeable gain in the camera head control system which was applied to object tracking. What is more, we assumed that in order to track objects moving on the ground, the UAV with the mounted camera would be used in this task. The main problem during the flight of the UAV was that various disturbances such as rapid rotations and shifts of the UAV might appear.

Furthermore, we assumed that the object was moving on the flat ground and the UAV's velocity was far greater than the maximum velocity of the tracked object. Owing to the fact that the UAV was equipped with the flight trajectory control system, it allowed to model the rotations and shifts of the UAV in the particular axes as independent control channels. In order to examine the legitimacy of using the changeable gain, the behaviour of the control system was examined for the different values and types of disturbances. We developed the function which defined the relationship between the error signal and gain in the control system. This function in turn let us track the object appropriately.

Next, we presented the examples which indicated on the fact that during major disturbances there were significant changes in the control error signal; therefore adjusting the gain to the value of the control error signal was the condition for limiting overshoots. Moreover, it was presented that using changable gains in the camera head control system, allowed to track the object for both, the small and major disturbances. Owing to the changeable gain, it was possible to eliminate rapid reactions of the control system during major disturbances. As a consequence, there was not any overshoot - the object was being tracked even during the high values of the disturbances.

However, one must bear in mind that prior to the proper selection of the function, which determined the relationship between the gain and error signal, it was necessary to conduct tests and experiments for the models of the particular devices.

Summing up, we may state that the modification of the control system by means of implementing the changeable gain, allowes to improve significantly the quality of control.

References

1. Davies, D., Palmer, P.L., Mirmehdi, M.: Detection and tracking of very small low contrast objects. In: Proceedings of the 9th British Machine Vision Conference (September 1998)
2. Zhang, S., Karim, M.A.: Automatic target tracking for video annotation. Op. Eng. 43, 1867–1873 (2004)
3. Irani, M., Peleg, S.: Improving resolution by image registration. CVGIP: Graph. Models and Image Process. 53, 231–239 (1991)
4. Chesnaud, C., Refegier, P., Boulet, V.: Statistical region snake-based segmentation adapted to different physical noise models. IEEE Trans. Patt. Anal. Mach. Intell. 21, 1145–1157 (1999)
5. Gordon, N., Ristic, B., Arulampalam, S.: Beyond the Kalman Filter: Particle Filters for Tracking Applications. Artech House, Boston (2004)
6. Sharp, C., Shakernia, O., Sastry, S.: A Vision System for Landing an Unmanned Aerial Vehicle. In: Proceedings of the 2001 IEEE International Conference on Robotics and Automation, vol. 2, pp. 1720–1727. IEEE, Los Alamitos (2001)
7. Casbeer, D., Li, S., Beard, R., Mehra, R., McLain, T.: Forest Fire Monitoring With Multiple Small UAVs, Porland, OR (April 2005)
8. Kuś, Z., Fraś, S.: Helicopter control algorithms from the set orientation to the set geographical Location. In: Nawrat, A., Simek, K., Świerniak, A. (eds.) Advanced Technologies for Intelligent Systems. SCI, vol. 440, pp. 3–14. Springer, Heidelberg (2013)
9. Nawrat, A.: Modelowanie i sterowanie bezzałogowych obiektów latających, Wydawnictwo Politechniki Śląskiej, Gliwice (2009)
10. Valavanis, K.P. (ed.): Advances In Unmanned Aerial Vehicles. Springer (2007)
11. Castillo, P., Lozano, R., Dzul, A.E.: Modelling and Control of Mini-Flying Machines. Springer (2005)
12. Padfield, G.D.: Helicopter Flight Dynamics, Backwell Science Ltd. (1996)
13. Manerowski, J.: Identyfikacja modeli dynamiki ruchu sterowanych obiektów latających. WN ASKON, Warszawa (1999)
14. Kearney, J.K., Thompson, W.B.: Optical flow estimation: An error analysis of gradient-based methods with local optimization. IEEE Transactions on Pattern Analysis and Machine Intelligence 9, 229–243 (1987)

15. Anandan, P.: A computational framework and an algorithm for the measurement of visual motion. International Journal Qf Computer Vision 2, 283–310 (1989)
16. Barman, H., Haglund, L., Knutsson, H., Granlund, G.: Estimation of velocity, acceleration, and disparity in time sequences. In: Proc. IEEE Workshop on Visual Motion, Princeton, NJ, pp. 44–51 (1991)
17. Bascle, B., Bouthemy, E., Deriche, N., Meyer, E.: Tracking complex primitives in an image sequence. In: Proc. 12th International Conference on Pattern Recognition, Jerusalem, pp. 426–431 (1994)
18. Butt, E., Yen, C., Xu, X.: Local correlation measures for motion analysis: A comparative study. In: Pattern Recognition and Image Processing Conference, Las Vegas, pp. 269–274 (1982)
19. Butt, E., Bergen, J.R., Hingorani, R., Kolczynski, R., Lee, W.A., Leung, A., Lubin, J., Shvayster, H.: Object tracking with a moving camera. In: Proc. IEEE Workshop on Visual Motion, Irving, pp. 2–12 (1989)
20. Buxton, B.E., Buxton, H.: Computation of optical flow from the motion of edges features in image sequences. Image and Vision Computing 2(2), 59–75 (1984)
21. Campani, M., Verri, A.: Computing optical flow from an overconstrained system of linear algebraic equations. In: Proc. 3rd International Conference on Computer Vision, Osaka, pp. 22–26 (1990)
22. Campani, M., Verri, A.: Motion analysis from firstorder properties of optical flow. CVGIP: Image Understanding 56(1), 90–107 (1992)
23. Carlsson, S.: Information in the geometric structure of retinal flow field. In: Proc. 2nd International Conference on Computer Vision, pp. 629–633 (1988)
24. Gessing, R.: Control Fundamentals. Silesian University of Technology, Gliwice (2004)
25. Kuś, Z., Nawrat, A.: Object Tracking in a Picture During Rapid Camera Movements. In: Nawrat, A., Kuś, Z. (eds.) Vision Based Systems for UAV Applications. SCI, vol. 481, pp. 77–91. Springer, Heidelberg (2013)
26. Kuś, Z., Nawrat, A.: Object Tracking for Rapid Camera Movements in 3D Space. In: Nawrat, A., Kuś, Z. (eds.) Vision Based Systems for UAV Applications. SCI, vol. 481, pp. 57–76. Springer, Heidelberg (2013)
27. The raport on the project O R 00 0132 12: "Design and implementation of innovative unmanned mobile platforms to monitor the state border" developed in Silesian Technical University, Gliwice (2010-2012)
28. Jędrasiak, K., Bereska, D., Nawrat, A.: The prototype of Gyro-Stabilized UAV Gimbal for Day-Night Surveillance. In: Nawrat, A., Simek, K., Świerniak, A. (eds.) Advanced Technologies for Intelligent Systems. SCI, vol. 440, pp. 107–116. Springer, Heidelberg (2013)

The Limitation for the Angular Velocity of the Camera Head during Object Tracking with the Use of the UAV

Zygmunt Kuś and Aleksander M. Nawrat

Silesian University of Technology, Institute of Automatic Control,
Akademicka 16 St., Gliwice, Poland
{zygmunt.kus,aleksander.nawrat}@polsl.pl

Abstract. The following paper examined the significance of the velocity limitations in the camera head control system. A prerequisite for the control system was that it should quarntee the tracking of an object moving on the ground, even in the case of disturbances such as the rapid rotations and shifts of the UAV. Equipping the UAV with the flight trajectory control system allowed to model the rotations and shifts of the UAV in the particular axes as independent control channels. What is more, the study examined behaviour of the control system for different values of the disturbances. The paper presented the comparison of the object tracking efficiency for the case with and without the limitations of the angular velocity of the camera head. The results showed that the limitation of the velocity of the camera head rotation are crucial in correct object tracking. There was the relationship between both, the value as well as the type of the disturbance and the requirements regarding the necessary velocity of the camera rotation. The paper showed that at the stage of designing the camera head, it was essential to adjust the velocities of the engines. The rationale for selecting the appropriate engines were the value and type of the disturbances to which the UAV might be exposed during operation. It was suggested in the paper that in the case of the disturbances, which require the velocities of rotation greater than the limit velocities of the engines, it was necessary to use additional information about the tracked object. For example, information about the velocity and direction of the tracked object might be used to track the object despite losing the object from the camera field of view for some time.

Keywords: Unmanned Aerial Vehicles, Object Tracking, Control System, Image Processing Methods, Velocity Limitation.

1 Introduction

Nowadays, the UAVs which are equipped with the camera, are frequently used to tack objects moving in the flat terrain. Hence, keeping the object within the camera field of view and keeping the object close to the UAV are the basic tasks for the control systems. In order to handle the aforementioned tasks, the image processing methods [1,2,3,4,5,6,7] can be used when the object is visible in the picture.

The proper turn of the camera towards the object is guaranteed by two control systems. There are independent control systems for the UAV and camera head. Therefore both systems should be able to compensate disturbances. When this requirement is not

A.M. Nawrat (ed.), *Innovative Control Systems for Tracked Vehicle Platforms*, Studies in Systems, Decision and Control 2,
DOI: 10.1007/978-3-319-04624-2_7,

met, the image processing methods [14] - [23] cannot be used in the object tracking task.

The dynamics of the camera head is faster than the dynamics of the whole UAV. Hence the return of the UAV to the set trajectory takes more time than turn of the camera head. Therefore the turn of the camera towards the object may be done faster than the UAV is able to return to the set trajectory. Consequently, counteracting the disturbances consists in compensating the changes of UAV's location or orientation by means of the proper camera head turn towards the object.

The aforementioned approach guarantees tracking the object without losing it from the picture when the object moves slowly or is stationary. It is the robust control problem.

The papers [25] and [26] assumed that the rapid and considerable disturbances are possible to be compensated by means of P regulators or by using PD compensator additionally. The angular velocities which were necessary to compensate disturbances were lower than maximum engines velocities [27], [28]. According to [29], it was examined whether it is possible to use the margin of the angular velocity to compensate far greater or more rapid disturbances. One must bear in mind that apparently there is the maximum limit of the angular velocity for the camera head engines. This problem has been discussed herein. Therefore the limits of the angular velocities were simulated and the influence of these limits on the control system operation was examined in this paper.

2 Modelling of the Dynamics of the UAV with the Flight Trajectory Control System

The unmanned helicopters are the examples of the UAVs. Several kinds of UAV's modelling, which consider the helicopters as an example, were presented in [8,9,10,11,12,13].

The various types of the UAVs require a different solutions for flight control systems which can guarantee reaching the set flight trajectory. Regardless of the UAV's type, we can define its state by means of the following variables: location in the coordinate system which is connected with the ground (x_h, y_h, z_h), the azimuth angle β_h and the angle α_h which describes the deviation from the perpendicular. The aforementioned angles, location and planes were presented in the figures 5 and 6.

We assume that the UAV's control system can properly steer the UAV's flight trajectory.

Therefore the UAV's model defines the influence of the disturbances and the set value on the UAV. The rapid gusts of wind can be treated as the disturbances which impact the UAV. Such disturbance causes rapid changes of UAV's location and orientation.

The model of the UAV's behaviour and its interaction with the environment should react promptly to disturbances and return slowly to the set flight trajectory. The aforementioned model has these dynamic properties. Despite the fact that the UAV is a multivariable controlled plant, due to control systems specific to a given the UAV's type, we can treat the UAV in such a way as there would not be any coupling channels.

For the discussed model of the plant, the output signals are location (x_h, y_h, z_h) and the UAV's rotation angles (α_h, β_h).

Owing to the UAV's own trajectory control system, it is possible to control each output separately not influencing other outputs.

This results in simplifying the model.

The abovementioned situation is the reason why we will present modelling only for the single control channel. For each control channel we will use the same type of the model.

On the basis of the UAV's properties, as was it presented in [8,9,10,11,12,13], we can conclude that the first order lag with the intergrator and the feedback loop guarantee the proper model of the dynamical UAV's properties. An example of the model for the control channel of the angle α_h is presented in the figure 1.

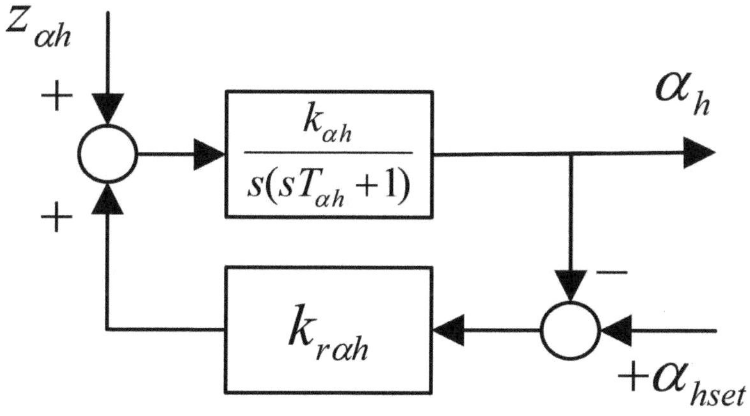

Fig. 1. The model of the disturbance effects and set value signal effects on a controlled plant for the control channel of the angle α_h

The signals marked in the figure 1, which presents a block diagram for a one-dimentional case, have the following meaning:

$y_{set} = \alpha_{hset}$ - the set point of signal $\alpha_h = y(t)$,
$z(t) = z_{\alpha h}$ - disturbance,
$y(t) = \alpha_h$ - a parameter of a UAV's flight (e.g. UAV's orientation or location - angle α_h in this case).

The synthesis of the model consists of the right selection of the values $k_{r\alpha h}$, $T_{\alpha h}$ and $k_{\alpha h}$. The simulation transients should correspond to the changes of orientation and location of a real UAV during test flights. The proper model of the UAV should be sensitive to disturbance and properly follow the signal w(t).

3 Modelling of the Dynamics of the Camera Head with a Mounted Camera

We have assumed herein a model of the camera head which depicts its dynamic properties. What is more, the same model is assumed for both rotation axes. This model is based on the known properties of the engines which are used in such applications. The parameters of the examnined camera head model were based on the description of the particular camera head [27], [28].

The figure 2 presents the spatial diagram of the variables defining the location and orientation of the camera head.

The figure 3 presents the control system of the camera head for the angle β taking into consideration the limitations of the rotation velocity.

The figure 4 presents the control system of the camera head for the angle α taking into account the limitations of the rotation velocity.

The first order lag with integrator, according to the figures 3 and 4, was a model of the behaviour of the camera head. Likewise, the same model was used in both, horizontal and vertical axes. The saturation blocks model the limitation of the angular velocities. The parameters of the transfer function which define the model of the engine may be interpreted as discussed below. For the engines of greater power, the angular velocities achieve bigger values. The time constants Talfa and Tbeta on the hand allow to model the load of the engines because of the weight of the camera and camera head elements. The greater time constant corresponds to the greater weight of these elements.

4 The Variables Defining the Location and the Orientation of the UAV, Camera and Tracked Object

The basis for computing the set values α_{kset} and β_{kset} are the geometrical relationships between the camera and tracked object. In this section we will define variables notations for the angles and distances which are crucial in control process.

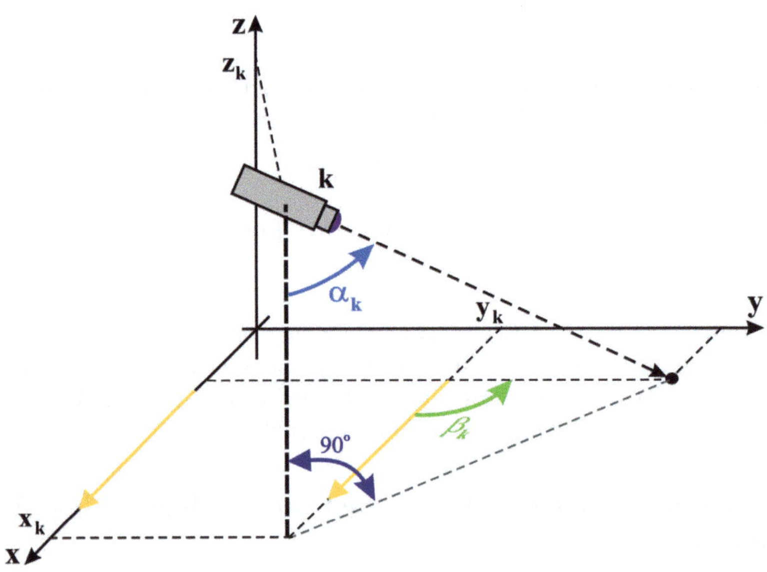

Fig. 2. The spatial diagram of the variables defining the location and orientation of the camera head

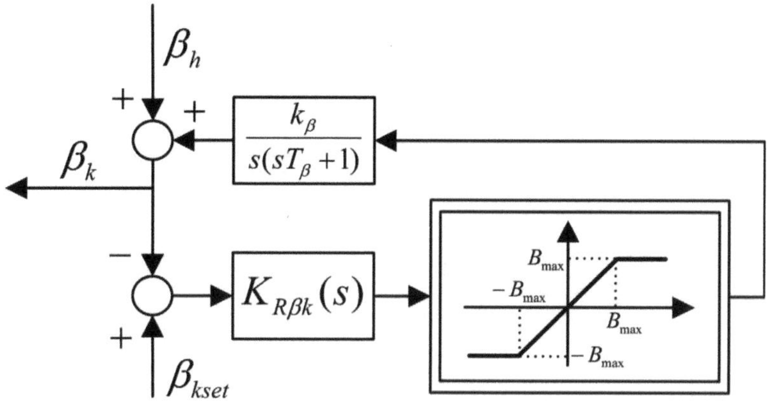

Fig. 3. The control system of the camera head for the angle β taking into consideration the limitations of the rotation velocity

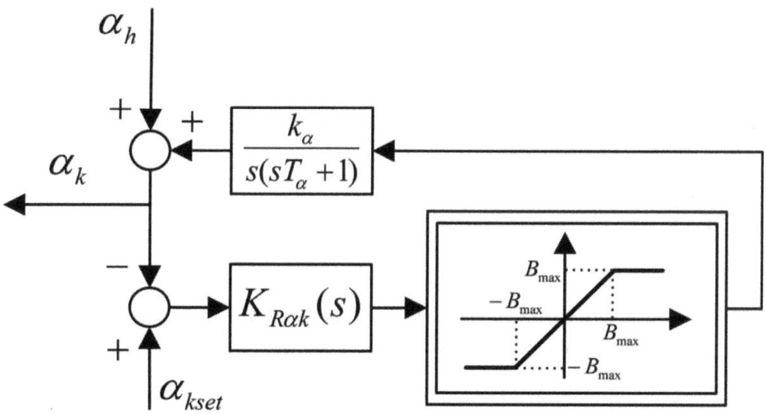

Fig. 4. The control system of the camera head for the angle α taking into account the limitations of the rotation velocity

In this study we use a helicopter as an example of a UAV.

Basic assumptions concerning geometrical relations for the examined problem were presented in [26].

Basic assumptions boil down to the fact that the helicopter position equals camera position and the tracked object moves on a flat ground ($z_o = 0$).

Both, camera head control and UAV's flight control require that we describe geometrical relationships between camera and tracked object's location. The camera head control algorithm is based on geometrical relationships between camera and tracked object locations.

Similarly, as it was presented in [26]. the current object position is calculated only in the moment when pattern recognition algorithms give information that the object is in the picture.

The figure 5 presents the angles and distances for the UAV, camera and tracked object in a projection on (x, y) plane.

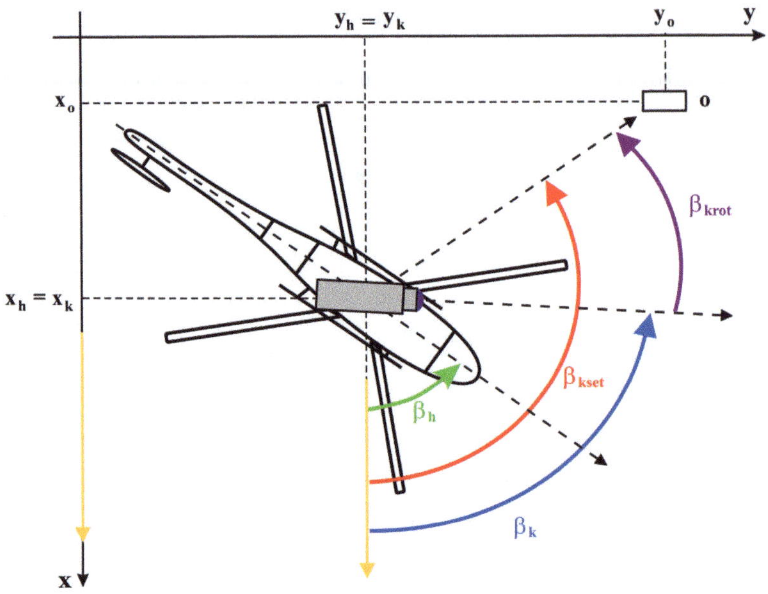

Fig. 5. The UAV, camera and tracked object as a projection on the (x, y) plane [29]

The figure 6 shows the angles and distances defining camera and tracked object's location and orientation in the vertical plane. The distance d_{xy} is the projection of the distance d on the plane (x, y).

The geometrical relationships between the angles and distances presented in figure 5 and figure 6 allow to calculate the set values α_{kset} and β_{kset} as well as to control the UAV's trajectory properly.

5 The Modelling of the Control Systems for the UAV and the Camera Head

In section 3 we have presented the camera head control systems for camera head's limited angular velocities. The saturation block, shown in figures 3 and 4, models the property of the angular velocity limitation. The limitation of the integrator's input value results in the limited velocity of the integrator's output. Selecting $k_{\alpha k}$ and B_{max} is necessary in order to achieve the compliance of the model and real propelling system which consists of an engine and gear. We have to take into account the load of the engine due to the camera's weight. Consequently, we obtain the model which meets the requirement $k_{\alpha k} \cdot B_{max} \leq \omega_{\alpha max}$. Where $\omega_{\alpha max}$ is the maximum angular velocity of the real propelling system.

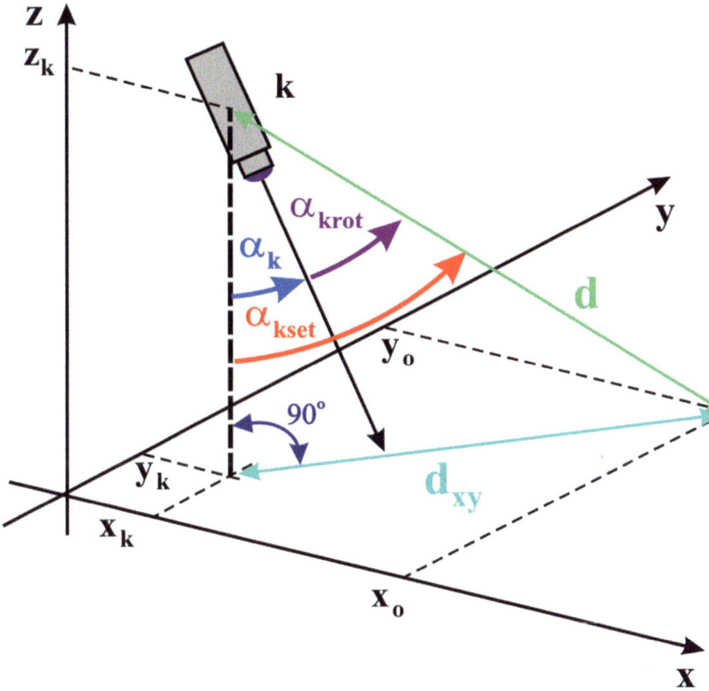

Fig. 6. The spatial scheme of the angles and distances for the camera and tracked object in the vertical plane

The same requirements have to be met for the control systems of the angles α_k and β_k.

The figure 7 represents the UAV and camera head's control system with the limitation of the camera's maximum rotation velocity for the control channel of the angle αk.

The same control system have to be met for the control channels of the angles α_k and β_k.

Generally speaking, we can describe the control algorithm as it is presented below.

Firstly, the control algorithm is used to correct rotation angle in a horizontal plane as it was presented in the figure 5. Secondly, the control system corrects rotation angles in a vertical plane as it is presented in the figure 6. The figure 6 presents relationships between the camera rotation angles in a vertical plane after correcting the rotation angles in the horizontal plane. Basing on the figure 6, the angle α_{krot} at which the camera has to be turned is calculated on the basis of the current camera rotation angle and the angle towards the object.

The first step of the control algorithm comprises of computing the object location when the object is visible in the picture.

The position of the object (x_o, y_o) is calculated on the basis of rangefinder data, GPS and IMU data as it was presented in (1).

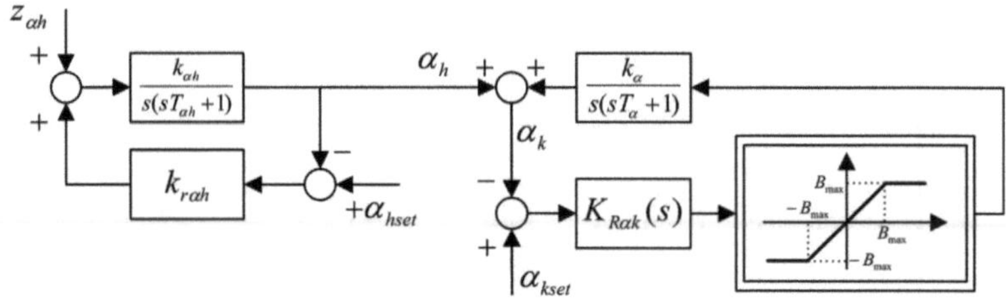

Fig. 7. The UAV and camera head's control system with the limitation of the camera's maximum rotation velocity

$$\begin{Bmatrix} x_o = x_k + d_{xy} \cdot \sin(90^o - \beta_k) \\ y_o = y_k + d_{xy} \cdot \cos(90^o - \beta_k) \end{Bmatrix} \quad \text{for} \quad \beta_k \in \langle 0^o, 180^o \rangle$$

$$\begin{Bmatrix} x_o = x_k + d_{xy} \cdot \sin(90^o + \beta_k) \\ y_o = y_k - d_{xy} \cdot \cos(90^o + \beta_k) \end{Bmatrix} \quad \text{for} \quad \beta_k \in \langle -180^o, 0^o \rangle \tag{1}$$

$$\text{where} \quad d_{xy} = z_k \cdot \tan(\alpha_k)$$

We can state that the distortion occurs when the object is lost from the picture or there are significant changes of the camera position or orientation. Then the helicopter is turned to the position and orientation calculated by the UAV's control system. The UAV's orientation should be almost horizontal and the distance between the UAV and the tracked object should not be too long. Afterwards $x_k, y_k, z_k, \alpha_k, \beta_k$ are measured. The target values of camera rotation angles α_{kset} and β_{kset} are calculated on the basis of the object and camera positions. These angles are calculated by the equations (2) stated below:

$$\alpha_{kset} = \arctan(\tfrac{d_{xy}}{z_k})$$

$$\text{where} \quad d_{xy} = \sqrt{(x_k - x_o)^2 + (y_k - y_o)^2} \tag{2}$$

$$\beta_{kset} = -90^o - \arctan(\tfrac{x_o - x_k}{y_o - y_k}) \quad \text{for} \quad y_k \geq y_o$$

$$\beta_{kset} = 90^o - \arctan(\tfrac{x_o - x_k}{y_o - y_k}) \quad \text{for} \quad y_k < y_o$$

In the next step the correction angles α_{krot} and β_{krot} are computed. The camera has to be turned at these angles. These angles are calculated as shown in the equation (3).

$$\begin{aligned} \alpha_{krot} &= \alpha_{kset} - \alpha_k \\ \beta_{krot} &= \beta_{kset} - \beta_k \end{aligned} \tag{3}$$

6 Examples of the Control System Operation at the Limited Angular Velocities of the Camera Head

The following examples illustrate that the limited angular velocities of the camera head may cause that the object is lost from the picture. The gains of the camera head control system were tuned in such a way as to obtain proper tracking of the object. This may result in high angular velocities in the camera head; moreover, the examples should show us whether it can disable object tracking.

Experiment 1

The low values of the disturbances were examined during the experiment 1.

Experiment 1a - the control system without limits for the angular velocities

The figure 8 presents the UAV's and tracked object's location. In the figure, we can clearly observe the moments of disturbances.

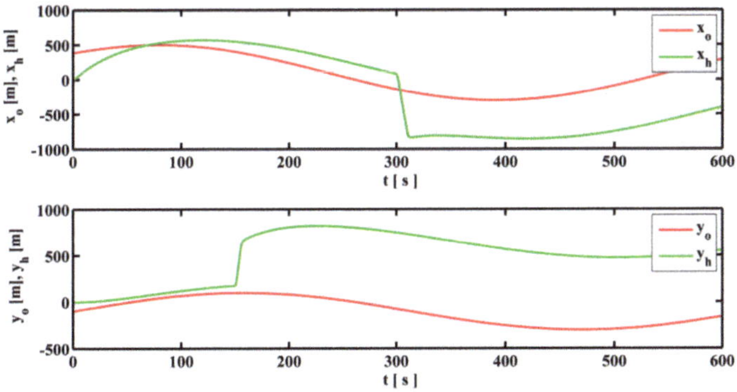

Fig. 8. The UAV's and tracked object's location in the experiment 1a

The figure 9 presents the transients of α_k, $e_{\alpha k}$, u_1, u_2 and $\omega_{\alpha k}$. In this particular case, $\omega_{\alpha k}$ reaches low values. The value of the signal u_1 equal 1 means that the object is out of the camera field of view. The value of the signal u_2 equal 1 means that the object is lost by the control system therefore it is necessary to track the object with a help of the human operator. This situation will be referred to as the "manual operation" further in this paper. u_1 equal 0 during all simulation denotes the correct operation of the control system.

The figure 10 presents the transients of β_k, $e_{\ betak}$, u_1, u_2 and $\omega_{\beta k}$.

We can observe the proper operation of the camera head control system. What is more, $\omega_{\alpha k}$ and $\omega_{\beta k}$ reach values significantly lower than the velocity limits.

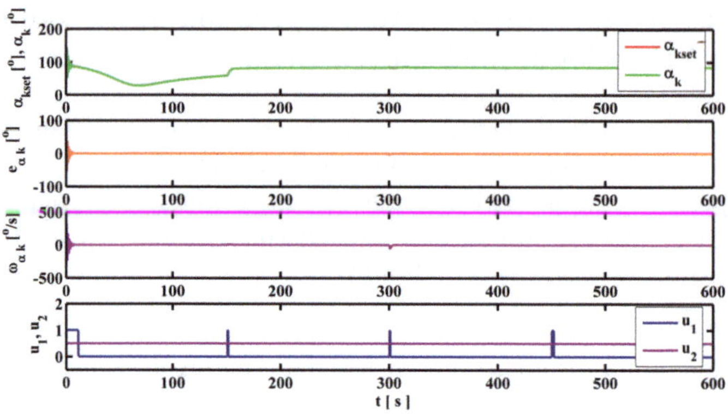

Fig. 9. The transients of α_k, $e_{\alpha k}$, u_1, u_2 and $\omega_{\alpha k}$ in the experiment 1a

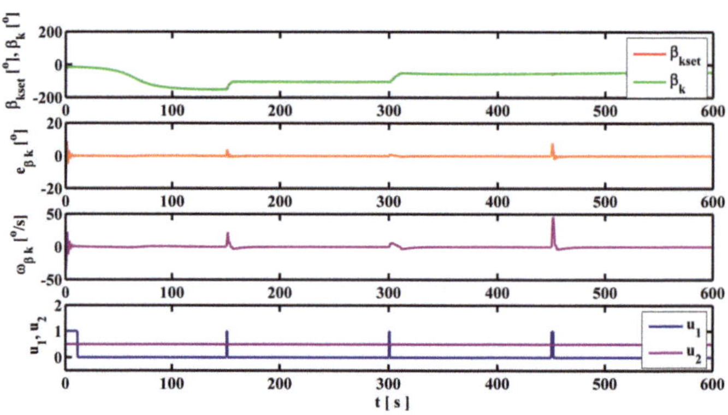

Fig. 10. The transients of β_k, $e_{\alpha k}$, u_1, u_2 and $\omega_{\beta k}$ in the experiment 1a

Experiment 1b - the control system with limits for the angular velocities

The figure 11 presents the UAV's and tracked object's location. In the figure, we can clearly observe the moments of disturbances.

The figure 12 presents the transients of α_k, $e_{\alpha k}$, u_1, u_2 and $\omega_{\alpha k}$.

u_1 equal 0 during all simulation denotes the correct operation of the control system.

The figure 13 presents the transients of β_k, e_{betak}, u_1, u_2 and $\omega_{\beta k}$.

u_1 equal 0 during all simulation denotes the correct operation of the control system.

Owing to the fact that $\omega_{\alpha k}$ and $\omega_{\beta k}$ reach low values in this particular case, the limitation of angular velocities does not influence the object tracking. We can observe the proper operation of the camera head control system.

Experiment 2

In the experiment 2, the examined values of the disturbances are greater than the disturbances in the experiment 1.

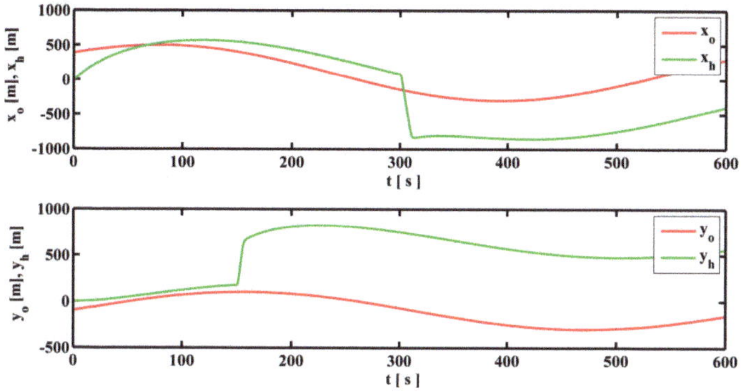

Fig. 11. The UAV's and tracked object's location in the experiment 1b

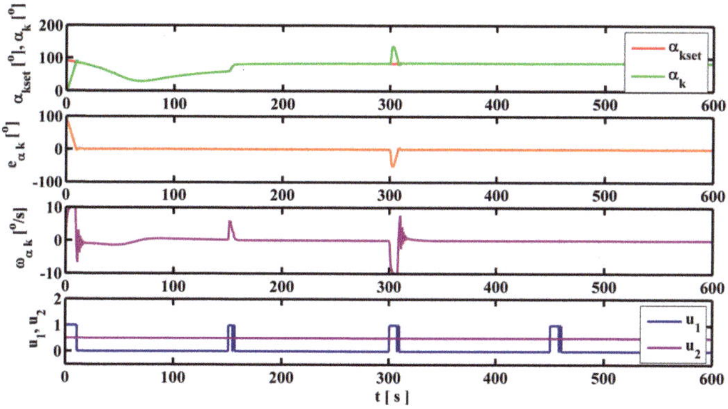

Fig. 12. The transients of α_k, $e_{\alpha k}$, u_1, u_2 and $\omega_{\alpha k}$ in the experiment 1b

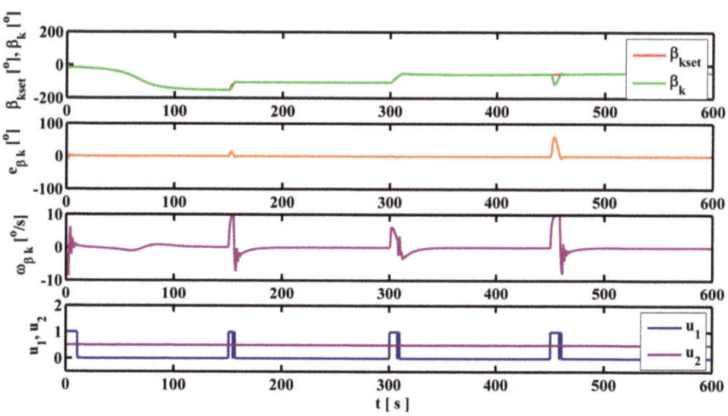

Fig. 13. The transients of β_k, e_{betak}, u_1, u_2 and $\omega_{\beta k}$ in the experiment 1b

Experiment 2a - the control system without limits for the angular velocities

The figure 14 presents the UAV's and tracked object's location. In the figure, we can clearly observe the moments of disturbances.

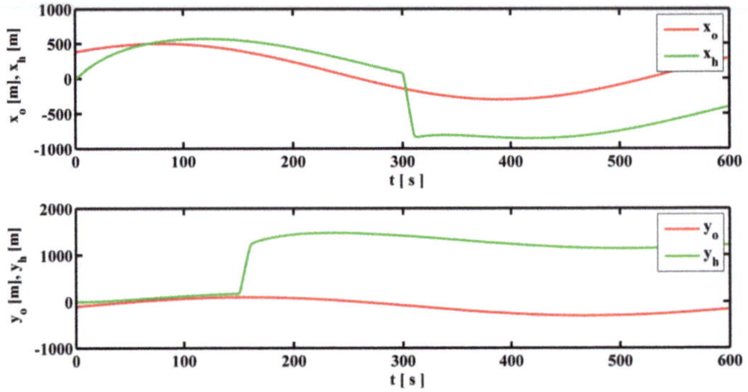

Fig. 14. The UAV's and tracked object's location in the experiment 2a

The figure 15 presents the transients of α_k, $e_{\alpha k}$, u_1, u_2 and $\omega_{\alpha k}$.

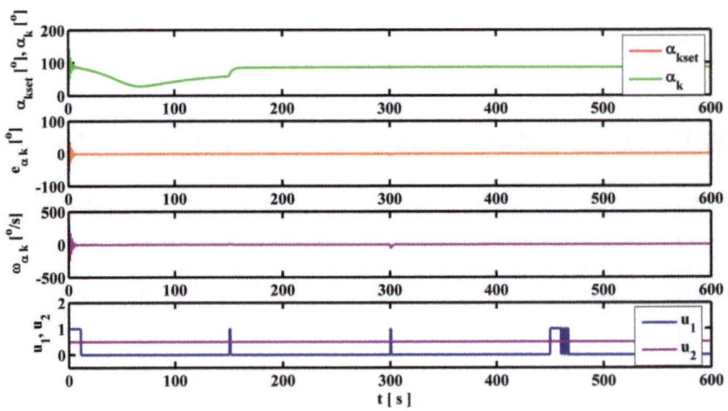

Fig. 15. The transients of α_k, $e_{\alpha k}$, u_1, u_2 and $\omega_{\alpha k}$ in the experiment 2a

The figure 16 presents the transients of β_k, $e_{\ betak}$, u_1, u_2 and $\omega_{\beta k}$.
In this particular case, $\omega_{\alpha k}$ reaches greater values than in the experiment 1.
u_1 equal 0 during all simulation denotes the correct operation of the control system.
We can observe the proper operation of the camera head control system despite the fact that $\omega_{\alpha k}$ and $\omega_{\beta k}$ reach greater values than in the experiment 1.

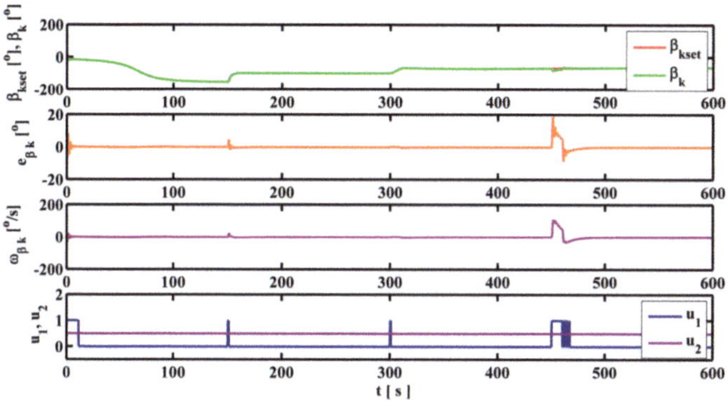

Fig. 16. The transients of β_k, $e_{\beta k}$, u_1, u_2 and $\omega_{\beta k}$ in the experiment 2a

Experiment 2b - the control system with limits for the angular velocities

The figure 17 presents the UAV's and tracked object's location. In the figure, we can clearly observe the moments of disturbances.

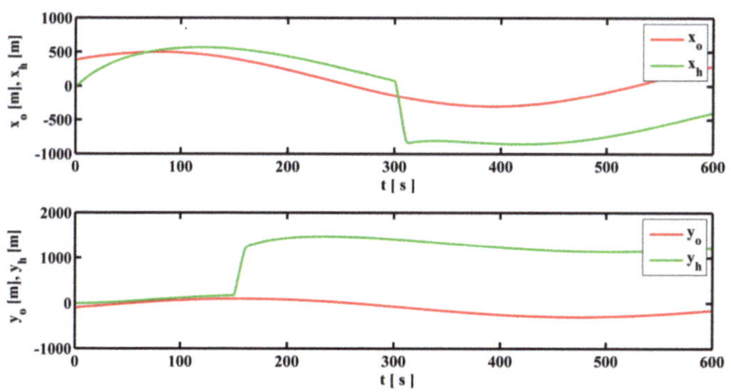

Fig. 17. The UAV's and tracked object's location in the experiment 2b

The figure 18 presents the transients of α_k, $e_{\alpha k}$, u_1, u_2 and $\omega_{\alpha k}$.

The figure 19 presents the transients of β_k, $e_{\ betak}$, u_1, u_2 and $\omega_{\beta k}$.

According to figures 18 and 19, we can notice that the angular velocities slightly exceed velocity limitation. This causes that u_1 equal 1 appears during the simulation. It means that the object is being lost from the picture; however, the control system can restore the correct camera's direction. It shows that the correct operation of the control system is possible when the disturbances are not too large.

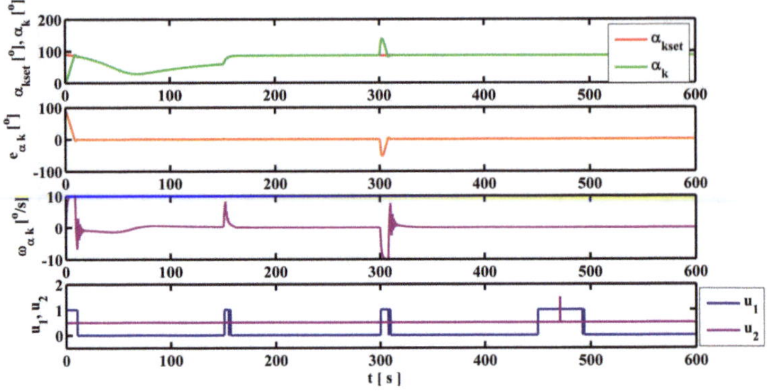

Fig. 18. The transients of α_k, $e_{\alpha k}$, u_1, u_2 and $\omega_{\alpha k}$ in the experiment 2b

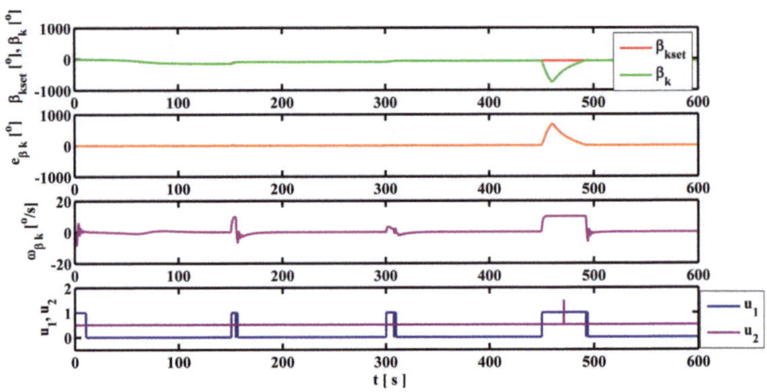

Fig. 19. The transients of β_k, $e_{\beta k}$, u_1, u_2 and $\omega_{\beta k}$ in the experiment 2b

Experiment 3

In the experiment 3, the examined values of the disturbances reach greater levels than the disturbances in the experiment 2.

Experiment 3a - the control system without limits for the angular velocities

The figure 20 presents the UAV's and tracked object's location. In the figure, we can clearly observe the moments of disturbances.

The figure 21 presents the transients of α_k, $e_{\alpha k}$, u_1, u_2 and $\omega_{\alpha k}$.

The figure 22 presents the transients of β_k, $e_{\ betak}$, u_1, u_2 and $\omega_{\beta k}$.

Due to the large values of disturbances, the great values of the angular velocities appear during the simulation. This causes that u_1 equal 1 appears during the simulation. Nevertheless, the control system is able to turn the camera head towards the tracked object.

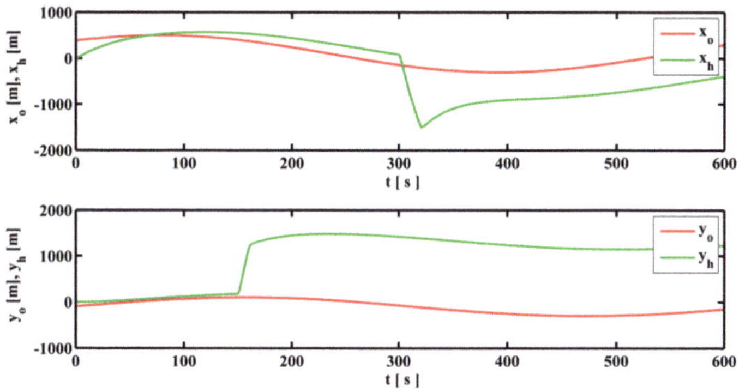

Fig. 20. The UAV's and tracked object's location in the experiment 3a

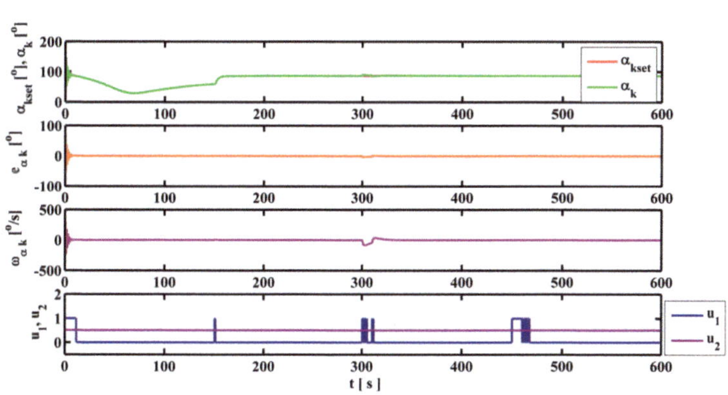

Fig. 21. The transients of α_k, $e_{\alpha k}$, u_1, u_2 and $\omega_{\alpha k}$ in the experiment 3a

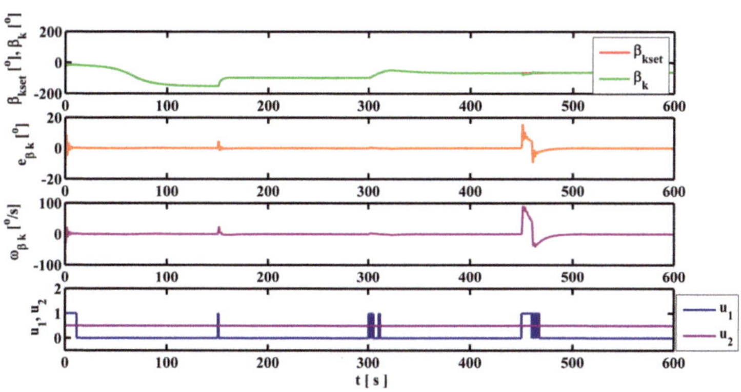

Fig. 22. The transients of β_k, $e_{\beta k}$, u_1, u_2 and $\omega_{\beta k}$ in the experiment 3a

Experiment 3b - the control system with limits for the angular velocities

The figure 23 presents the UAV's and tracked object's location. In the figure, we can clearly observe the moments of disturbances.

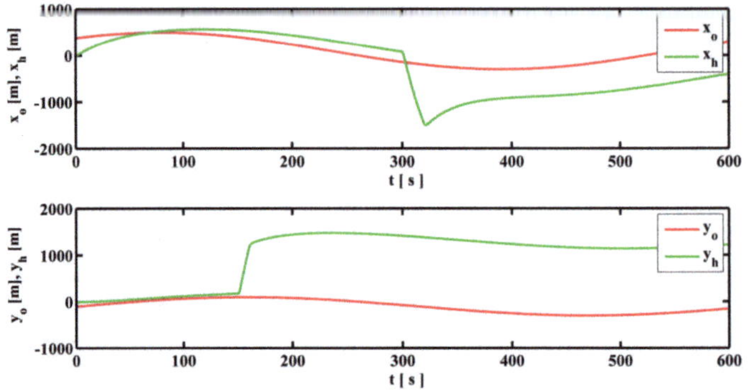

Fig. 23. The UAV's and tracked object's location in the experiment 3b

The figure 24 presents the transients of α_k, $e_{\alpha k}$, u_1, u_2 and $\omega_{\alpha k}$.

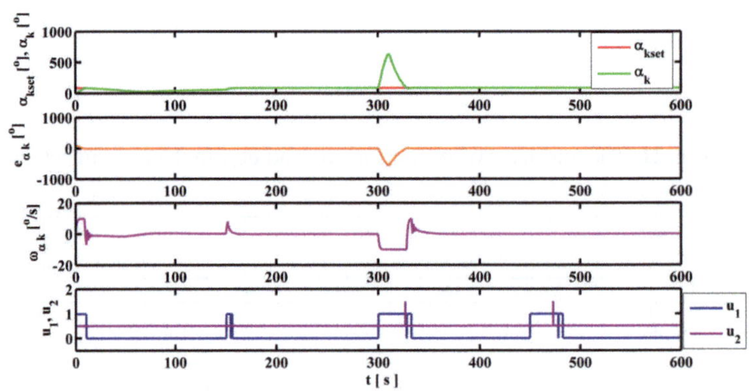

Fig. 24. The transients of α_k, $e_{\alpha k}$, u_1, u_2 and $\omega_{\alpha k}$ in the experiment 3b

The figure 25 presents the transients of β_k, $e_{\ betak}$, u_1, u_2 and $\omega_{\beta k}$.

According to figures 21 and 22, we can notice that for large disturbances the angular velocities greatly exceed velocity limitation. In the pictures 24 and 25 the signal u_1 equal 1 appears during the simulation and it lasts for most of the time. It indicates on the fact that the control system cannot correct the direction of the camera head. The abovementioned situation is caused by too slow resoponse of the control system to the disturbance.

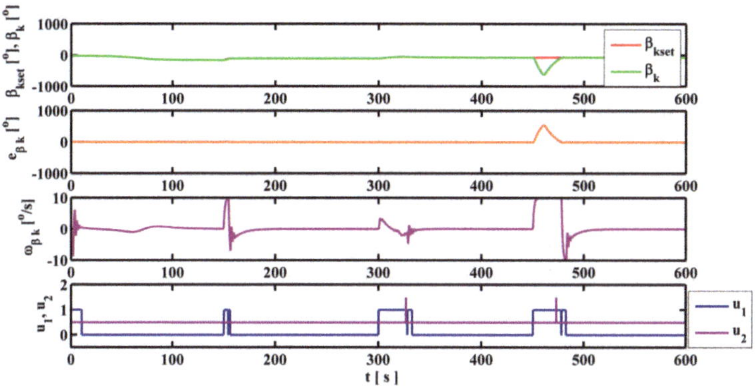

Fig. 25. The transients of β_k, $e_{\beta k}$, u_1, u_2 and $\omega_{\beta k}$ in the experiment 3b

Owing to the presented examples, we can state that the velocity limitations do not disable object tracking in case of minor disturbances. On the other hand, in case of large disturbances the control system has to react promptly. At the stage of design, the camera head engines should be selected according to the predictable weather conditions. These conditions influence the requirements which the UAV and camera head control system have to meet.

7 Conclusions

This study focused on testing the significance of the velocity limitations in the camera head control system. We assumed that the control system should quarntee the tracking of an object moving on the ground, even in the case of the major disturbances. Additionally, the rapid rotations and shifts of the UAV were treated as the disturbances. Furthermore, it was assumed that the object moves on the ground with the velocity lower than the maximum speed of the UAV.

Crucially, the UAV equipped with the flight trajectory control system was considered in the paper. This fact allowed to model the rotations and shifts of the UAV in the particular axes as independent control channels. The study examined the behaviour of the control system for different values of the disturbances. There was presented the comparison of the object tracking efficiency for the case with and without the limitations of the angular velocity of the camera head.

The results showed that in some cases the limitations of the velocity of the camera head rotation are crucial in correct object tracking. However, the abovementioned situation does not always occur. If the disturbances are small or the engines of the camera head have high maximum velocities, than the influence of the velocity limitation can be negligible. In the paper we presented the relationship between both, the value as well as the type of the disturbance and the requirements regarding the necessary velocity of the camera rotation. Another conclusion was that in the case of the disturbances which rotated the UAV in the horizontal or vertical plane, the limitations of the rotation velocity caused far greater worsening of the object tracking than in the case of the disturbances shifting the UAV in the plane (x, y).

The main conclusion of this study was that at the stage of designing the camera head, it was essential to adjust the velocities of the engines to the predictable working conditions - including weather and speed related conditions. What is more, it was suggested in the paper that in the case of the disturbances, which require the velocities of rotation greater than the limit velocities of the engines, it was necessary to use additional information about the tracked object. The velocity and direction of the tracked object might be used to track the object despite losing the object from the camera field of view for some time.

To sum up, the results showed that the limitation of the angular velocity of the camera head is the important parameter of the design of the object tracking system.

References

1. Davies, D., Palmer, P.L., Mirmehdi, M.: Detection and tracking of very small low contrast objects. In: Proceedings of the 9th British Machine Vision Conference (September 1998)
2. Zhang, S., Karim, M.A.: Automatic target tracking for video annotation. Op. Eng. 43, 1867–1873 (2004)
3. Irani, M., Peleg, S.: Improving resolution by image registration. CVGIP: Graph. Models and Image Process. 53, 231–239 (1991)
4. Chesnaud, C., Refegier, P., Boulet, V.: Statistical region snake-based segmentation adapted to different physical noise models. IEEE Trans. Patt. Anal. Mach. Intell. 21, 1145–1157 (1999)
5. Gordon, N., Ristic, B., Arulampalam, S.: Beyond the Kalman Filter: Particle Filters for Tracking Applications. Artech House, Boston (2004)
6. Sharp, C., Shakernia, O., Sastry, S.: A Vision System for Landing an Unmanned Aerial Vehicle. In: Proceedings of the 2001 IEEE International Conference on Robotics and Automation, vol. 2, pp. 1720–1727. IEEE, Los Alamitos (2001)
7. Casbeer, D., Li, S., Beard, R., Mehra, R., McLain, T.: Forest Fire Monitoring With Multiple Small UAVs, Porland, OR (April 2005)
8. Kuś, Z., Fraś, S.: Helicopter control algorithms from the set orientation to the set geographical Location. In: Nawrat, A., Simek, K., Świerniak, A. (eds.) Advanced Technologies for Intelligent Systems. SCI, vol. 440, pp. 3–14. Springer, Heidelberg (2013)
9. Nawrat, A.: Modelowanie i sterowanie bezzałogowych obiektów latających, Wydawnictwo Politechniki Śląskiej, Gliwice (2009)
10. Valavanis, K.P. (ed.): Advances In Unmanned Aerial Vehicles. Springer (2007)
11. Castillo, P., Lozano, R., Dzul, A.E.: Modelling and Control of Mini-Flying Machines. Springer (2005)
12. Padfield, G.D.: Helicopter Flight Dynamics, Backwell Science Ltd. (1996)
13. Manerowski, J.: Identyfikacja modeli dynamiki ruchu sterowanych obiektów latających. WN ASKON, Warszawa (1999)
14. Kearney, J.K., Thompson, W.B.: Optical flow estimation: An error analysis of gradient-based methods with local optimization. IEEE Transactions on Pattern Analysis and Machine Intelligence 9, 229–243 (1987)
15. Anandan, P.: A computational framework and an algorithm for the measurement of visual motion. International Journal Qf Computer Vision 2, 283–310 (1989)
16. Barman, H., Haglund, L., Knutsson, H., Granlund, G.: Estimation of velocity, acceleration, and disparity in time sequences. In: Proc. IEEE Workshop on Visual Motion, Princeton, NJ, pp. 44–51 (1991)

17. Bascle, B., Bouthemy, E., Deriche, N., Meyer, E.: Tracking complex primitives in an image sequence. In: Proc. 12th International Conference on Pattern Recognition, Jerusalem, pp. 426–431 (1994)
18. Butt, E., Yen, C., Xu, X.: Local correlation measures for motion analysis: A comparative study. In: Pattern Recognition and Image Processing Conference, Las Vegas, pp. 269–274 (1982)
19. Butt, E., Bergen, J.R., Hingorani, R., Kolczynski, R., Lee, W.A., Leung, A., Lubin, J., Shvayster, H.: Object tracking with a moving camera. In: Proc. IEEE Workshop on Visual Motion, Irving, pp. 2–12 (1989)
20. Buxton, B.E., Buxton, H.: Computation of optical flow from the motion of edges features in image sequences. Image and Vision Computing 2(2), 59–75 (1984)
21. Campani, M., Verri, A.: Computing optical flow from an overconstrained system of linear algebraic equations. In: Proc. 3rd International Conference on Computer Vision, Osaka, pp. 22–26 (1990)
22. Campani, M., Verri, A.: Motion analysis from firstorder properties of optical flow. CVGIP: Image Understanding 56(1), 90–107 (1992)
23. Carlsson, S.: Information in the geometric structure of retinal flow field. In: Proc. 2nd International Conference on Computer Vision, pp. 629–633 (1988)
24. Gessing, R.: Control Fundamentals. Silesian University of Technology, Gliwice (2004)
25. Kuś, Z., Nawrat, A.: Object tracking in a picture during rapid camera movements. In: Nawrat, A., Kuś, Z. (eds.) Vision Based Systems for UAV Applications. SCI, vol. 481, pp. 77–91. Springer, Heidelberg (2013)
26. Kuś, Z., Nawrat, A.: Object Tracking for Rapid Camera Movements in 3D Space. In: Nawrat, A., Kuś, Z. (eds.) Vision Based Systems for UAV Applications. SCI, vol. 481, pp. 57–76. Springer, Heidelberg (2013)
27. The raport on the project O R 00 0132 12: "Design and implementation of innovative unmanned mobile platforms to monitor the state border" developed in Silesian Technical University, Gliwice (2010-2012)
28. Jędrasiak, K., Bereska, D., Nawrat, A.: The prototype of Gyro-Stabilized UAV Gimbal for Day-Night Surveillance. In: Nawrat, A., Simek, K., Świerniak, A. (eds.) Advanced Technologies for Intelligent Systems. SCI, vol. 440, pp. 107–116. Springer, Heidelberg (2013)
29. Kuś, Z., Nawrat, A.: Camera head control system with a changeable gain in a proportional regulator for object tracking

Camera Calibration for Tracked Vehicles Augmented Reality Applications

Dawid Sobel[1], Karol Jędrasiak[1], Krzysztof Daniec[1], Józef Wrona[2,3], Piotr Jurgaś[1], and Aleksander M. Nawrat[1]

[1] Institute of Automation, Silesian University of Technology, Gliwice, Poland
{dawid.sobel,karol.jedrasiak,krzysztof.daniec,piotr.jurgas, aleksander.nawrat}@polsl.pl
[2] Military University of Technology, ul. gen. Sylwestra Kaliskiego 2, 00-908 Warsaw, Poland
jwrona@wat.edu.pl
[3] Industrial Research Institute for Automation and Measurements, Aleje Jerozolimskie 202, 02-486 Warsaw, Poland
jwrona@piap.pl

Abstract. The article presents a method for camera calibration for tracked vehicles augmented reality applications. The method does not require any additional optical markers or symbols placed within the observed scene. The approach can be used regardless the quality of placement of optical markers, which could be disturbed due to object's damage or contamination. The presented method can be applied to every object of known dimensions and is immune to light conditions, objects contamination and known deformations due to e.g. age or damage. Moreover it can be used both in real-time as well as in post-processing using object's schematics. It allows implementation of Augmented Reality applications and systems in various places like tracked vehicles that previously were thought to be challenging or impossible. Therefore, the presented approach can be used to improve the quality of usage, diagnostics and repairs of tracked vehicles. In the article there are presented algorithms and their quality evaluation and an examples of applications.

Keywords: camera calibration, augmented reality, image processing.

1 Introduction

Augmented reality is slowly becoming popular area of everyday life. It is justified by the fact that it is possible to find unlimited number of useful AR solutions. Various applications can be applied in almost every field of science and industry like e.g. military, medicine, education technology. Augmented Reality connects like a bridge the real world with a virtual, generated by the computer, world. It allows augmentation of human senses in particular our sense of sight. Especially e.g. detected and identified objects by the pilot, telemetry information about aerial vehicles or psychophysical condition of the patient. The augmented information is

A.M. Nawrat (ed.), *Innovative Control Systems for Tracked Vehicle Platforms*, Studies in Systems, Decision and Control 2,
DOI: 10.1007/978-3-319-04624-2_8, © Springer International Publishing Switzerland 2015

merged with our natural perception of world therefore it is visually acceptable. In order to use AR a specific device like AR helmet is required.

There are two main problems to solve before the widespread usage of AR. First, the problem of rendering and merging virtual objects along with the high quality video stream [13]. Second problem can be stated as a problem of finding a transformation of virtual scene into the perceived by the human scene (video camera). It requires that virtual objects are placed in the correct 3D position, orientation and comply with the human eyes scale factor. For every frame of video stream the virtual objects have to be placed in a correct position and orientation and rendered again. It is a complex and time consuming process. The process of finding the geometric transformation between the coordinate systems is called camera calibration. There are multiple methods of camera calibration the most common are based on detection of specified optical markers, known pattern like checker board or tracking visual features called self-calibration. However, the research is constantly being carried out in order to find more reliable, easier or faster methods of calibration.

2 Augmented Reality

The concept of Augmented Reality is nothing new. However the term and the possibilities for applications were born recently. The probably first use of the term is dated 1997 [3]. Due to rapid development of computer technologies, GPS receivers and inertial measurement units it was possible to create more immerse technology for human users. The unlimited possibilities of improving the perception of the world of given by AR to the scientists and developers increased the number of AR devices and applications. A good example of an application is a system consisting of computer software and paperboards with optical markers printed on them. The system allows checking how virtual furnishings would fit into a real room [4]. The user is responsible for placing the markers within the room. Next, using the camera the scene of the room is perceived. The computer software is responsible for augmenting the detected and identified markers with 3D models of furniture. Rotating the paperboards with markers rotate the virtual furniture allows adjustment of them along the vision of the end user of the room. Such system allows quick visualization of results and saves time of both architects and end users. A similar application is existing in some high-schools where pupils during chemistry classes are able to observe three dimensional models of water particles after placing the optical markers of hydrogen and oxygen particles in front of AR camera [5].

Augmented reality applications are also widely used by producers of software and devices in the navigational field. Such devices or program display virtual map module with marked trajectory of the vehicle and real-time video stream from the video camera mounted on the vehicle. The system is responsible for detection and identification of traffic signs, speed limits and any additional information that might be useful for driver. Therefore the system increases safety of drivers by placing one single display in front of eyes of the driver in a way the he/she does not need to waste time looking around or in the maps and can concentrate on safe driving [6]. Another scientist deducted that it might be useful for a driver to display an arrow pointing where to drive, the speed, the distance traveled or incoming voice transmissions along with a speaker's avatar [7].

Another type of AR devices is a head mounted display (HMD). It is a device with a small screen and lenses in front of user's eyes [8]. The video signal obtained from a camera mounted to the helmet is processed by a microprocessor and finally displayed to the user using HMD in such a way that the user is observing at the same time real world and augmented information. Currently used solutions are large and disturbs normal functioning. However, minimalizing the volume of the helmet or glasses [9] is an important problem and numerous scientists try to solve it in order to provide a device that can be carried by everyone without a shame. A concept of glasses presented by Google company utilizes functionalities like: informing about incoming transmissions, messages, weather broadcasts, personal schedule and so on. Nowadays for all the functionalities the smartphones are used. However due to the AR technology it is possible to simplify their usage and allow applications in e.g. tracked vehicles [16].

3 Camera Calibration

Camera calibration relies on finding the internal and external camera parameters. Internal parameters are related to the optical and electronic properties of the camera and they include: the focal length, radial distortion and tangent coefficients and the actual coordinates of the camera image center. External parameter are those that describe the transformation, that is the rotation and transformation of the system associated with the observed 3D scene to camera system (Fig.1). With this transformation, we are able to find the points registered on the camera image corresponding to the 3D points. Camera calibration procedure usually involves targeting it to an object of known shape and known pattern or an object with known 3D coordinates of its characteristic points.

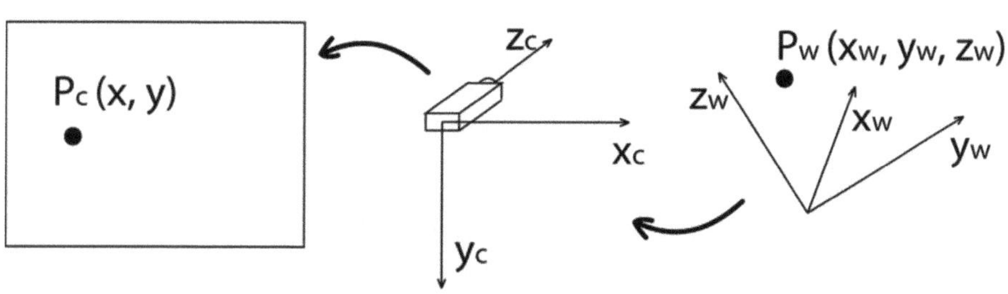

Fig. 1. Transformating 3D coordinates to 2D image coordinates

Various calibration methods often differ in terms of precision and requirements for computing power. According to preferences a fast but prone to error methods are used like DLT [11]. The method allows to find a linear transformation between points in coordinate system of 3D scene and image coordinate system. The transformation does not include the internal camera parameters. Therefore, it is only required to provide a solution for a system of linear equations that significantly increases the performance of the solution. Unfortunately, most of currently used video cameras suffer from various imperfections that should be taken into account during computations in order to acquire results within an acceptable error margin.

$$\begin{bmatrix} x \\ y \end{bmatrix}_{camera} = A \begin{bmatrix} x \\ y \\ z \end{bmatrix}_{word}, \tag{1}$$

Unlike the DLT a method focus on quality was presented by Fig [12]. The method requires high computing power due to expanded camera model which takes into account both internal camera parameters as well as video camera acquisition system characteristics. In the method the initial guess values are further improved using nonlinear optimization. Such method may be used in systems that require the highest precision and are designed for offline utilization.

The method that combines the advantages of both of the methods was presented by Tsai [1]. In order to neutralize those two problems a concept of split the algorithm into two parts was suggested. First, the linear method is used in order to find an initial guess for both internal and external video camera parameters. Second, the values are improved using nonlinear optimization procedure (eq. 2, eq. 3). Because of the precise first step usually only one or two iterations are sufficient in order to find the solution within an acceptable margin.

$$s_x^{-1} d'_x X + s_x^{-1} d'_x X k_1 r^2 = f \frac{r_1 x_w + r_2 y_w + r_3 z_w + T_x}{r_7 x_w + r_8 y_w + r_9 z_w + T_z}, \tag{2}$$

$$d'_y Y + d'_y Y k_1 r^2 = f \frac{r_4 x_w + r_5 y_w + r_6 z_w + T_y}{r_7 x_w + r_8 y_w + r_9 z_w + T_z}, \tag{3}$$

where sx, dx, dy parameters are the parameters of CDD matrix. Points X and Y are respectively the coordinates of the pixel in the image vertically and horizontally and where xw, yw, zw are the their corresponding coordinates in the observed 3D scene. Radial distortion coefficient k1 and facal length f represent the camera internal parameters and where rotating matrix elements r1… r9 and the shift vector elements Tx, Ty, Tz are the external parameters.

An interesting propos al of eliminating the negative effects of the distortion of the camera's optical system is a method introduced in [2]. Using simple formulas (Eq. 4, Eq. 5) it is possible to calculate the actual location of the distorted point in the image (xd, yd).

$$x_d = x(1 - \gamma) + x_c,$$
$$y_d = y(1 - \gamma) + y_c \quad (4)$$

$$\gamma = \lambda \frac{x^2 + y^2}{x_c y_c}, \quad (5)$$

$$x = x_r - x_c, y = y_r - y_c, \quad (6)$$

where x_c, y_c are the coordinates of the image's centre, parameter λ is selected by the user with an experimental method in the field (0.001, 0.1) depending on the level of distortion. Coefficients x_r, y_r are the coordinates of pixels on the recorded image to a particular point.

4 Camera Calibration Algorithm

The presented camera algorithm assumes that both 3D coordinates in a 3D scene coordinates system and 2D coordinates in the image coordinate system (pixels) are known.

The first step is to collect the video frame, on which the largest number of calibration points is seen.

These points are either in the local coordinates system of the object or in the external circuit of the observed 3D scene. After collecting the frame, detecting the pixels coordinates in an image (Xfi, Yfi) corresponding to those points is needed. The number of points must be greater than 7. Then, using the information from the manufacturer we determine the value of each parameter: Ncx – number of sensors in the CCD level, Nfx – width of the collected frame in pixels, dx – distance between the centers of succeeding sensors on the X, d'x = dxNcx/Nfx, dy distance between the centers of the sensors on the Y, sx – the scale factor calculated in the calibration process at the beginning equal to 1. If we do not know the values of these parameters, all are assumed as 1. Then the image center coordinates Cx and Cy are set. Most often this is done by dividing the width and height of the image in half, however, it is recommended to check if the edges of the image do not contain blank lines of pixels. If such exist, they are not counted to the width of height of an image and then the coordinates of the frame, reduced by the blank pixel lines, are determined. Next we trans form the selected calibration points (Eq. 7, Eq. 8) as follows. We perform this procedure for each selected point.

$$X_{di} = s_x^{-1} d'_x (X_{fi} - C_x), \quad (7)$$

$$Y_{di} = d_y (Y_{fi} - C_y). \quad (8)$$

The next step is to calculate the 7 unknown points: $T_y^{-1} s_x r_1$, $T_y^{-1} s_x r_2$, $T_y^{-1} s_x r_3$, $T_y^{-1} s_x T_x$, $T_y^{-1} r_4$, , $T_y^{-1} r_6$. For this purpose, an formed overdetermined linear system is formed, in which the unknowns are the above- mentioned parameters and each equation corresponds to one calibration point. (Eq. 9). This system can be solved, for ex ample, with the method of east squares.

$$[Y'_{di}x_{wi} \quad Y'_{di}y_{wi} \quad Y'_{di}z_{wi} \quad Y'_{di} \quad -X'_{di}x_{wi} \quad -X'_{di}y_{wi} \quad -X'_{di}z_{wi}] \begin{bmatrix} T_y^{-1}s_x\, r_1 \\ T_y^{-1}s_x\, r_2 \\ T_y^{-1}s_x\, r_3 \\ T_y^{-1}s_x\, T_x \\ T_y^{-1}r_4 \\ T_y^{-1}r_5 \\ T_y^{-1}r_6 \end{bmatrix} = X'_{di}. \tag{9}$$

After determining the unknown parameters, a calculation is needed ($r_1, \ldots, r_9, T_x, T_y$). For this purpose, we set $|T_y|$ using the (Eq. 10) formula. To simplify, the following signs are taken:

$$a_1 = T_y^{-1}s_x\, r_1, a_2 = T_y^{-1}s_x\, r_2, a_3 = T_y^{-1}s_x\, r_3, a_4 = T_y^{-1}s_x\, T_x,\; a_5 = T_y^{-1}r_4,\; a_6 = T_y^{-1}r_5,\; a_7 = T_y^{-1}r_6. \tag{10}$$

Having $|T_y|$ the sign T_y has to be determined. A point, which coordinates in the computer image are the further away from the center of the(C_x, C_y) is selected. The T_y sign is set as positive and following factors are calculated:

$$r_1 = (T_y^{-1}s_x\, r_1)T_y, r_2 = (T_y^{-1}s_x\, r_2)T_y, r_4 = (T_y^{-1}r_4)T_y, r_5 = (T_y^{-1}r_5)T_y, x = r_1x_w + r_2y_w + T_x, \quad y = r_4x_w + r_5y_w + T_y, T_x = (T_y^{-1}s_x\, T_x)T_y.$$

Values $T_y^{-1}s_x\, r_1T_y$, $T_y^{-1}s_x\, r_2$, , $T_y^{-1}r_5$, $T_y^{-1}s_x\, T_x$ were determined from the system of equations (Eq. 9). Parameters x_w and y_w are the coordinates in the 3D space of the selected point and X and Y are the coordinates of the point in the image. If the calculated x and X have the same sign, y and Y have the same sign, then sign T_y is positive . In other case, sign T_y is negative. Using the previously calculated absolute value $|T_y|$ (Eq. 10) the value of the scale factor is calculated s_x (Eq. 11).

$$s_x = \sqrt{a_1^2 + a_2^2 + a_3^2}\,|T_y|. \tag{11}$$

Knowing s_x parameters $r_1, \ldots, r_7$ and T_x displacement are calculated.

$$r_1 = a_1 T_y / s_x, r_2 = a_2 T_y / s_x, r_3 = a_3 T_y / s_x$$
$$r_4 = a_5 T_y, r_5 = a_6 T_y, r_6 = a_7 T_y \tag{12}$$
$$r_7 = r_2 r_6 - r_3 r_5, T_x = a_4 T_y.$$

Taking advantage of the fact that the matrix R is orthonormal and using the right-hand rule we determine the three rotation θ, ϕ, ψ (Eq. 13) which determine the rotations towards axis x, y, z.

$$\psi = atan2(r_4, r_1),$$
$$\phi = atan2(r_3 \sin\psi - r_6 \cos\psi,\ r_5 \cos\psi - r_2 \sin\psi), \tag{13}$$
$$\theta = atan2(-r_7,\ r_1 \cos\psi + r_4 \sin\psi).$$

We calculate the rotation elements of the matrix R (Eq. 14):

$$R = \begin{bmatrix} \cos\psi\cos\theta & \sin\psi\cos\theta & -\sin\theta \\ -\sin\psi\cos\phi + \cos\psi\sin\theta\cos\phi & \cos\psi\cos\phi + \sin\psi\sin\theta\sin\phi & \cos\theta\sin\phi \\ \sin\psi\sin\phi + \cos\psi\sin\theta\cos\phi & -\cos\psi\sin\phi + \sin\psi\sin\theta\cos\phi & \cos\theta\cos\phi \end{bmatrix}. \tag{14}$$

At this stage, the complete rotation matrix R and two elements of the displacement vectors T_x *and* T_y are known. The next step is to determine the approximate values of the focal length f and T_z displacement. To do that, a system of equations (Eq. 15) is created, which size depends on the number of calibration points and where the unknown are the parameters mentioned above. Again, a method of least squares can be used to solve this formula:

$$[y_i \quad -d_y Y_i], \begin{bmatrix} f \\ T_z \end{bmatrix} = w_i d_y Y_i. \tag{15}$$

where y_i and w_i for succeeding points are described by formulas (Eq. 16).

$$y_i = r_4 x_{wi} + r_5 y_{wi} + T_y,$$
$$w_i = r_7 x_{wi} + r_8 y_{wi}. \tag{16}$$

The calculation of the approximate values f, T_z and k_1= 0 should be used as initial conditions for non-linear optimization. The equation on the basis of which we optimize was already presented by (Eq. 3). On its basis, the quality index is created which allows to find the optimal solution. Method that can be used here is a method of steepest descent.

The nonlinear optimization is the last step of the presented calibration method. After all the internal and external parameters are established. Using them it is possible to rectify the distorted images or render the virtual objects in a correct position, orientation and scale. In order to register two images that vary in terms of image distortion it is required to find the transformation or to undistorted both images. In order to render 3D virtual object into image from video camera multiple steps are required to pass. First, the 3D mesh of vertices of the object has to be created. Next, each of the points of the mesh has to be transformed from 3D scene coordinates system into 2D image coordinates system. The transformation is described by eq. 17:

$$\begin{bmatrix} x_{camera} \\ y_{camera} \\ z_{camera} \end{bmatrix} = R \begin{bmatrix} x_{word} \\ y_{word} \\ z_{word} \end{bmatrix} + \begin{bmatrix} T_x \\ T_y \\ T_z \end{bmatrix}. \tag{17}$$

The transformation from 3D into image 2D is performer in two steps. First, a conversion from an origin of the 3D coordinates system into camera coordinates system. Then from 3D to 2D using eq. 18:

$$x = f(x_{camera}/z_{camera}), \quad y = f(y_{camera}/z_{camera}). \tag{18}$$

Then in the second step the position of (x, y) of our point is distorted to fit it into the naturally distorted image (Eq. 19):

$$x_d = x(1 - \gamma) + C_x, \quad y_d = y(1 - \gamma) + C_y, \tag{19}$$

where γ is defined by (Eq. 5). Knowing the position of an object in an image we are able to draw it.

5 Augmented Reality Calibration Usage

Camera calibration allows combination of virtual world and real world objects in a single display. Augmented reality applications all are based on a process camera calibration. Regardless the method goal is to find a transformation from one coordinates system into another. Most of calibration methods used for AR are based on paper boards with unique markers printed on them. By observing such markers it is possible to identify the rotation and scale of the marker, therefore rendering a virtual object on top of it. Unfortunately, such paper or printed markers can be easily damaged making them useless in most of military applications where the used devices are often slightly damaged or contaminated during battle operations.

There is a request from industry for a solution that is immune to such conditions. The presented in the article calibration method requires only that observed in the video stream objects are of known dimensions. It can be easily achieved by using schematics. All military devices comply to strict norms, therefore their dimensions are well known. Such method can be easily used both in designed for it vehicles as well in old, already existing ones. It is worth to mention that it is completely immune to light conditions and objects contamination. At the same time any damage to objects can be measured or simply estimated by the system operator and sufficient corrections can be applied. Such flexibility is novelty of practical importance. If a remote connection to the HQ is active it is possible even for a person without tools or knowledge of the object to use the calibration method thanks to communication with an expert. Another useful feature of the method is the possibility to track the changes of external camera parameters using for e.g. inertial measurement units or optical flow. That way it is not required to find corresponding points in every frame but only in one and then track the changes in external camera parameters matrix.

Using the calibration method it is easy to augment the image with a text or 3D models in order to aid vehicles diagnostics or remote repairs. In a crisis situation a remote expert can successfully guide the user signaling the breakdown of the vehicle.

6 Software

In order to verify and visualize the capabilities of the algorithms a dedicated application was designed and implemented. The software main functionalities are loading and zooming images and then augmenting them. The process is distributed into few steps. First, user chooses a photograph or a movie that will be used for augmentation. During next step it is required to point and click an origin of the 3D coordinates system and then to select corresponding points in 2D along with their distances in real world (3D). After selection of a sufficient number of corresponding points a calibration is performed. When external and internal parameters are computed by the software it is possible to render additional text, lines or scaling, rotating and translating virtual objects rendered into the image. The software can be used for calibration into any existing devices of a known dimensions and for augmenting and configured text, lines or 3D models therefore its potential for future applications is infinite. The main areas for future applications are: telemedicine, experiments with goniometer, military applications and all already recorded movies that should be augmented offline.

Visual verification of correctness of the calibration method and the capabilities of the software for image augmentation is presented in fig. 2. The process of selection corresponding points in 3D and 2D can be automated by one of existing methods especially in military applications. However it is beyond the scope of this article.

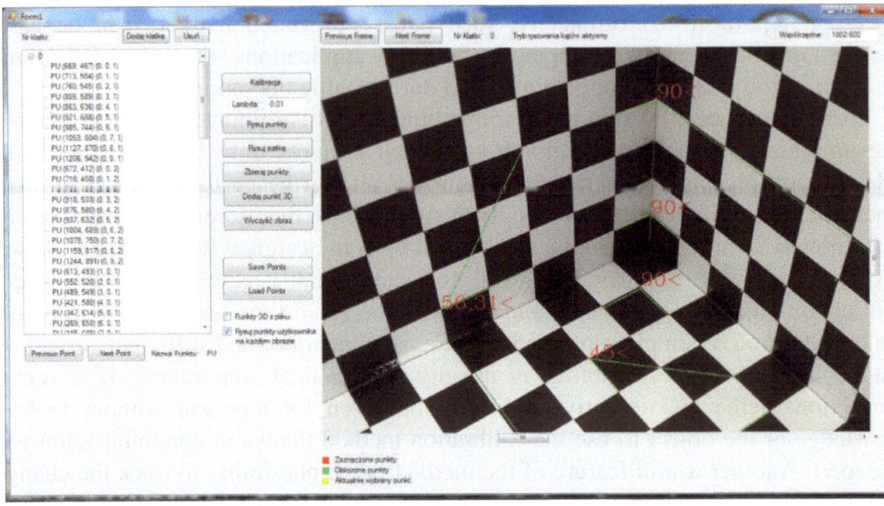

Fig. 2. The graphical user interface presenting the calibration pattern augmented with various virtual elements

7 Tests

The presented camera calibration algorithm has been tested using the described software. The machine used during tests was a HP Pavilion dv6 1308 laptop. Aim of the tests was checking the quality of the calibration depending on the amount of the selected points and the ability to calibrate on any object of known dimensions. Two various objects were used during the tests. The first, shown in the screenshot from the application (fig. 1) is composed of three vertically stacked together checker boards. Size of each checker board is 20 cm x 20 cm. Each square side length was 2 cm. The use of this pattern allowed to determine the 3D coordinates of the calibration points that were chosen at the corners of the squares of the same color. Selection of the corresponding image coordinates formed the basics for further processing. As a second object we used an old monitor with size 39 cm x 33 cm x 40 cm. During tests there was performed an attempt to camera calibration based on knowledge of only the observed object's real world dimensions.

8 Measuring the Reprojection Error

For this test a spatial pattern with checker boards was used. During the initial test a maximum number of visible characteristic points (corners) were selected and during experiments some of them were randomly omitted during the camera calibration process. The reprojection error was computed using the 3D coordinates of the omitted points and their computed 2D position in the image. The difference between the real position and the calculated was treated as an error. The total error was a sum of all point errors divided by the number n (eq. 20):

$$Error = \sqrt{\frac{\sum_{0}^{n}[(X_{comp}^{i} - X_{calib}^{i})^2 + (Y_{comp}^{i} - Y_{calib}^{i})^2]}{n}} \quad (20)$$

The acquired results are presented in the fig. 3. and fig. 4. It can be seen that the number of points affects the process of calibration. The application allows initializing the process of calibration with 8 points or more. For a small number of points, the error is significant due to insufficient information about camera parameters within them. With the increase of the amount of used points the error begins to decrease until it reaches a minimum for 36 points in the case of spatial checker pattern. This is because the number of points is sufficient to properly define the space but due to the small number of points only in the area closely associated with calibration points. It is known that reprojection error increases with distance of the objects from the camera [13]. It can be observed that using more or equal to 36 points the error begins to grow. However it is still relatively small and sufficient for AR applications. It is an effect of inaccurate indication of the points due to imperfections of the checker board pattern and selection of corners. For a significant amount of points like over 75 the error decreases due to significant coverage of the image which allows more accurate estimation of camera parameters in the case of optical distortion. In the fig. 4 the whole data was cropped to the interval starting from 30 points.

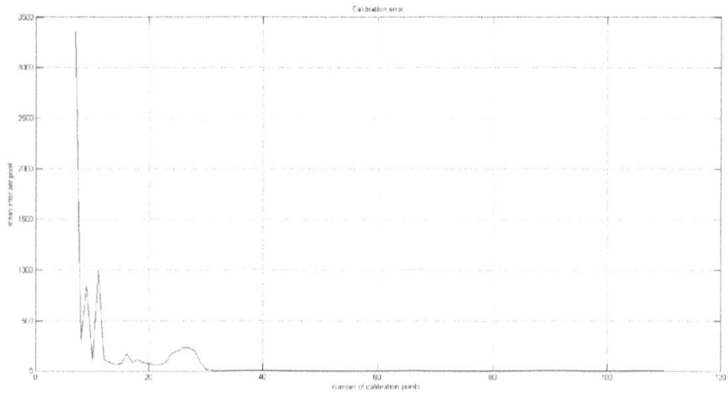

Fig. 3. A plot of the change of the reprojection error depending on the number of calibration points

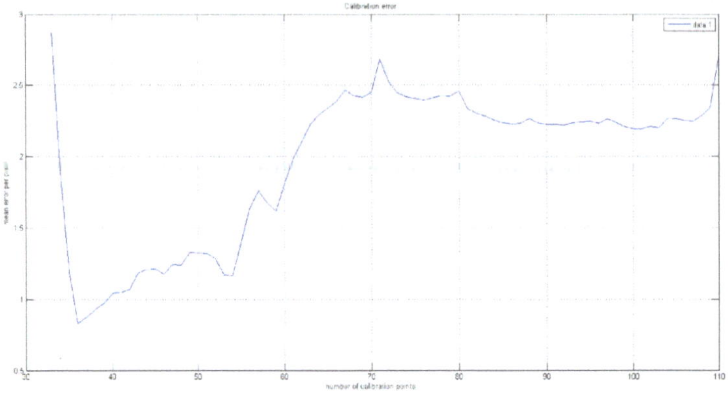

Fig. 4. A plot of the change of the reprojection error depending on the number of calibration points

9 Augmentation of the Object of Only Known Dimensions

In order to verify the practical capabilities of the presented method and software an augmentation test was designed. The goal of the test was to take image of the same scene from two different viewports and use the developed software in order to find the camera parameters (calibrate it) and use them for AR application. The simplified virtual model of one of Silesian University buildings along with the "POLSL" text was used. During the experiment the visible in the image table, monitor and notebook were measured and used as a source for corresponding points. The origin of the

Fig. 5. An image augmented with the predefined 3D models

Fig. 6. An image augmented with the predefined 3D models from other perspective

coordinates system was placed in a lower left corner of the table (right next to the blank a4). The axes were selected as follows: x axis – depth, y axis – width, z axis – height. The acquired results are presented in fig. 5 and fig. 6. It can be observed that they are both visually acceptable. The presented result was for 27 corresponding points.

10 Conclusions

Nowadays Augmented Reality applications are becoming more and more popular due to their simplicity for presenting complex data to the end user. In the article it was presented a novel method for camera calibration dedicated for AR applications. The method can be applied to augmenting the cockpit of tracked vehicles and aiding their diagnostics and repairs. Especially in battlefield conditions when some of the elements may by damaged and contaminated. Authors present as well a designed and implemented software for camera calibration and image and video augmentation as well. It is dedicated for tracked vehicle applications. However due to its abstract design it can be tuned for other applications like telemedicine or medicine, especially for aiding orthopedists.

The presented algorithm can be treated as an another step into building useful bridge between the real and virtual reality worlds. It was presented that basic knowledge and information such as objects dimensions are sufficient for calibrating the camera that was used to perceive the images in the test. Such method can be used for both real-time video camera stream and offline already recorded movie. The software allows for inserting virtual objects into the 3D scene with a visually acceptable quality.

The potential application of the software and method are limited only by the creativity of the end user. It is worth to mention that the method is immune to light conditions, objects contamination or known deformations due to age or damage.

References

[1] Tsai, R.Y.: A Versatile Camera Calibration Technique for High-Accuracy 3D Machine Vision Metrology Using Off-the-Shelf TV Cameras and Lenses. IEE Journal of Robotics and Automation RA-3(4) (August 1987)

[2] Babiarz, A., Bieda, R., Jaskot, K.: Vision System for Group of Mobile Robots. In: Nawrat, A., Kuś, Z. (eds.) Vision Based Systems for UAV Applications. SCI, vol. 481, pp. 139–156. Springer, Heidelberg (2013)

[3] Azuma, R.: A Survey of Augmented Reality, Presence: Teleoperators and Virtual Environments 6(4), 355–385 (August 1997)

[4] Kato, H., et al.: Virtual Object Manipulation of a Table-Top AR Environment. In: Proc. Int'l Symp. Augmented Reality 2000 (ISAR 2000). Munich, 5-6, pp. 111–119 (October2000)

[5] Information on: http://www.pdxstudio.com/tecnologia/realidad-aumentada

[6] Navab, N., Bani-Hashemi, A., Mitschke, M.: Merging visible and invisible: Two camera-augmented mobile C-arm (CAMC) applications. In: IWAR 1999: Proc. 2nd Int'l Workshop on Augmented Reality, San Francisco, CA, USA, October 20-21, pp. 134–141. IEEE CS Press (1999) ISBN 0-7695-0359-4

[7] Information on, http://www.telegraph.co.uk/motoring/motoringvideo/9915073/Augmented-reality-car-windscreens-to-display-phone-calls-and-GPS-information.html

[8] Thompson, J.I.: A Three Dimensional Helmet Mounted Primary Flight Reference for Para-troopers, AFIT/GCS/ENG/05-18, Department of the Air Force, Ohio (2005)

[9] Information on, http://www.google.com/glass/start/

[10] Bradski, G., Kaehler, A.: Learning OpenCV: Computer Vision with the OpenCV Library. O'Reilly Maedia (2008)

[11] Hartley, R., Zisserman, A.: Multiple View Geometry in computer vision. Cambridge University Press

[12] Faig, W.: Calibration of close-range photogrammetry systems: Mathematical formulation. Photogrammetric Eng. Remote Sensing 41, 1479–1486 (1975)

[13] Topór-Kamiñski, T., Krupanek, B., Homa, J.: Delays Models of Measurement and Control Data Transmission Network. In: Nawrat, A., Simek, K., Świerniak, A. (eds.) Advanced Technologies for Intelligent Systems. SCI, vol. 440, pp. 257–278. Springer, Heidelberg (2013)

[14] Babiarz, A., Jaskot, K.: The Concept of Collision-Free Path Planning of UAV Objects. In: Nawrat, A., Simek, K., Świerniak, A. (eds.) Advanced Technologies for Intelligent Systems. SCI, vol. 440, pp. 81–94. Springer, Heidelberg (2013)

[15] Babiarz, A., Bieda, R., Jaskot, K.: Vision System for Group of Mobile Robots. In: Nawrat, A., Kuś, Z. (eds.) Vision Based Systems for UAV Applications. SCI, vol. 481, pp. 139–156. Springer, Heidelberg (2013)

[16] Czapla, T., Wrona, J.: Technology Development of Military Applications of Unmanned Ground Vehicles, Vision Based Systems for UAV Applications. In: Nawrat, A., Kuś, Z. (eds.) Vision Based Systems for UAV Applications. SCI, vol. 481, pp. 293–309. Springer, Heidelberg (2013)
[17] Orkisz, M., Swiech, L., Zacharzewski, J.: Fatigue tests of motor glider wing's composite spar. Eksploatacjai Niezawodnosc 14(3), 228–232 (2012)
[18] Martynowicz, P., Kciuk, S., Mezyk, A.: Rotary-shock-absorber with magnetorheological valves. Advanced Materials Research 628, 505–511 (2013)
[19] Kciuk, S., Kciuk, M., Turczyn, R., Martynowicz, P.: An Investigation of the Behaviour of Mag-netorheological Fluids in The Rotary Shock-Absorber. Advanced Materials Research 628, 512–517 (2013)
[20] Barnat, W., Panowicz, R., Niezgoda, T.: Numerical and Experimental Comparison of Combined Multilayer Protective Panels. ActaMechanica et Automatica 6(1), 13–16 (2012)
[21] Barnat, W., Panowicz, R., Niezgoda, T., Dybcio, P.: Numerical analysis of IED detonation effect on steel plate. Acta Mechanica et Automatica 6(1), 10–12 (2012)
[22] Panowicz, R., Barnat, W., Niezgoda, T., Szymanczyk, L., Grzymkowski, J.: Acta Mechanica et Automatica 6 (1), 49-52 (2012)
[23] Barnat, W., Niezgoda, T., Panowicz, R., Sybilski, K.: The Influence of Conical Composite Fill-ing on Energy Absorbtion During The Progressive Fracture Process. WIT Transactions on Modelling and Simulation 51, 625–633 (2011)
[24] Niezgoda, T., Panowicz, R., Sybilski, K., Barnat, W.: Numerical Analysis of Missile Impact Be-ing Shot By Rocket Propelled Grenades with Rod Armour. WIT Transactions on Modelling and Simulation 51, 625–633 (2011)
[25] Daniec, K., Jedrasiak, K., Koteras, R., Nawrat, A.: Embedded Micro Inertial Navigation System. Applied Mechanics and Materials 249, 1234–1246 (2013)
[26] Jędrasiak, K., Bereska, D., Nawrat, A.: The Prototype of Gyro-Stabilized UAV Gimbal for Day-Night Surveillance. In: Nawrat, A., Simek, K., Świerniak, A. (eds.) Advanced Technologies for Intelligent Systems. SCI, vol. 440, pp. 107–116. Springer, Heidelberg (2013)
[27] Jędrasiak, K., Nawrat, A., Wydmańska, K.: SETh-link The Distributed Management System for Unmanned Mobile Vehicles. In: Nawrat, A., Simek, K., Świerniak, A. (eds.) Advanced Technologies for Intelligent Systems of National Border Security. SCI, vol. 440, pp. 247–256. Springer, Heidelberg (2013)
[28] Jędrasiak, K., Nawrat, A.: Image Recognition Technique for Unmanned Aerial Vehicles. In: Bolc, L., Kulikowski, J.L., Wojciechowski, K. (eds.) ICCVG 2008. LNCS, vol. 5337, pp. 391–399. Springer, Heidelberg (2009)
[29] Nawrat, A., Jedrasiak, K.: SEThSystem Spatio-Temporal Object Tracking Using Combined Color and Motion Features. In: WSEAS International Conference, Proceedings, Mathematics and Computers in Science and Engineering, vol. 9 (2009)
[30] Jedrasiak, K., Daniec, K., Nawrat, A.: The Low Cost Micro Inertial Measurement Unit. In: Industrial Electronics and Applications (ICIEA), pp. 403–408 (2013)
[31] Bereska, D., Jędrasiak, K., Daniec, K., Nawrat, A.: Gyro-Stabilized Platform for Multispectral Image Acquisition. In: Nawrat, A., Kuś, Z. (eds.) Vision Based Systems for UAV Applications. SCI, vol. 481, pp. 115–121. Springer, Heidelberg (2013)
[32] Daniec, K., Iwaneczko, P., Jędrasiak, K., Nawrat, A.: Prototyping the Autonomous Flight Algo-rithms Using the Prepar3D Simulator. In: Nawrat, A., Kuś, Z. (eds.) Vision Based Systems for UAV Applications. SCI, vol. 481, pp. 219–232. Springer, Heidelberg (2013)

[33] Babiarz, A., Bieda, R., Jędrasiak, K., Nawrat, A.: Machine Vision in Autonomous Systems of Detection and Location of Objects in Digital Images. In: Nawrat, A., Kuś, Z. (eds.) Vision Based Systems for UAV Applications. SCI, vol. 481, pp. 3–25. Springer, Heidelberg (2013)
[34] Bereska, D., Daniec, K., Fraś, S., Jędrasiak, K., Malinowski, M., Nawrat, A.: System for Multi-Axial Mechanical Stabilization of Digital Camera. In: Nawrat, A., Kuś, Z. (eds.) Vision Based Systems for UAV Applications. SCI, vol. 481, pp. 177–189. Springer, Heidelberg (2013)

Document-Oriented Data Stores of Vision Objects

Małgorzata Bach and Aleksandra Werner

Silesian University of Technology,
Akademicka 16 44-100 Gliwice, Poland
{Malgorzata.Bach,Aleksandra.Werner}@polsl.pl

Abstract. Nowadays vision systems are widely used in many areas of human life. They are applied to industrial quality assurance, burglar alarms, intelligent control systems of traffic control, car parks monitoring or e.g. to the border crossings control. The activity of such systems is concerned with the need of large amounts of LOB data types storage. In order to satisfy efficient access to such data it's necessary to develop adequate mechanisms of data storage and processing. Because of the fact that traditional relational systems may be inefficient, the possibilities of using NoSQL database that may speed up the video content management are described in the paper.

Keywords: NoSQL, key–value databases, vision systems, LOB.

1 Introduction

Electronic information plays an important role in modern systems of public safety. It is difficult to imagine modern and well-organized state organs, working for public safety and order, without a well-functioning IT systems. Thanks to the access to modern computer solutions public services can effectively and efficiently accomplish their tasks. Nowadays the preventive activities of the services are increasingly based on the collection, analysis and the usage of information of various human activity areas. In crime-prevention campaigns, it is essential to use the most modern and effective tools for sharing information – and thus – tools for fast data processing and its analysis. Due to this fact, many terabytes of data stored over the years in different locations must be reorganized in a way that can provide the most optimal access to them.

The physical architecture of the crime detection or policy violations system consists of a set of cameras that monitor the area and detect an intrusion attempt which are coupled with their recorders as well as a central visualization server, data repository and client terminals (Figure 1). Data from cameras' recorders that is compressed with the standard e.g. H.264 or MPEG-4 is transmitted via Internet to a visualization server equipped with some tools of monitoring devices' integration, where data is processed. Then data is sent to a storage device – data repository and, finally, it is usually analyzed in a form of reports or visualizations by such clients as police, council services or customs officers.

A.M. Nawrat (ed.), *Innovative Control Systems for Tracked Vehicle Platforms*,
Studies in Systems, Decision and Control 2,
DOI: 10.1007/978-3-319-04624-2_9, © Springer International Publishing Switzerland 2015

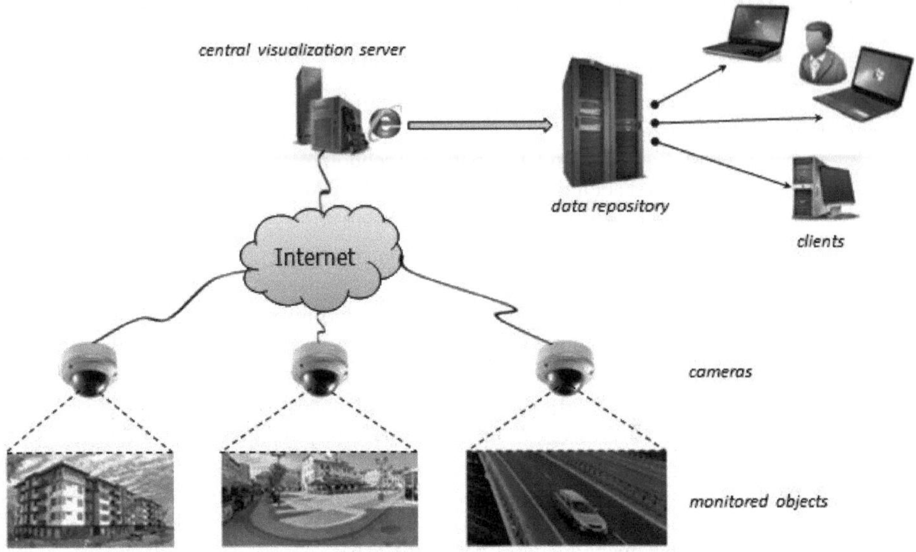

Fig. 1. The physical architecture of a typical monitoring system

It should be emphasized that recorded information, related to the observed events, is consuming extensive storage capacity. For example, for one during object tracking [10-12] using e.g. Sanyo VCC-HD 5600 camera, the amount of data stored within 1 day equals on average 2-4 tens of gigabytes [7]. Depending of the camera's resolution only one frame needs approximately 1,2MB [8]. Taking into account the fact that vision objects are usually stored in a relational database tables as fields of specialized data types such as BLOBS (Binary Large Objects) corresponding to the images, it is obvious that some techniques should be applied in order to counteract the system overloading. However this solution allows to store the image data it misses the possibility to dig inside and extract the useful information that is stored.

In [4] the new conception was introduced that proposed to separate the set of metadata from the acquisitioned images corresponding to the stream of film frames. The idea was based on the observation that single frame has a lot of data that is uninteresting for anyone. As an example the situation presented in the Figure 2 will be analyzed, where violating security policies occurred. There is a part of the city center, observed by a camera, and unlawful movement of a vehicle is detected on a pedestrian part of the street. The only interesting picture area is marked with black rectangle (a pedestrian part of the street), so the rest background can be ignored. Even if the required fragment is approximated by a camera, it is still a relatively large area of unusable information.

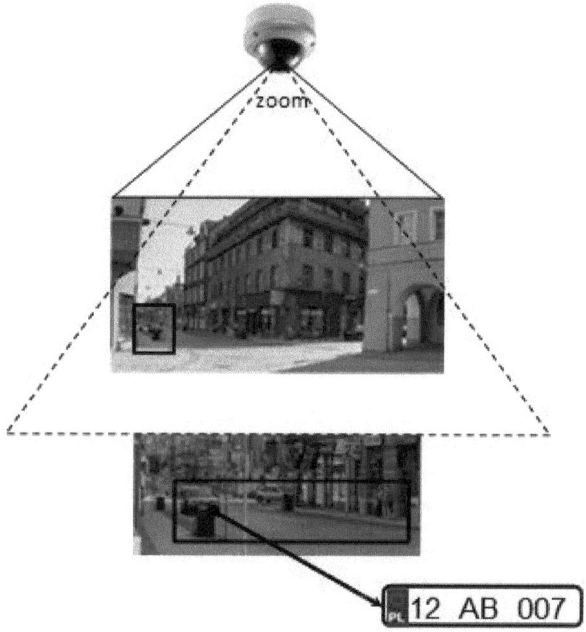

Fig. 2. Detection of violating security policies

Besides, it must be remembered that a camera raw file contains unprocessed and uncompressed picture data from a digital camera's image sensor, along with information about how the image was captured (i.e. metadata). This additional data, relating to a vision device, includes camera intrinsic and extrinsic parameters – such as: focal length, resolution, the position of the camera center, its rotation, etc. Moreover every image has its own attributes that must be stored in a conjunction with BLOB object namely size, compression, distance from the center of the image to the center of the field of view of the camera. Thus, the full description of one image can require the tens or hundreds of additional data generation.

The traditional relational databases are not efficient enough and do not ensure the required performance, so the reasonable solution seems to be the usage of a NoSQL database.

2 NoSQL

For over the last 40 years, the relational database model has been the dominant in a database management, however the volumes of data that is being stored have increased massively for over the last decade. RDBMS's which were designed are not able to cope with the requirements of modern applications.

NoSQL (Not only SQL) databases were developed to deal with limitations of relational databases, particularly concerning scale, replication and unstructured data storage. Non-relational database provides the mechanisms for storage and retrieval of the data that use consistency models that are not so strong in comparision with the traditional relational databases, in order to achieve horizontal scaling and higher availability [9].

NoSQL encompasses a wide variety of different database technologies but generally all NoSQL databases have some features in common. Generally, these databases include the following designs:

- **Document databases** pair each key with a complex data structure known as a document. Documents can contain many different key-value pairs, or key-array pairs, or even nested documents.
- **Graph stores** are used to store information about networks, such as social connections.
- **Key-value stores** are the simplest NoSQL databases. Every single item in the database is stored as an attribute name, or a key, together with its value.
- **Wide-column stores** are optimized for queries over large datasets, and store columns of data together, instead of rows.

Because of the fact that traditional relational systems may be inefficient, the possibility of using MongoDB, which is document-oriented data store, that may speed up the video content management, is proposed in this paper.

2.1 MongoDB

MongoDB is widely recognized as one of NoSQL database leader. Instead of storing data in rows and columns as it is in a relational database MongoDB stores a binary form of JSON documents (BSON). Data in MongoDB has the flexible schema, so the collections do not enforce the document structure.

For storing and retrieving files that exceed the BSON-document size limit of 16MB, GridFS is necessary to be used. Instead of storing a file in a single document, GridFS divides the file into chunks and stores each of those chunks as a separate document. By default GridFS limits a chunk size to 256k. GridFS uses two collections to store files. These collections are placed in a common bucket by prefixing each ones with the bucket name. By default, GridFS uses two collections with names prefixed by fs bucket: fs.chunks and fs.files. The first one stores the file chunks, the second stores file metadata.

GridFS is useful not only for storing files that exceed 16MB but also for storing any files for which an access without having to load the entire file into memory is required. Range queries can be performed on files stored through GridFS. Information from arbitrary sections of files can also be accessed, which allows to "skip" into the middle of a video or audio file. It can be very useful for monitoring systems [6].

3 Solution Description

In the earlier research presented in [4] and [5] the structure of the relational database for video content management, such as in Fig. 3, was proposed.

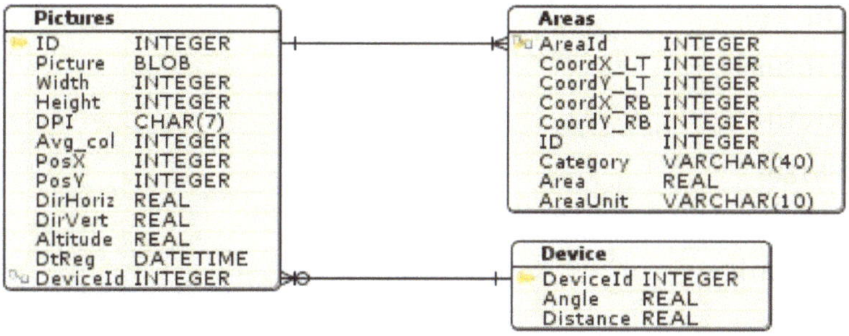

Fig. 3. The schema of tested relational database

In the assumption the photos will not only be analyzed, but also will be processed in order to distinguish some objects interesting for the observer (AREAS) [1]. Therefore one picture is related to a lot of areas, as it is shown in Figure 3 (1:N relationship). Images are recorded by a camera (table DEVICE). The attributes of PICTURES table are mostly the photos parameters e.g. image resolution (DPI), width, height of image, etc. There is also another data such as the location of the camera, it means GPS data (PosX, PosY) and extra information about the direction of the device, at the moment of taking the photo (DirHoriz, DirVert). The angle and distance of camera's view are recorded in the DEVICE table.

The schema of tested database was implemented in the Oracle database system and a sample data was loaded.

Oracle allows manipulating big binary data that is generally represented as stream of bytes or buffers, in two ways. The first method is to store a pointer to the image data (BFILE), while the second – to store data in the same row in a BLOB element (the amount of held data must be up to maximum 4GB). The algorithm that load an external file into the table comprises the following stages [9]:

1. Create a table to store the BLOBs.
2. Create a logical directory in the database to the physical file system.
3. Create a procedure to load the BLOBs from the file system using the logical directory and the BFILE locator variable.
4. Create a procedure to display the image (optionally).

Once physical files are associated with records using SQL DML, subsequent read operations on the BFILE can be performed using a PL/SQL DBMS_LOB package. The DBMS_LOB package provides subprograms to operate on BLOBs, and BFILENAME is a built-in function that can be used to initialize a BFILE column to point to an external file:

BFILENAME('*directory_name*', '*image_filename*').

In contrary to a relational database data, MongoDB has a flexible schema. That's why the collection of documents that will store information about PICTURES, AREAS and DEVICES must be created [2].

For the study purposes the following structure of the documents was defined:

```
{  _id :…,
   picture: …,
   width: …,
   height:…,
   DPI:…,
   Avg_col:…,
   PosX:…,
   PosY:…,
   DirHoriz:…  ,
   DirVert:…,
   Altitude:…,
   DtReg:…,
   Device: {
           Angle:…,
           Distance:…},
   Areas: [
           {CoordX_LT:…,
           CoordY_LT:…,
           CoordX_RB:…,
           CoordY_RB:…,
           Category:…,
           Area:…,
           AreaUnit:…},
             { CoordX_LT:…,
           CoordY_LT:…,
           CoordX_RB:…,
           CoordY_RB:…,
           Category:…,
           Area:…,
           AreaUnit: …},
       …]
   }
```

In these documents the picture attribute is an identifier of the file stored in fs.files collection, that was mentioned in the previous chapter.

4 Test Results

The test environment consisted of two 2.5 GHz PC computers, with 8GB of RAM and Intel Core i5-2520M CPU.

In order to verify the usefulness of NoSQL usage in monitoring data processing, the set of variety of queries was considered. Finally, queries with/without tables joining, with/without aggregate functions, with GROUP BY and HAVING clauses, etc., were performed in the discussed research and sample tasks were as follows:

1. Find the ID of all the camera shots that recorded the traced (e.g. suspicious) object.
2. Find all objects recorded on 3rd of May 2013 between 3 p.m. and 7 p.m.
3. Find all objects registered by the device with a known ID.
4. Find all objects that were recorded later than the specific object.

For research purposes, not only different types of queries were performed, but also the number of table's rows varied from 2,000 up to 100,000 in Oracle database and respectively – 2,000 up to 100,000 documents in MongoDB. These figures refer only to AREAS and PICTURES tables, because the size of DEVICE one was negligibly small.

The implementation of chosen queries for MongoDB and Oracle is given in the table below.

	Find all objects recorded in 3rd of May 2013 between 3 p.m. and 7 p.m.
MongoDB	`db.MonitoringSytem.find({DtReg : {$gte: ISODate('2013-05-03T03:00:00.000Z'), $lt: ISODate('2013-05-03T07:00:00.000Z')}});`
Oracla	`SELECT *FROM PICTURES WHERE DtReg > '2013-05-03 03:00:00.00' AND DtReg < '2013-05-03 07:00:00.00' ;`
	Find all objects registered by the device with ID=1
MongoDB	`db.Publikacje.find({DeviceId : 1});`
Oracle	`SELECT * FROM PICTURES P, DEVICE D, AREAS A WHERE P.DEVICEID = D.DEVICEID AND A.ID=P.ID AND DEVICEID=1;`

For better visualization the differences in speed of query execution in Oracle and MongoDB database systems, the percentage scale on Y axis was used, as it is shown in Fig. 4 and Fig. 5.

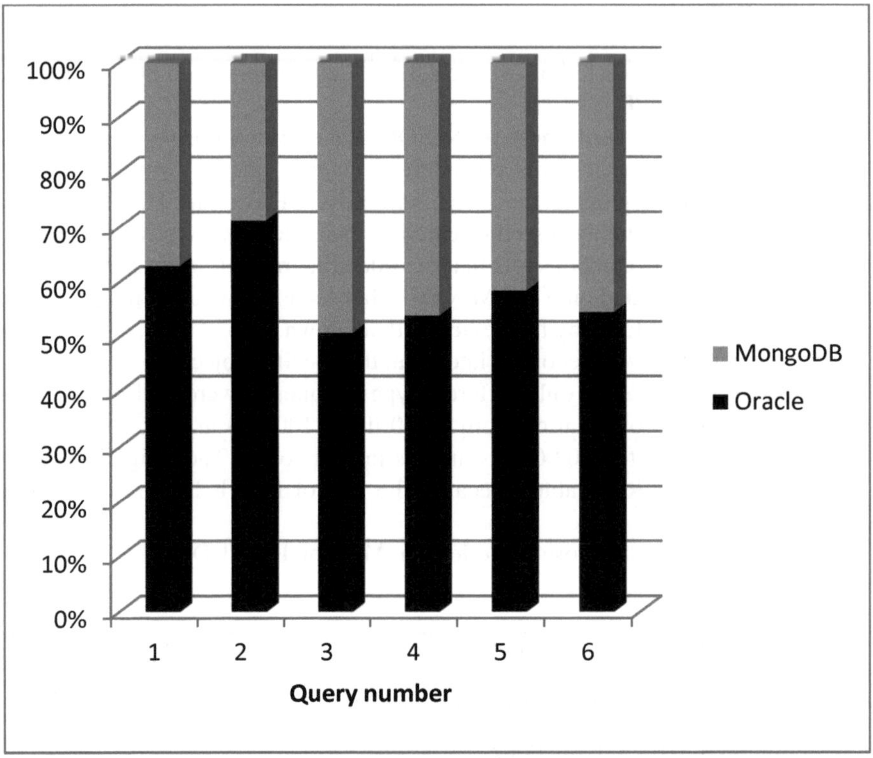

Fig. 4. The cumulative execution time of sample queries (for 2000 rows/documents)

The bars of the chart in Fig. 4 show that in nearly every type of executed query recorded time was longer in Oracle system than in NoSQL database. Especially it is visible with regard to query 2, where the observed time was considerably – about 2 times – longer in the classic, relational database system, than in the document database.

It can be easily noticed that the performance increase depends on the number of processed rows/documents. More satisfactory speedup than observed in Fig. 4 was achieved for the larger number of rows. For example the time of MongoDB record searching for 100,000 documents was about 1.8-3.5 times shorter than it was in Oracle, while for 2,000 rows/documents it was only approximately 1.2 - 2.5 times.

It can be also noticed that MongoDB much better copes with the scalability.

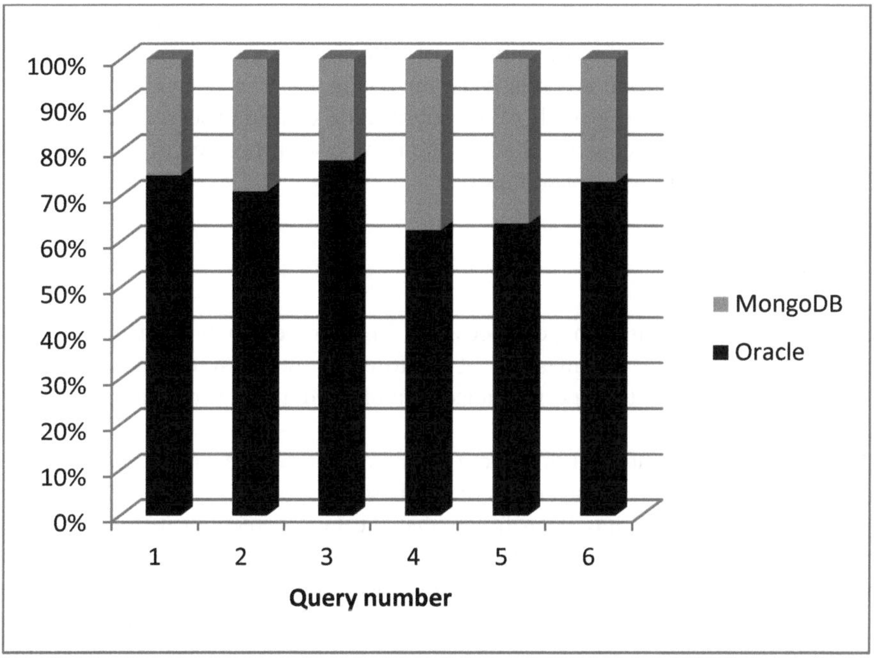

Fig. 5. The cumulative execution time of sample queries (for 100 000 rows/documents)

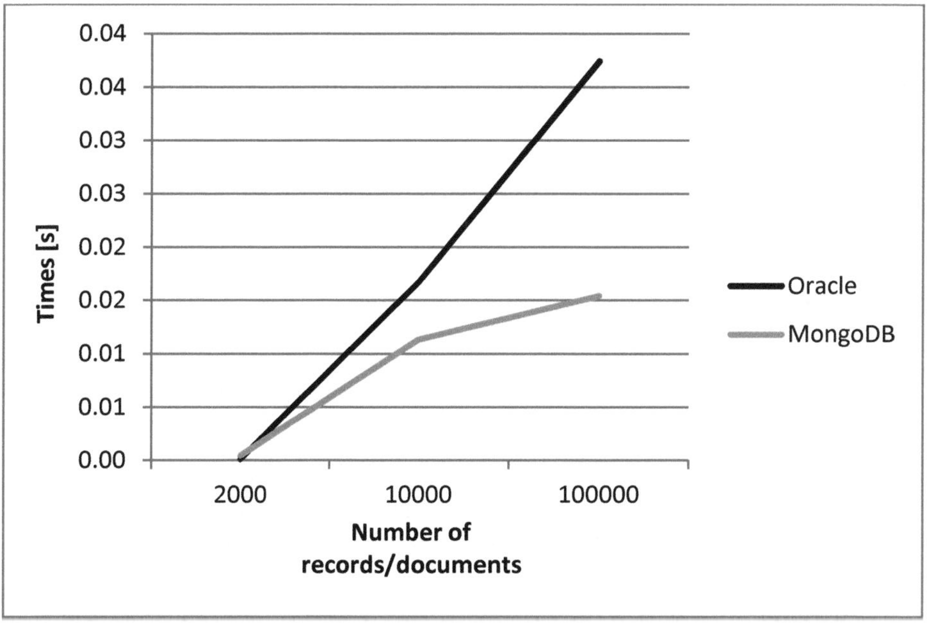

Fig. 6. The dependence of query execution time on the number of processed data

It can be seen that the time needed to execute the query in Oracle system sharply increased with the increase in the number of tables' rows (records), while in the case of MongoDB – the increase was relatively small (Fig. 6).

In MongoDB the steep growth of execution time can be observed between 2,000 and 10,000 processed documents, but for the number of processed documents greater than 10,000 the growth was not so rapid.

5 Summary

Desired decrease of digital image processing time can be achieved, inter alia, by the use of alternative data organization. Therefore the research on the influence of document organization at the performance of multimedia data processing is described in this paper.

The test results indicate that document oriented data storage significantly speeds up the search operations for various types of database queries.

In our observations it has been found that the tasks executed in MongoDB take approximately 1,5-2 times less than in the Oracle system. In MongoDB the speed of time increase at the beginning was noticeably faster, but it dropped for the number of the processed documents greater than 10,000. The graph in Fig. 6 shows that on the contrary to the NoSQL solution the line of time growth in the case of Oracle system was significantly sharper and after reaching a point of 10,000 documents – even became steeper.

As it was mentioned previously, in general, the document database was significantly faster but required the more complicated code to implement each task and it took longer to tune. So the only inconvenience in the proposed solution was the applied notation. Queries in a high-level languages, such as SQL, are easier to write, easier to modify, and what is important much easier to understand for a unskilled user. Meanwhile, diverse and often very unintuitive languages which are to learn to use NoSQL databases with queries that are usually created at a very low level of abstraction, can cause a serious problem for a large number of potential users.

References

[1] Duszeńko, A., Maćkowski, M., Nurzyńska, K., Skoroniak, K.: Wykorzystanie baz danych do zarzadzania multimedialnymi metadanymi w systemach monitoringu, Bazy danych nowe technologie, Red.In: Kozielski, S., Małysiak-Mrozek, B., Kasprowski, P. Mrozek, D.: WKiŁ, pp.219-228. Warszawa (2007)

[2] G. Miller, S. Fels and S. Oldridge: A Conceptual Structure for Computer Vision. In: Conference on Computer and Robot Vision (2011), web page (2013), http://www.freeviewpointvideo.co.uk/Publications/Papers/Miller11CRV.pdf

[3] Werner, A., Bach, M., Duszeńko, A.: Vision Object Access And Management with Usage of the In-Memory Databases. In: Nawrat, A., Simek, K., Świerniak, A. (eds.) Advanced Technologies for Intelligent Systems. SCI, vol. 440, pp. 231–246. Springer, Heidelberg (2013)

[4] Bach, M., Duszenko, A., Werner, A.: Column-oriented metadata organization of vision objects. In: Kapczyński, A., Tkacz, E., Rostanski, M. (eds.) Internet - Technical Developments and Applications 2. AISC, vol. 118, pp. 51–58. Springer, Heidelberg (2012)
[5] GridFS (2013), http://docs.mongodb.org/manual/core/gridfs/
[6] Wereszczyński, K.: Inteligentne i interaktywne gromadzenie danych wizyjnych w systemach wielokamerowych. Finalbachelor'sproject, promoter: A. Świtoński (PhD), PJWSTK, Bytom (2012)
[7] Camera specifications (2013), http://www.dipol.com.pl
[8] Oracle® Database PL/SQL Packages and Types Reference (2013), http://docs.oracle.com/cd/B19306_01/appdev.102/b14258/d_lob.htm
[9] Redmond, E., Wilson, J.: Seven Databases in Seven Weeks, A Guide to Modern Databases and the NoSQL Movement The Pragmatic Bookshelf, USA (2012) ISBN-13: 978-1-93435-692-0
[10] Kuś, Z., Nawrat, A.: Object Tracking for Rapid Camera Movements in 3D Space. In: Nawrat, A., Kuś, Z. (eds.) Vision Based Systems for UAV Applications. SCI, vol. 481, pp. 57–76. Springer, Heidelberg (2013)
[11] Kuś, Z., Nawrat, A.: Object Tracking for Rapid Camera Movements in 3D Space. In: Nawrat, A., Kuś, Z. (eds.) Vision Based Systems for UAV Applications. SCI, vol. 481, pp. 77–91. Springer, Heidelberg (2013)
[12] Babiarz, A., Bieda, R., Jędrasiak, K., Nawrat, A.: Machine Vision in Autonomous Systems of Detection and Location of Objects in Digital Images. In: Nawrat, A., Kuś, Z. (eds.) Vision Based Systems for UAV Applications. SCI, vol. 481, pp. 3–25. Springer, Heidelberg (2013)
[13] Babiarz, A., Bieda, R., Jaskot, K.: Vision system for group of mobile robots. In: Nawrat, A., Kuś, Z. (eds.) Vision Based Systems for UAV Applications. SCI, vol. 481, pp. 139–156. Springer, Heidelberg (2013)
[14] Bieda, R., Jaskot, K., Jędrasiak, K., Nawrat, A.: Recognition and location of objects in the visual field of a UAV vision system. In: Nawrat, A., Kuś, Z. (eds.) Vision Based Systems for UAV Applications. Studies in Computational Intelligence, vol. 481, pp. 27–45. Springer, Heidelberg (2013)
[15] Kciuk, S., Martynowicz, P.: Special application magnetorheological valve numerical and experimental analysis. Diffusion and Defect Data Pt. B Solid State Phenomena 177, 102–115 (2011)
[16] Orkisz, M., Swiech, L., Zacharzewski, J.: Fatigue tests of motor glider wing's composite spar. EksploatacjaiNiezawodnosc 14(3), 228–232 (2012)
[17] Martynowicz, P., Kciuk, S., Mezyk, A.: Rotary-shock-absorber with magnetorheological valves. Advanced Materials Research 628, 505–511 (2013)
[18] Mezyk, A.: Minimization of Transient Forces in an Electromechanical System. Structural Optimization 8(4), 251–256 (1994)
[19] Jedrasiak, K., Daniec, K., Nawrat, A.: The Low Cost Micro Inertial Measurement Unit. In: Industrial Electronics and Applications (ICIEA), pp. 403–408 (2013)
[20] Bereska, D., Jędrasiak, K., Daniec, K., Nawrat, A.: Gyro-Stabilized Platform for Multispectral Image Acquisition. In: Nawrat, A., Kuś, Z. (eds.) Vision Based Systems for UAV Applications. SCI, vol. 481, pp. 115–121. Springer, Heidelberg (2013)

[21] Daniec, K., Iwaneczko, P., Jędrasiak, K., Nawrat, A.: Prototyping the Autonomous Flight Algorithms Using the Prepar3D® Simulator. In: Nawrat, A., Kuś, Z. (eds.) Vision Based Systems for UAV Applications. SCI, vol. 481, pp. 219–232. Springer, Heidelberg (2013)
[22] Babiarz, A., Bieda, R., Jędrasiak, K., Nawrat, A.: Machine Vision in Autonomous Systems of Detection and Location of Objects in Digital Images. In: Nawrat, A., Kuś, Z. (eds.) Vision Based Systems for UAV Applications. SCI, vol. 481, pp. 3–25. Springer, Heidelberg (2013)
[23] Bereska, D., Daniec, K., Fraś, S., Jędrasiak, K., Malinowski, M., Nawrat, A.: System for Multi-Axial Mechanical Stabilization of Digital Camera. In: Nawrat, A., Kuś, Z. (eds.) Vision Based Systems for UAV Applications. SCI, vol. 481, pp. 177–189. Springer, Heidelberg (2013)
[24] Barnat, W., Panowicz, R., Niezgoda, T.: Numerical and Experimental Comparison of Combined Multilayer Protective Panels. Acta Mechanica et Automatica 6(1), 13–16 (2012)
[25] Barnat, W., Panowicz, R., Niezgoda, T., Dybcio, P.: Numerical analysis of IED detonation effect on steel plate. Acta Mechanica et Automatica 6(1), 10–12 (2012)
[26] Panowicz, R., Barnat, W., Niezgoda, T., Szymanczyk, L., Grzymkowski, J.: Acta Mechanica et Automatica 6(1), 49–52 (2012)
[27] Barnat, W., Niezgoda, T., Panowicz, R., Sybilski, K.: The Influence of Conical Composite Filling on Energy Absorbtion During The Progressive Fracture Process. WIT Transactions on Modelling and Simulation 51, 625–633 (2011)
[28] Niezgoda, T., Panowicz, R., Sybilski, K., Barnat, W.: Numerical Analysis of Missile Impact Being Shot By Rocket Propelled Grenades with Rod Armour. WIT Transactions on Modelling and Simulation 51, 625–633 (2011)

Simulations and Analysis of Performance Algorithms Used in Wireless Networks and Measurement Systems

Beata Krupanek

Institute of Measurement Science, Electronics and Control,
Silesian University of Technology, Gliwice, Poland
beata.krupanek@polsl.pl

Abstract. Paper presents a study under the Quality of Service in modern wireless sensor networks. Such a networks are characterized by small amount of data transmitted in fixed periods. Very often this data must by transmitted in real time so data transmission delays should be well known. This article shows two networks (point-to-point and star topology) simulated in packet OPNET Modeler.

Keywords: delays, WiFi, OPNET, Quality of Service.

1 Introduction

Nowadays, one of the popular wireless technologies, use to transfer different types of data, is the IEEE 802.11 standard. This popular "WiFi" technology provides low-cost wireless Internet capability for end users and also is used in modern measurement and control distributed systems. Wireless medium is a shared medium, so as more and more devices demand the bandwidth, concentrating on performance becomes critical. As the wireless LANs are giving a competition to the Ethernet, apart from speed it will also have to assure the Quality of Service, especially when dealing with voice, video [10,11] and other real time services.

In this situation, reliable data transport with Quality of Service criteria, related to throughput or latency [9], is needed. To analyze the QoS parameters, especially referring to wireless communication, is not easy because it is affected by a number of parameters like: channel access method, physical and environmental conditions, the number of nodes, distance etc. The WLAN media is error prone and the bit error rate (BER) is very high compared to the BER of wired networks. In addition, Carrier Sensing is difficult in wireless networks because a station is incapable of listening to its own transmissions in order to detect a collision. The Hidden Terminal problem also decreases the performance of a WLAN.

A proper selection of protocols and topology helps in maintaining and improving the QoS support of a wireless network. IEEE 802.11 standard has two sublayers MAC protocols: Distributed Coordination Function (DCF) and Point Coordination Function (PCF). The DCF supports best effort delivery of packets while the PCF is built on the top of DCF and provides services for time-bounded traffic. The 802.11 MAC layer

A.M. Nawrat (ed.), *Innovative Control Systems for Tracked Vehicle Platforms*, 175
Studies in Systems, Decision and Control 2,
DOI: 10.1007/978-3-319-04624-2_10, © Springer International Publishing Switzerland 2015

aims to provide access control functions to the wireless medium such as access coordination, addressing or frame check sequence generation.

2 Quality of Services in Wireless Measurement Systems

The IEEE 802.11 standard for wireless LANs covers the Medium Access Control (MAC) and the physical (PHY) layers. Three different Physical layers specifications exist in the 2.4 GHz range of the unlicensed spectrum, but commonly used hardware is based on the Direct Sequence Spread Spectrum (DSSS) technique. The MAC protocol operates in the DCF and PCF modes. The DCF is the fundamental access method. It uses Carrier Sense Multiple Access with Collision Avoidance (CSMA/CA) for channel access, Request-To-Send and Clear-To-Send (RTS/CTS) messages for reservations and Acknowledgements (ACK) for reliability. The PCF mode is used only in infrastructure networks. It relies on a Point Coordinator (PC) that resides in the Access Point to perform polling, so that polled stations may have contention-free access to the channel [1].

In this paper is modeled the basic DCF (CSMA/CA) protocol, the DCF protocol with RTS/CTS, and the PCF protocol for medium large office network. Both these mechanisms are not equipped with full-fledged built in system for supporting real time services makes it very difficult to provide the required QoS[7]. The main requirement of IEEE 802.11 WLAN is to guarantee the QoS requirements due to unaware functions such as dedicated bandwidth, controlled jitter and latency.

2.1 Distributed Coordination Function DCF

DCF implements the Carrier Sense Multiple Access with Collision Avoidance (CSMA/CA) algorithm with binary exponential backoff algorithm for accessing the medium. Carrier sensing is done through physical and virtual mechanisms. A station senses the medium to check if another station is transmitting in the physical mechanism. The virtual carrier-sense is done by distributing reservation information with RTS/CTS mechanism.

DCF includes random backoff mechanism to ensure low collision probability. This mechanism provides best effort type of service for transfer of data. DCF uses either two-way handshaking or four-way handshaking technique Chile transmitting the data. While two-way handshaking uses explicit acknowledgement (ACK) for recept confirmation, RTS and CTS frames are used by four-way handshaking technique [2].

DCF is the basic medium access mechanism for both ad-hoc and infrastructure modes. In DCF mode, a station should ensure that the medium is free before it starts transmitting data over the channel. In the case of a busy channel, the node has to wait until the transmission by other node is complete [2]. In addition, the node has to wait for another amount of time to provide sufficient gap between subsequent frame transmissions.

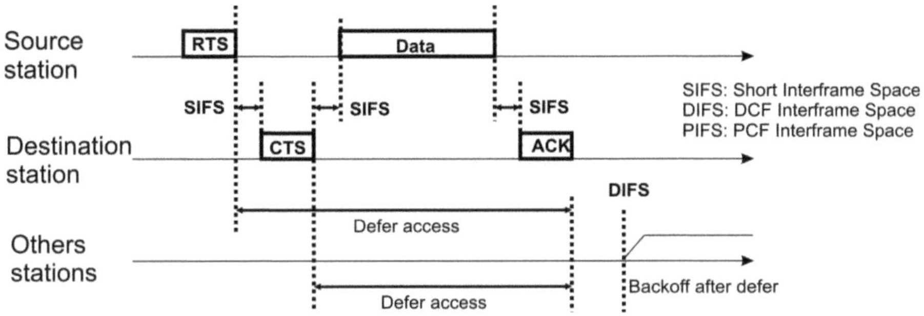

Fig. 1. Distributed Coordination Function DCF

There are two basic rules that need to be followed by every sensor node operating in the DCF mode. Firstly every station that has data to transmit should confirm that the channel was idle for at least DIFS (Distributed Inter-Frame Space) amount of time. Secondly, it should wait a random amount of time when it has another datagram to transmit after a successful transmission or it has a datagram to transmit but it was denied channel access, as channel was busy (Fig.1). Each time the carrier sense operation detects the medium to be idle, the backoff timer decrements the backoff time. After the collision detect transmitter retransmits the data.

When a sensor station receives a data frame, it waits for a SIFS (Short IFS) duration and transmits an ACK to the sender. There is no carrier sensing or backing off for ACK frames. The transmitter expects an acknowledgement within SIFS time frame. Only after receiving an ACK frame correctly, the transmitter assumes that the data frame was delivered successfully.

The SIFS and DIFS time frames enable the communicating pairs to complete the frame exchange sequence without collision.

2.2 Point Coordination Function PCF

Point Coordination Function (PCF) is a centralized, polling-based access mecha-nism which requires the presence of an AP that acts as Point Coordinator (PC). In the PCF mode, time is divided into superframes. Each superframe consists of a contention period where DCF is used and a contention-free period (CFP) where PCF is used. The CFP is started by a beacon frame sent by the PC using DCF. The CFP may vary from superframe to superframe, as the base station has to contend for the medium. Once the CFP starts, the PC polls each station in its polling list (the high priority stations) when they can access the medium. To ensure that no DCF stations are able to interrupt this mode of operation, the inter frame space between PCF data frames (PIFS) is shorter than the DIFS.

To start the CFP, PC gains the control of the medium by waiting PIFS period of time and then sends a beacon frame (Fig.2). It is a control frame that has all the attributes used in the CFP period. Point Coordinator builds a polling list of all the PCF enabled nodes. This list is organized according to the nodes MAC addresses.

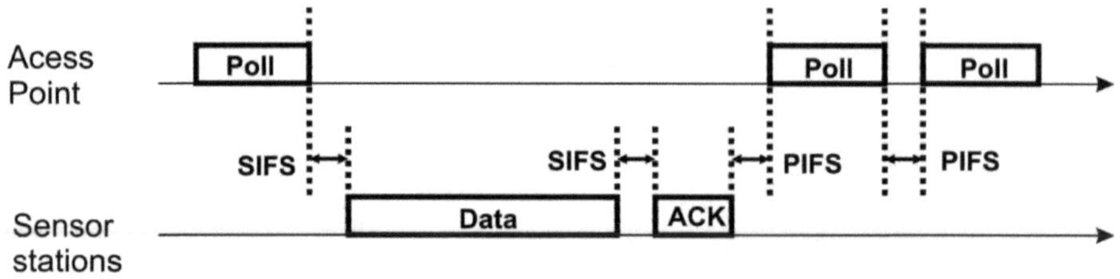

Fig. 2. Point Coordination Function PCF

Upon receiving CFP, stations wait for SIFS period of time and then either respond with data frame along with the acknowledgement of data received or if they don't have any data to send then they will send a null frame in order to release the medium. This way PC will not wait for the PCF sensor stations that don't have any data to send. After wait-ing for SIFS period, if PC does not receive the data or null frame then it will repeat the polling of the same node. It keeps repeating this until maximum number of poll failures occurs. It then drops that node in that CFP period and moves to the next station [2].

The PCF has not been studied as much as the DCF. Existing studies of PCF have concentrated on the ability of the PCF mode to service real time audio traffic and on improving its efficiency using new polling mechanisms. This is because sequential polling wastes bandwidth and causes increased delays when the network is lightly loaded.

3 Simulations and Simulation Set Up

To compare DCF and PCF mechanisms in wireless sensing networks (usually small amounts of data transmitted in fixed time intervals) the simulations were done in modern packed called OPNET Modeler. The simulations were done for two different networks. One of them is small point-to-point network and the second consists of couple of elements. This shows main differences between transmissions in such a networks.

3.1 OPNET Modeler Environment

In this research, the performance of DCF and PCF channel access methods were compared with respect to their support for real-time (measured data) traffic [3]. The research was carried out using a specialized software environment - OPNET Modeler network simulator. Modeler is one of the most widely used network simulators in both academic and industrial communities. It supports most of the networking protocols, including wireless protocols. PCF and DCF modules are included in the basic OPNET as part of its wireless.

Software developed by OPNET Technologies Inc. is a virtual environment that allows professionals to simulate the work of each network infrastructure as a whole, including technologies, protocols, specific equipment and applications used by end users. The main element of the environment is a package of OPNET Modeler (Fig.3), which relies on mapping simulations and running of the network using the graphical user interface (GUI) witch allows the creation of models, data collection and their subsequent processing. The program has a built-in library containing most of the models currently, used protocols and the TCP / IP applications, as well as models of various types of connections currently used for connecting nodes in the network [4].

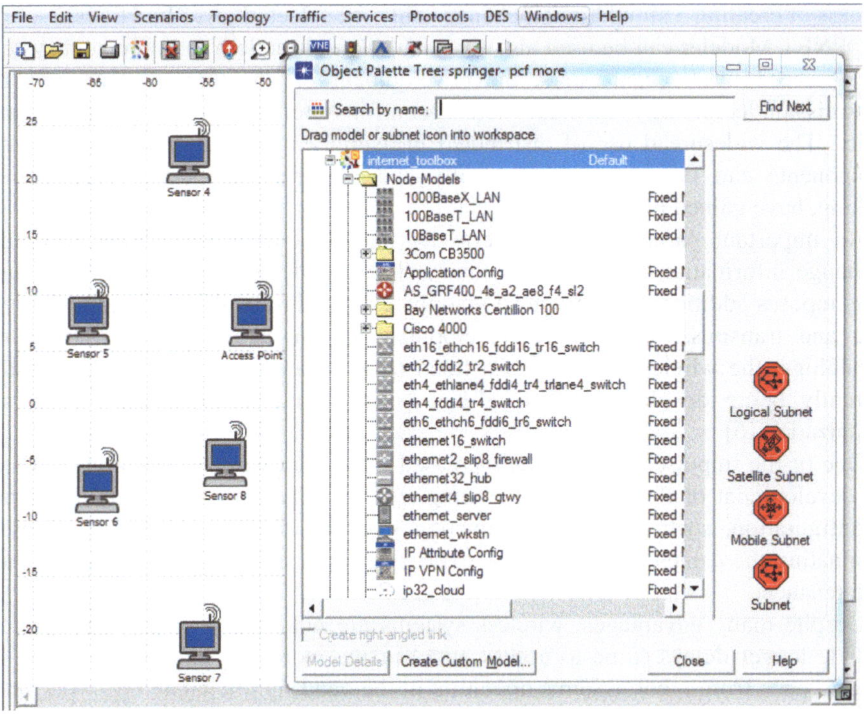

Fig. 3. Main window of OPNET Modeler and library of models and objects

Modeler environment takes into account the relationship between network elements and provides a wealth of information necessary to evaluate the performance of installed applications and the work of the network as a whole. With this information it is possible to complete the planning of information systems, their configuration, development and diagnostics, and above all optimize the structure, in terms of cost and quality, measurement and computer networks [5]. OPNET Modeler software also allows the implementation of planning applications on the network, a network analysis of the configuration and centralized management of data from various sources and the planning of the existing structure of migration to new technologies.

Modeling in OPNET Modeler environment takes place at three hierarchical levels [7]. The highest of these is the level of the network and the most widely used. Allows the connection of network nodes such as switches, amplifiers and routers through appropriate communication links.

The lowest level in the hierarchy is the level of the node on which it is possible to change the internal structure of the network created by the level of node objects, ie. sender, recipient, queue and a set of protocols. The level of the process is characterized by very high detail. The third level, allows the changing of the parameters of your own network using the C + +. Such a division simplifies the process of creating a simulation, it changes only the selected layer functions .

OPNET Modeler can be used also for simulation of modern measurement systems. As observed over recent years, the development of computer systems and data transmission technology has made modern control systems cease to exist as a separate entity. The widespread use of distributed measurement and control systems, whose components can be separated physically and preserve independence in decision making, have gained popularity.

An important parameter of modern distributed systems is technology used to exchange information between elements. Industry and industry control - measuring here imposes additional constraints due to the expected reliability and indivisibility of data and transmission security. Development in this area has come from wire technology, the wireless systems of a fixed network topology, and wireless systems currently being developed or an ad-hoc basis with a certain capacity for self-organization [6].

One of the impulses causing the increased interest in wireless solutions in data is the development of flexible manufacturing systems. There is the possibility of rapid reconfiguration, such as on the production line. In addition, wireless systems, by eliminating the cables are clearly less expensive: both in terms of installation and in maintenance.

Despite many advantages, wireless systems are not free of defects, which could include longer delays (time to access node) or the increased likelihood of data loss (interference from other systems operating in the band, multichanneling, work on the border of the range). It should be noted that these deficiencies can be compensated for by the appropriate choice of technology, data transmission and appropriate design of the system.

OPNET allows the simulation of distributed measurement systems based on the standard simulation of WLAN with QoS.

3.2 Measurement System

Two different scenarios were considered for this research work. In the first scenario, both PCF and DCF were deployed in the wireless domain with two measurement stations (Fig.4).

The workstations shown in fig.3.2.1 represent two sensors. Sensor_1 is responsible for the measurement of appropriate quantities such as temperature or pressure, and then sends the measurement data to Sensor_0, which eventually will be connected to the server responsible for the acquisition and retention of data.

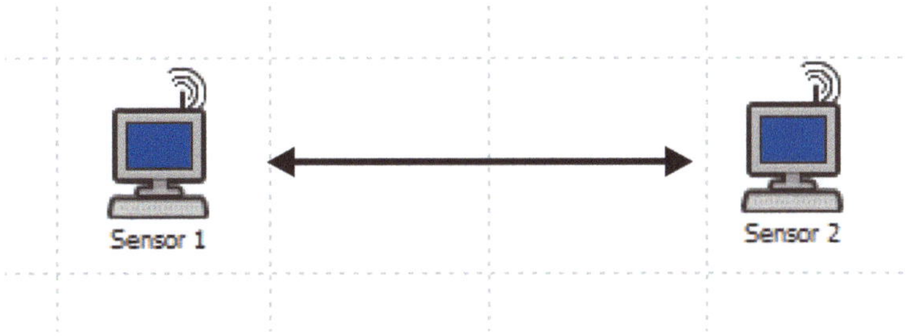

Fig. 4. First measurement system

The distance between workstations was changed during the simulation. Three separate scenarios in the OPNET Modeler software were created where the sensors were split apart by 500m, 750m and 1000m. The project did not take into account confounding factors such as the extent of the station, other wireless networks, walls, barriers and weather conditions.

For each sensor station the parameters defining the generation of network traffic was set. These parameters included the time when the packets started and stopped generating and also the duration of the broadcast transmission. The start time which specified the distribution name and arguments used for generating the random start time across different nodes was constant and set to 2 seconds (Fig.5).

Attribute	Value
name	Sensor_1
Destination Address	0
⊟ Traffic Generation Parameters	(...)
Start Time (seconds)	constant (2)
ON State Time (seconds)	exponential (4)
OFF State Time (seconds)	exponential (4)
⊞ Packet Generation Arguments	(...)
Stop Time (seconds)	Never
Traffic Type of Service	Best Effort (0)
⊞ Wireless LAN	

Fig. 5. Traffic generation parameters for Sensor_1

'ON State Time' specifies the distribution name and arguments to be used for generating random outcomes for time spent in the 'ON' state. Packets are generated only in the 'ON' state. The 'OFF State Time' was the distribution name and arguments used for generating random outcomes for time spent in the 'OFF' state. No packets are generated in the 'OFF' state [8]. Baud rate was fixed at 11Mbps.

'Load' in this measurement system which represents the total load (in bits/sec) submitted to wireless LAN layers by all higher layers in all WLAN nodes of the network is shown in Fig.6. The load is almost identical for all scenarios.

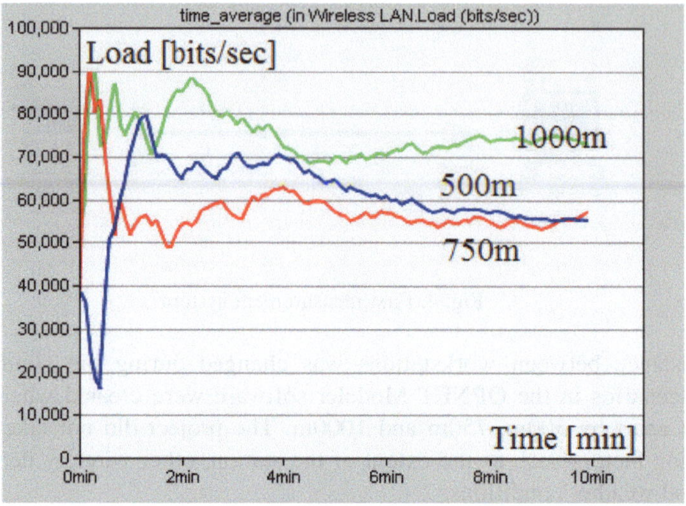

Fig. 6. Load of packets in seconds

Packet delay through the transmission for DCF mechanism by increasing the distance between the measuring stations is in Fig.7.

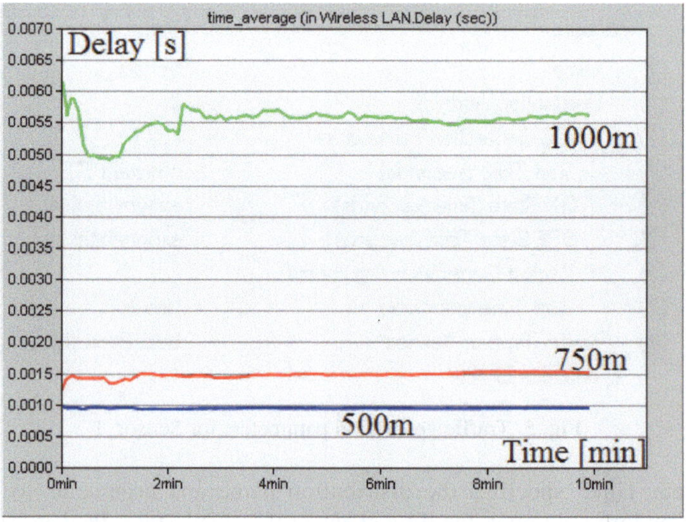

Fig. 7. Delay of packets in seconds for DCF scenario

It is clear that increasing the distance positively affects the packet's delay time. Up to 500 meters, this time is almost identical to the later increases and almost exponentially increases the amount of packets lost too. The PCF mechanism performed the same simulation time delay of packets (Fig.8).

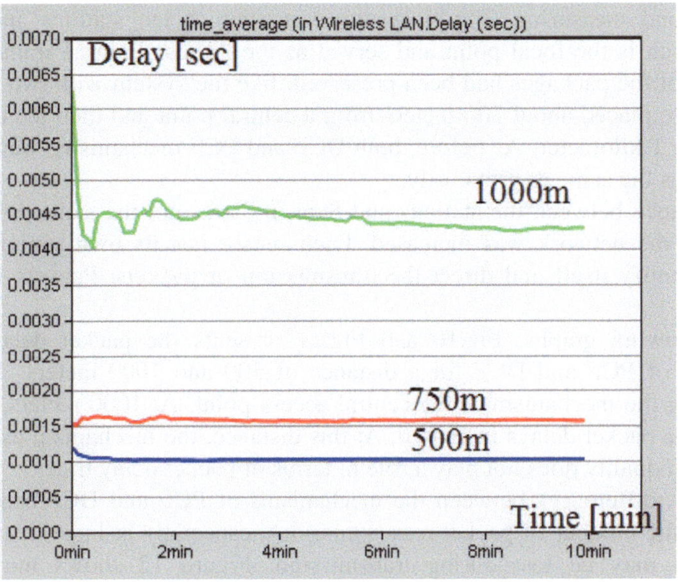

Fig. 8. Delay of packets in seconds for PCF scenario

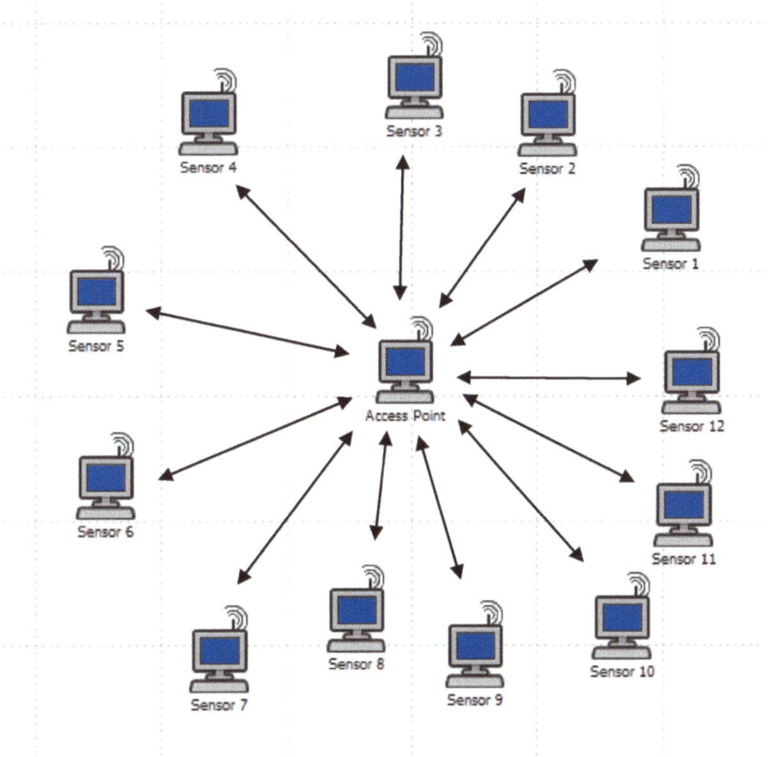

Fig. 9. Second measuring system

The second measuring (Fig.9) system consisted of ten stations and a sensing system, which is the focal point and served as the Access Point. Parameters for the generation of the packages had been preserved, like the system with two stations. All stations were placed about 500 meters from a central point and then the distance was increased to 1 kilometer. As before, both DCF and PCF mechanisms were simulated. The load was the same as previously.

The distance between the stations and Sensor_0 was not the same. Distance from the axis of the network was increased. Each sensor has its own station address in order to identify itself and direct the transmission of packets. Packets were sent at random.

The following graphs: Fig.10 and Fig.11 presents the packet delay using the mechanism of PCF and DCF for a distance of 500 and 1000 meters. The delay is higher using the mechanism of the central access point. At 1000 meters, the number of very large packet delays increased. At this distance, the mechanism used to ensure transmission quality does not play a role in terms of packet delay time.

However, differences between the mechanisms of PCF and DCF can be seen by comparing the number of packet retransmissions, especially at large distances where the packets may be lost during transmission. Figure 12 shows the number of retransmissions for station Sensor_1.

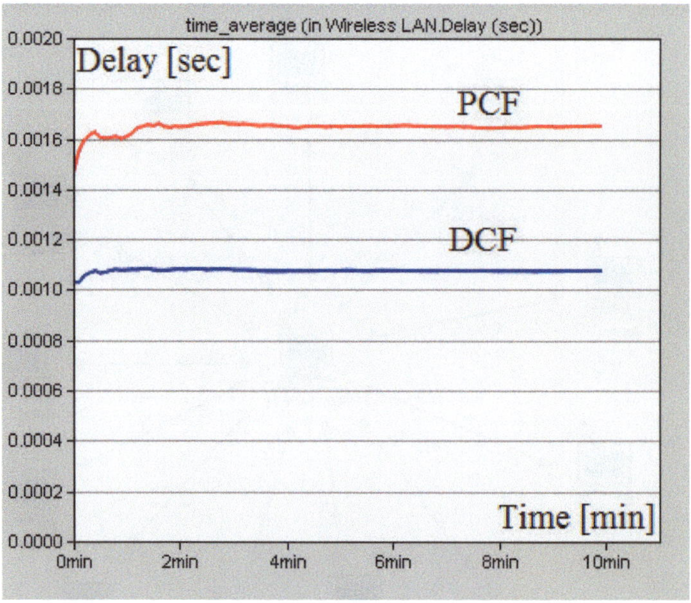

Fig. 10. Delay of packets for 500 meters

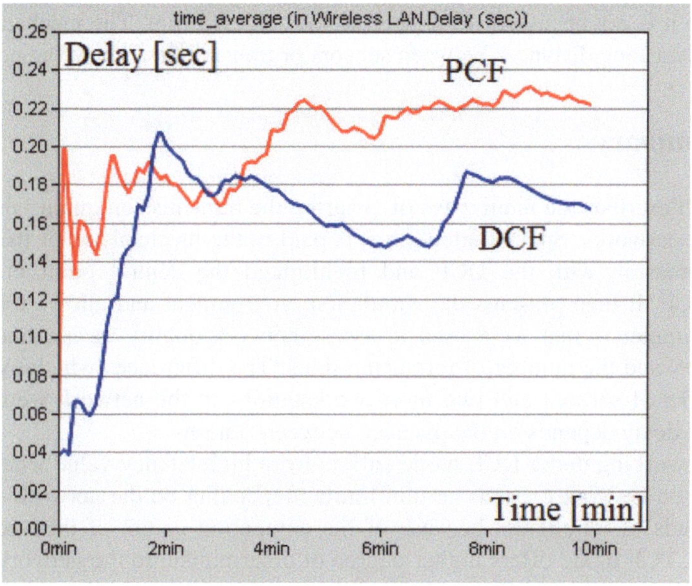

Fig. 11. Delay of packets for 1000 meters

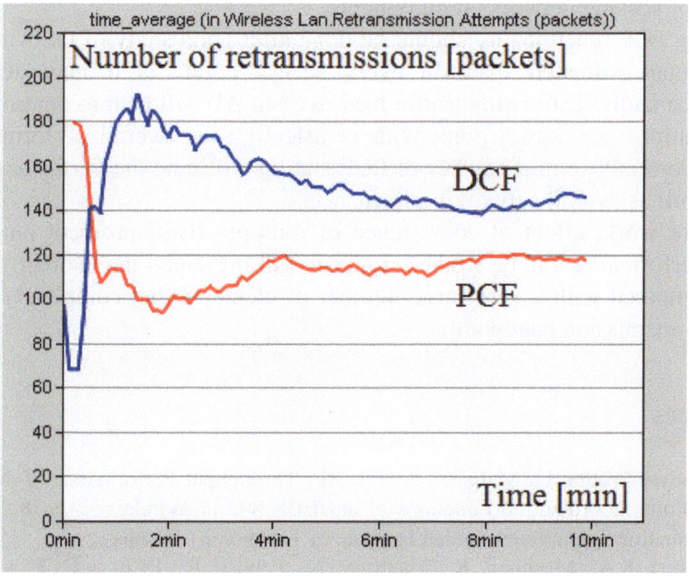

Fig. 12. Number of retransmissions of packets for 1000 meters

The number of retransmitted packets is greater than the distributed transmission mechanism DCF, while using the mechanism of PCF, not such a large amount of data, was lost so it is not necessary to re-send to the Access Point. The mechanism of PCF was used over long distances between sensors or transmitters, where the power is low.

4 Summary

This paper describes the main ways of ensuring the transmission quality measurement in wireless networks. Special attention was paid to the mechanisms of the distributed data transmission with the DCF and highlighted the central point in the sensor network PCF. It then presents the simulation environment and allows for the design and test parameters that were used to assess network quality measurement, such as packet delay and the number of retransmissions. The differences which exist between the networks of sensors and two more workstations in the network were presented. The packet delay depends on the distance between stations.

WLAN working under DCF mode suffers from high latency value when the traffic offered to the network exceeds certain limits (depending on the network size and the load demands of its clients) because of the contention nature of this mode. On the other hand, PCF mode offers higher degree of determinism to the network clients but suffers from the negative effect of polling overhead in the case of low load conditions. The wise selection of PCF parameters must takes into consideration several aspects such as the load expected to/from the nodes, number of nodes and the level of service provided for sensing networks requirements.

Enabling PCF functionality should be done after good study of the whole network and the total estimated load on every access point, as it may affect overall performance badly. Estimating traffic load over an AP will help to determine number of PCF stations per access point without affecting the overall performance of the network. Correctly setting number of PCF station will tune the performance of those nodes as well as overall network performance.

At future work, effect of coexistence of multiple IEEE protocol on the sensing network performance can be studied. Later it will presented the another network that will be compared with a very large number of elements that completely occupy the available transmission bandwidth.

References

[1] Chandran-Wadia, L., Mahajan, S., Iyer, S.: Throughput Performance of the Distributed and Point Coordination Functions of an IEEE 802.11 Wireless LAN, K. R. School of Information Technology, Indian Institute of Technology Bombay

[2] Rasheed, S.A., Masnoon, K., Thanthry, N., Pendse, R.: PCF vs DCF: a performance comparison, Department of Electrical and Computer Engineering Wichita State University. IEEE (2007)

[3] Kurytnik, I.P., Karpiński, M.: Bezprzewodowa transmisja informacji, Wydawnictwo PAK, Warszawa (2008)

[4] Krupanek, B., TOPór-Kamiński, T.: Wykorzystanie środowiska OPNET Modeler w symulacjach sieci komputerowych, Metrologia dziś i jutro, materiały konferencyjne MKM Gdańsk (2009)

[5] Krupanek, B.: Badania symulacyjne bezprzewodowego systemu eksperymentalnego opartego na standardzie ZigBee przy wykorzystaniu środowiska Opnet Modeler; Monografia XI Międzynarodowych Warsztatów Doktoranckich OWD (2009)
[6] Topór-Kamiński, T., Krupanek, B.: Niezawodność przesyłu danych w rozproszonych systemach pomiarowych. Pomiary Automatyka Kontrola (October 2009)
[7] Fitzgerald, J., Dennis, A.: OPNET Lab Manual. OPNET Technologies, Inc. (2009)
[8] http://www.opnet.com/university_program/teaching_with_opnet/textbooks_and_materials/materials/OPNET_Modeler_Manual.pdf
[9] Topór-Kamiñski, T., Krupanek, B., Homa, J.: Delays Models of Measurement and Control Data Transmission Network. In: Nawrat, A., Simek, K., Świerniak, A. (eds.) Advanced Technologies for Intelligent Systems of National Border Security. SCI, vol. 440, pp. 257–278. Springer, Heidelberg (2013)
[10] Kuś, Z., Nawrat, A.: Object Tracking for Rapid Camera Movements in 3D Space. In: Nawrat, A., Kuś, Z. (eds.) Vision Based Systems for UAV Applications. SCI, vol. 481, pp. 57–76. Springer, Heidelberg (2013)
[11] Kuś, Z., Nawrat, A.: Object Tracking for Rapid Camera Movements in 3D Space. In: Nawrat, A., Kuś, Z. (eds.) Vision Based Systems for UAV Applications. SCI, vol. 481, pp. 77–91. Springer, Heidelberg (2013)
[12] Werner, A., Bach, M., Duszeňko, A.: Vision Object Access and Management with Usage of the In-Memory Databases. In: Nawrat, A., Simek, K., Świerniak, A. (eds.) Advanced Technologies for Intelligent Systems. SCI, vol. 440, pp. 231–246. Springer, Heidelberg (2013)
[13] Bach, M., Duszenko, A., Werner, A.: Column-Oriented Metadata Organization of Vision Objects. In: Kapczyński, A., Tkacz, E., Rostanski, M. (eds.) Internet - Technical Developments and Applications 2. AISC, vol. 118, pp. 51–58. Springer, Heidelberg (2012)
[14] Babiarz, A., Bieda, R., Jędrasiak, K., Nawrat, A.: Machine Vision in Autonomous Systems of Detection and Location of Objects in Digital Images. In: Nawrat, A., Kuś, Z. (eds.) Vision Based Systems for UAV Applications. SCI, vol. 481, pp. 3–25. Springer, Heidelberg (2013)
[15] Orkisz, M., Swiech, L., Zacharzewski, J.: Fatigue tests of motor glider wing's composite spar. Eksploatacja i Niezawodnosc 14(3), 228–232 (2012)
[16] Kciuk, S., Martynowicz, P.: Special application magnetorheological valve numerical and experimental analysis. Diffusion and Defect Data Pt. B Solid State Phenomena 177, 102–115 (2011)
[17] Bieda, R., Jaskot, K., Jędrasiak, K., Nawrat, A.: Recognition and Location of Objects in the Visual Field of a UAV Vision System. In: Nawrat, A., Kuś, Z. (eds.) Vision Based Systems for UAV Applications. SCI, vol. 481, pp. 27–45. Springer, Heidelberg (2013)
[18] Kciuk, S., Kciuk, M., Turczyn, R., Martynowicz, P.: An Investigation of the Behaviour of Magnetorheological Fluids in The Rotary Shock-Absorber. Advanced Materials Research 628, 512–517 (2013)
[19] Gasiorek, D., Mezyk, A., Switonski, E.: Modelling of Electromechanical Drive Systems With The Use of Modal Condensation. Transport 20(1), 23–31 (2005)
[20] Mezyk, A.: Minimization of Transient Forces in an Electromechanical System. Structural Optimization 8(4), 251–256 (1994)

Application of Telemedical Technologies in Remote Evaluation of Soldiers' Vital Signs during Training and in Combat Conditions

Ireneusz Wochlik[1], Jarosław Bułka[1], Łukasz Folwarczny[2], Krzysztof Daniec[3], Karol Jędrasiak[3], Roman Koteras[3], and Aleksander M. Nawrat[3]

[1] AGH University of Science and Technology, 30-059 Krakow, Poland
[2] Silvermedia Sp. z o.o., 30-415 Krakow, Poland
[3] Institute of Automatic Control, Silesian University of Technology, Akademicka 16, 44-100 Gliwice, Poland

Abstract. The article presents the potential use of computer systems for remote evaluation of soldiers' vital signs and the transfer of results to the Command and Control Center or the Medical Support Centre. This article also suggests a number of possible solutions that can be successfully adopted both in training and in combat conditions. Furthermore, it indicates new sources of development to this rapidly progressing field of technology.

Keywords: remote control systems, teleoperations, UCEV, UGV, ARV.

1 Introduction

For many years, the remote evaluation of general condition has been growing in popularity and recognition amongst physicians and has been widely accepted by patients. Prototype products and telemedical services are introduced in general applications more frequently, in compliance with the dual-use philosophy, not only in military services but also in civil applications. Analysis of the problem domain clearly suggests choosing telemonitoring as the utmost attractive and highly desirable method for command support. Thanks to constant observation and automated analysis of vital signs, telemonitoring systems enable, among other things, an improvement of current control of soldiers' general conditions, adjusting modes and training programs to each soldier individually, increasing soldiers' sense of safety in the battlefield and optimizing human resources in military conditions and stabilization missions. Diagnostic mobile devices supported and remotely managed by modern computer systems, transmitting data via encrypted communication channels, can be used not only to monitor soldiers' general conditions on the battlefield, but also to supervise the general condition of garrisons in the battle vehicles. This chapter contains information about the use of the computer-based telemonitoring system designed and

A.M. Nawrat (ed.), *Innovative Control Systems for Tracked Vehicle Platforms*,
Studies in Systems, Decision and Control 2,
DOI: 10.1007/978-3-319-04624-2_11, © Springer International Publishing Switzerland 2015

tested by the authors. It is equipped with dedicated diagnostic devices used to remotely monitor selected vital signs of soldiers during training and the conduct of combat operations.

2 Usage Scenarios

Telemonitoring of soldiers' vital signs can be applied in many areas where an immediate health assessment is crucial in the context of decision-making and the medical diagnostics and health monitoring are much impeded. The following section presents a detailed description of an IT platform used to carry out the tasks of telemedicine. It introduces exemplary measuring devices integrated with the platform and discusses two examples of soldier telemonitoring that differ in purpose: the telemonitoring of soldiers during training and simulated combat and the telemonitoring of soldiers during true combat conditions.

2.1 Soldier Telemonitoring Conducted during Training and Simulated Combat

Adequate training of soldiers is the key element for ensuring optimal performance in combat conditions. Comprehensive, real-time remote assessment of the state of health of soldiers can serve to match the mode and intensity of exercise to the individual shape of each of the soldiers. This makes it possible to increase efficiency and reduce the risk of overtraining or serious injury. In addition, telemonitoring can be used during individual workouts to determine the physical fitness of each of the soldiers.

Continuous recording and analysis of vital signs such as ECG, heart rate, body temperature, level of relaxation, concentration or stress makes it possible to assess the health of the soldiers and their mental condition during various training scenarios; such as, training carried out in buildings, in the woods, during operations with the use of combat vehicles or in open space. Modern telemedicine systems, in such cases, enable us to observe the influence of changing weather conditions and the influence that the characteristics of the environment have on the behavior and psyche of soldiers.

To facilitate the process of command and training management the IT system should have implemented algorithms for the analysis of recorded ECG signals and a system of notifications informing the individual monitoring of anomalies which may endanger the life or health of one of the soldiers. This will allow for rapid intervention to reduce the intensity of the workout, or, if necessary, to remove a soldier from the firing range or training grounds. The measured vital signs should have diagnostic value and be able to be used by physicians to analyze the state of health of the soldiers.

The software should also take into account the archiving of measured vital signs, as they may be helpful for further analysis of the conducted training or comparative analysis of a series of training sessions. Observation of both the behavior of the soldiers during training and their vital signs may allow one to estimate with high probability how they will react to various stimuli in combat.

2.2 Telemonitoring in Combat Conditions

The contact between soldiers and the headquarters or the center of medical support during combat activities is, for obvious reasons, much inhibited. In addition, sometimes commanders on the battlefield have no current knowledge about the state of health of their soldiers.

As is generally known, all attempts to communicate information about the state of health of the members of the sub-division, verbally or in writing, may reveal the physical and psychological state of the soldier or the group to the enemy. This may lead to the rapid identification of the weak point by the enemy, who may use this information during combat operations. On the other hand, limiting all contact between soldiers and the command center can lead to a situation in which the commander supervising the combat operations loses control over the state of his troops. This could eventually lead to unnecessary delays in providing aid to the injured in combat operations or, at best, to inadequate disposition of human resources to perform the assigned functions. As a result, this may lead to the weakening of the combat capability of the specific unit.

In order to avoid this type of situation, the general condition of the soldiers on the battlefield can be continuously monitored. The best way to do this is the remote, real-time monitoring of biomedical measurements. Unquestionable benefits of such monitoring are:

- Enhancing the situational awareness of the commander by increasing the control of the state of health of the soldiers – Thanks to monitoring, the occurrence of incorrect records obtained from measurements of vital signs is registered by the headquarters or the center of medical support and allows them to immediately take the necessary action. This may include sending a medical support vehicle to the scene or informing another soldier who is on the battlefield of the need to provide assistance immediately. Any action of this type must be done while maintaining the highest level of security, without risking one's own life or the lives of other members of the group.
- Increasing the security of conducted actions – reducing the occurrence of situations that are dangerous for the soldiers while carrying out orders in the field.
- Increasing the troops' sense of security on the battlefield – Having an electronic system constantly checking basic vital signs, such as ECG and heart rate, provides soldiers with the confidence that in case of any health problems the

center of medical support will be informed immediately and the appropriate rescue unit will be sent in order to provide the necessary help.

- The assessment of situations on the battlefield without having to communicate with the members of the crew or team by means of registration of the psychophysical response of soldiers. This allows commanders to delegate tasks to the soldiers with optimal capacity at a given time.
- An improvement of warfare conduct – thanks to the monitoring of soldiers during training, the command center is able to predict and plan the course of combat operations, taking into account the results obtained by each of the members of the crew.

A very important aspect of monitoring soldiers in combat conditions is the monitoring of vital signs of the crews who are in combat vehicles. Software allowing for visualization of measurements should allow control of collective, aggregated data presenting (in an easy and quick way to analyze) the indicators exemplifying the state of health of individual crew members and of a group of combat vehicles altogether. The software should for this purpose make a preliminary analysis of the measurements and properly visualize the data. An IT system that informs the commander in which of the vehicles irregularities among the crew were detected by autonomous operating system algorithms would significantly help him to make optimal decisions, and in extreme cases allow a rapid intervention aiming at the evacuation of the wounded.

3 Ways of Measuring Vital Signs

The ways of taking measurements, together with possibilities arising from the measurements will be presented in the chapter below. Including: measurements of heart activity, mental state of the soldiers and additional medical measurements facilitating diagnostics, such as, body temperature and level of blood oxygenation.

3.1 Measurements of the Activity and Quality of Soldier's Heart Rate

The measurement of heart muscle activity is the crucial parameters depicting both general conditions of a soldier, as well as, the heart's reaction to external stimuli. Monitoring heart rate activity is useful during both training and combat conditions.

It allows for early detection of cardio-vascular problems. The measurement of heart muscle activity can help uncover irregularities that would not be seen with routine medical examinations. This can help prevent deploying a soldier to combat when a potential medical issue exists.

Thanks to early diagnosis, it is possible to start a required and consistent treatment in advance.

Another advantage of monitoring heart rate is the possibility to thoroughly recognize the opportunities of subordinates by their commanding officer. Registering the course of the heart rate during particular tasks allows one to observe soldiers' reaction through the noticeable change of heart rate. This reflects in the process of ECG and pulse cues. It is then easy to objectify the evaluation of such reactions as surprise, pain, stress or fear.

Monitoring heart rate in combat conditions aims at observing the general condition of a soldier in real-time. It provides the commanding officer with information regarding the possible need to send medical assistance to the operational areas. It has a direct impact on the decision to transport or evacuate a soldier to the base camp, or even to stop combat operations and retreat from fighting.

The soldier, who is being monitored, due to the need of maintaining adequate freedom of movement, has to be monitored with a device that has the following properties: the highest possible current efficiency, the smallest mass and the minimum number of electrodes, while at the same time, providing the best possible quality and accuracy of measurements taken.

These devices can be, for example, a 1-electrode ECG devices, which provide crucial information, such as, whether the soldier is alive and able to continue his mission. The device can be designed as a wristband which simplifies its use. Apart from cleansing the place of electrode's fastening with the appropriate means of disinfection, there is no additional means, such as, disposing of excessive hair, which would be necessary if electrodes were applied to the chest. Special attention should be paid to the mechanical properties of the device, regardless of its diagnostic quality. It is important to remember its resistance to disturbances and energy efficiency for obvious reasons. Complementing power supplies in combat conditions is extremely difficult.

Possibilities to choose ECG devices are greater in training conditions. Due to the fact that in this case it is more important to take a more specific heart rate measurement which will allow for the comparison of a lot of leads, it is possible to use ECG devices that are 3-, 4-, 5-, 6-, 7-, or even 12-leads. In this case, ECG devices can have a form of electrodes clipped to the body and connected with a small ECG device pinned to the soldier's belt, or electrodes embedded in suits worn like a shirt, directly on the body.

3.2 Monitoring of Soldiers' Mental State

Monitoring mental state allows one to determine soldier's alertness, focus or stress. It is of invaluable help when analyzing the capabilities of particular soldiers, as well as, the engagement in performing certain tasks in training conditions. Retrospective examinations of this type enable one to make precise plans of combat operations in the field.

Monitoring soldiers' mental state is possible by tracking the course of electric activity in the brain with the help of brain wave measurement collected from the frontal area (point FP1). The device is incredibly lightweight and its firm design fits entirely under a soldier's helmet. What is more, the device is wireless, thus it does not impair a soldier's performance during training.

3.3 Monitoring Additional Medical Parameters

In order to make a thorough assessment of a soldier's general condition it may be required to make additional measurements such as:

- Body temperature
- Saturation

Body temperature is repeatedly the first sign which informs us about various infections and health problems. This parameter has a wide range of applications in diagnostics. Low blood oxygenation is of similar importance. Additionally, it is a significant diagnostic parameter, necessary in the evaluation of heart muscle efficiency. Measurements of saturation can be taken with a device that uses infrared waves.

4 Methods of Transferring Medical Data

Depending on the use, the transfer of medical data can be carried out in many ways. The choice of technology and the way of potential combination depends largely on whether it is necessary to monitor in real time, or if the measurements can be sent in the so-called packages such as the 20-second ECG waveform records. If the measurements are made during a workout, a good solution is to use data transfer via Bluetooth or Wi-Fi. Medical data transmitted via Bluetooth can be collected in transceiving station and sent over via Wi-Fi or radio waves. All data transferred via GSM or GPRS can be delivered directly to the system visualizing the measured vital signs. Telemonitoring, carried in combat conditions (and training conditions), requires transmission methods which guarantee the desired level of security of the data transmitted through the use of transmission encryption and the use of key agreement mechanisms.

5 Suggested Solutions

This section presents the architecture of the proposed solution with regard to the means of transmitting medical data between different devices. Next, it will present the measuring device constructed by a team of researchers from the Silesian University of Technology and a program designed by a team of experts from the Academy of Science and Technology and Silvermedia Company.

5.1 Architecture of the Suggested Solutions

The proposed solution assumes the existence of two different models of communication between the measurement devices and the application installed on the computer in the monitoring base. Depending on the conditions in which the monitoring system operates, different means of data transfer will be used.

The first model is the monitoring of soldiers during training. In this model, the results of the measurements of vital signs can be sent directly from each measurement device via Wi-Fi or through a Bluetooth module.

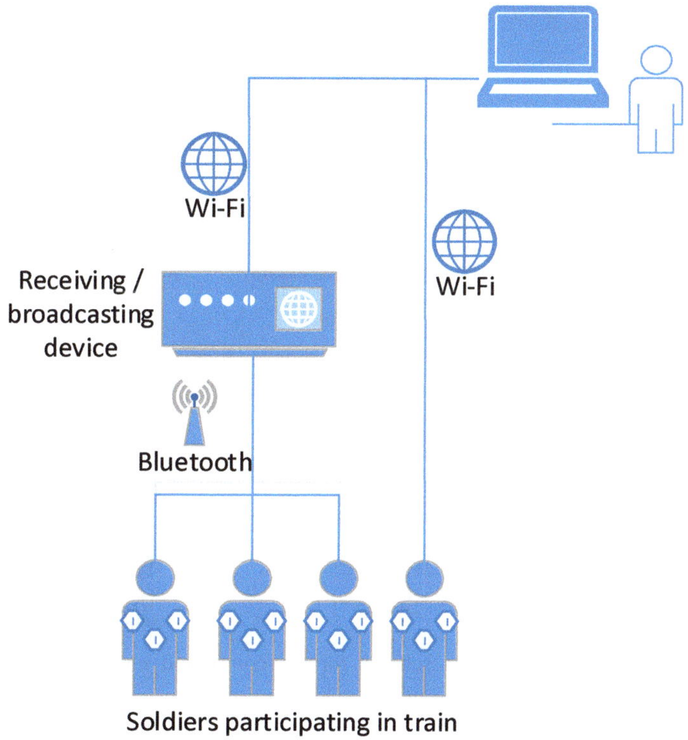

Fig. 1. Medical data transmission scheme for the considered models of communication in training conditions

The second model assumes monitoring of vital signs (heart rate, ECG) of soldiers during warfare. In this case, it is possible to transfer data directly from the measuring device using GSM/GPRS communication or radio waves to the receiving application in the monitoring base. The same data can also be sent in the same manner to the unmanned flying object which then communicates with the receiving application via the satellite link.

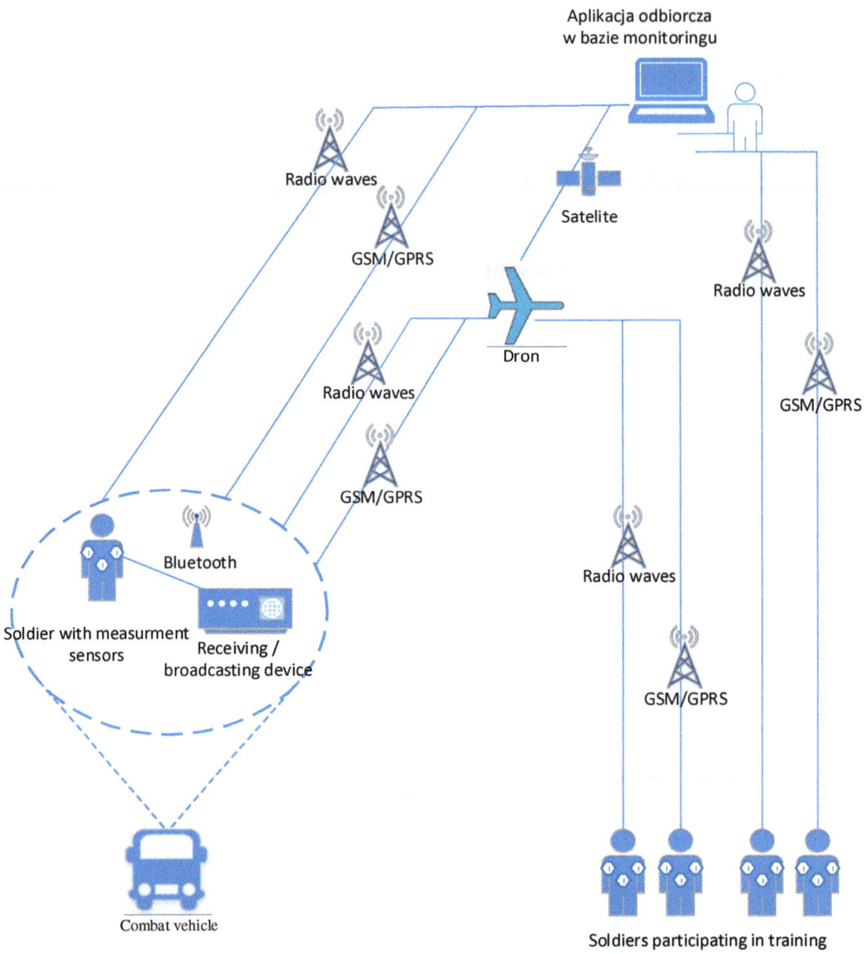

Fig. 2. Medical data transmission scheme for the considered models of communication in combat conditions

In both of the above cases, you should be aware that the majority of commercially available measurement devices designed to monitor performance during training first perform the measurement and stores it in memory and only after the signal acquisition, sends it to the receiving station.

5.2 Device for Measuring Vital Signs

A mobile device for measuring vital signs has been designed and produced for the purpose of this project.

One of its advantages is the small size and weight. The modular design of the device is flexible when it comes to the modification of measuring and communication system elements.

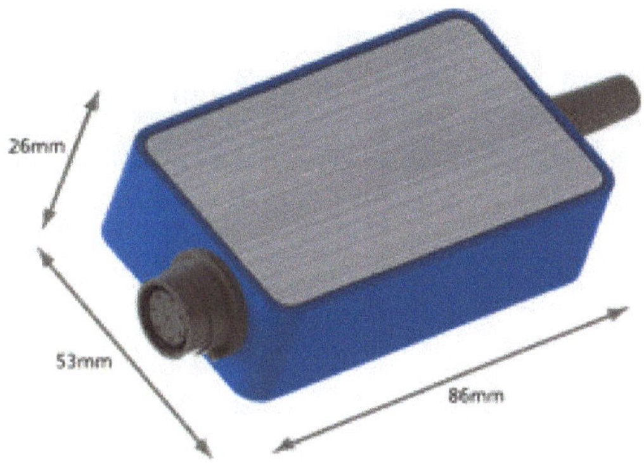

Fig. 3. Device for measuring vital signs of the soldiers

It can be easily modified depending on its intended use. The device enables measurement multi-lead ECG. The number of supported lead can be modified accordingly with the intended use. The device can be adjusted to measure 3-, 4-, 5-, 6-, 7-, 12- lead ECG while monitoring soldiers taking part in training. In case of adopting it to battle conditions, the device measures only 1- lead ECG signal based upon whether it is possible to specify if the monitored soldier is alive or not.

Thanks to the system of detecting unattached electrodes (for the device having more than one electrode) the person monitoring the situation can contact the soldier with a request to re-attach them.

Additionally, the device has an embedded sensor of body temperature. Its design allows integration with systems equipped with CAN bus. The device can cooperate with simulation systems in training mode. The simulations include tracking movements or shoot down.

Modular design of the device facilitates easy replacement of the communication module, to adjust the means of data transfer to an intended use. Telemonitoring in combat conditions can require data transfer other than in training conditions. Systems of encryption provide security of data transfer.

5.3 System for the Acquisition, Evaluation and Visualization of Medical Data

A dedicated, modular, on-line system has been designed for the purpose of this project. Functionalities of the system have been consulted with military experts and in each of the scenarios provide: the acquisition and archiving of biomedical data, automatic quality evaluation of the parameters, visualization of critical data and support for comparative research.

Visualization and analysis of measured vital signs are the crucial elements necessary to evaluate the general condition of the soldiers. A dedicated software has been designed for the purpose of this project. It enables individuals in charge of monitoring soldiers' general condition to control their health condition right after taking the measurement.

Thanks to innovative algorithms, biomedical signals sent in real-time are automatically analyzed for the occurrence of any abnormalities or disorders.

Two modes of the software are provided:

- Module designed to monitor soldiers vital signs during training and in combat operations in the field

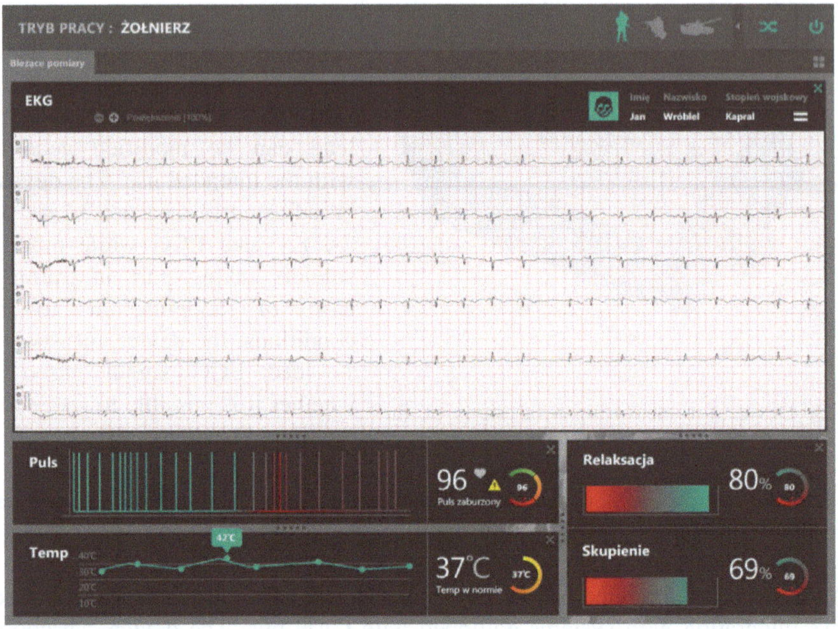

Fig. 4. Module designed to monitor soldiers' vital signs during training and in combat operations in the field

- Module designed to monitor vital signs of combat vehicle garrisons

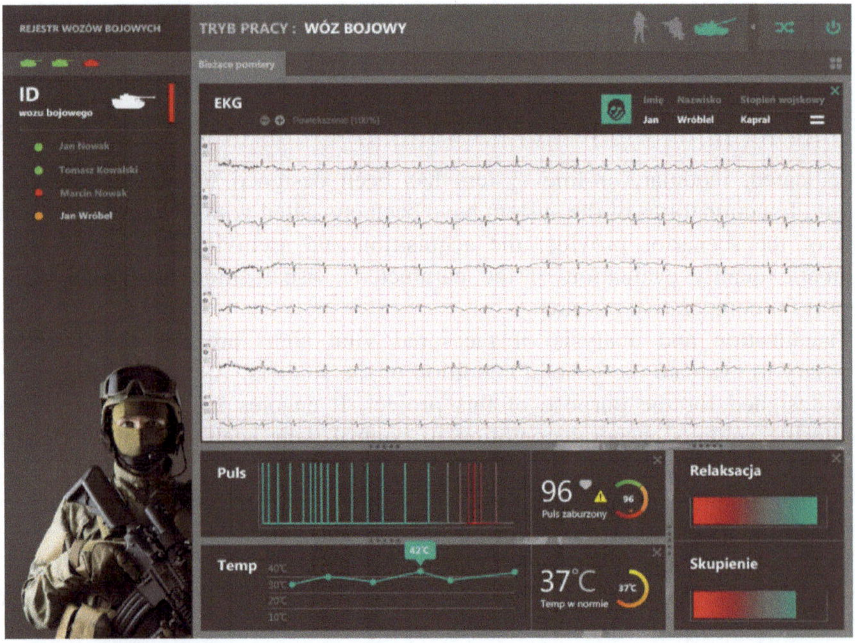

Fig. 5. Module designed to monitor vital signs of combat vehicle garrisons

Thanks to the modular design and extensibility of the system, further development is possible. The system has the ability to add further vital signs to be monitored. The way of presenting specific vital signs can also be configured, for example, the number of displayed ECG leads.

While using the software, the user is able to freely switch between the modes. It will not affect the functioning of the program and algorithms. Embedded algorithms combined with alarm functions provide vital assistance to the the individuals monitoring the soldiers. Notification systems make constant observation of all soldiers' vital signs unnecessary.

Special control panels simplify handling of the system. A change in the control color from green to orange or red is an alarm signal.

In case of monitoring members of the combat vehicle garrison, the soldiers are assigned to specific vehicles that are subject to monitoring. All vehicles, as well as the soldiers, are marked with a special alarm control. It changes color from green to orange or red whenever an algorithm embedded in the system detects any disorder or abnormality in any of the soldiers in a specific vehicle.

Archiving measured vital signs is crucial for the analysis of training sessions that measure physical efficiency, both group and individual. One of the possible solutions to archive the measurements is integrating with the central system.

It is possible to configure for which soldiers the parameters should be archived after training.

In the application designed for visualization of measured parameters in real-time, it is possible to configure for which soldiers the parameters should be archived after completing the training. The user can also choose which of the measured physiological parameters should be archived. A comprehensive archiving of all the data is not necessary.

6 New Directions of Development

The continuous development of biomedical technology allows for the continued expansion of possibilities of medical diagnosis support with the latest solutions in the field of bio cybernetics; within which lies the measurement of signals collected from the human body, their processing, automated analysis and visualization of the results obtained thereby.

The IT system designed as part of the present project is also ready for acquisition and analysis of images. Another desirable direction of development in this area is the solution providing information about the soldier's mental and physical condition by analyzing the image obtained from the surface of the eye. Analysis of pupil diameter and its changes may indicate the occurrence of a strong emotional reaction to the sudden appearance of life threatening danger or to soldier's remaining under the influence of drugs or poisons sprayed by the enemy on the combat area. Another parameter, which is increasingly indicated as a parameter supporting selection of soldiers when planning actions in the combat field, is the measurement of blood glucose levels. Abnormal levels of blood glucose significantly affect the performance of the peripheral nervous system, while the value of the concentration of glucose in the blood changes in this parameter provide a basis for assessment of the exercise capacity of soldiers.

7 Summary

Telemedicine systems, as the authors attempted to prove in this chapter, can be a very valuable source of information about the health and fitness of soldiers in all conditions of their service and therefore provide the commander with invaluable and objective support to make key decisions when conducting training or warfare. Designed and manufactured IT systems together with developed, dedicated diagnostic equipment create a consistent and automatically evaluating key parameter biomedical platform that can be used as a tool for command support.

References

[1] Izworski, A., Koleszyńska, J., Tadeusiewicz, R., Bułka, J., Wochlik, I.: Artificial intelligence algorithms in computer aided diabetes therapy. In: Arabnia, H.R., Yang, M.Q., Deng, Y., et al. (associate eds.) BIOCOMP 2009: Proceedings of the 2009 International Conference on Bioinformatics & Computational Biology, Las Vegas, Nevada, USA, July 13-16, vol. 1, pp. S.364–S.369. Bibliogr. s. 368–369, Abstr. CSREA Press, USA (2009) ISBN: 1-60132-093-0, 1-60132-094-9 (1-60132-095-7)

[2] Bułka, J., Izworski, A., Wochlik, I.: Wykorzystanie słuchowych potencjałów wywołanych z pnia mózgu w diagnostyce systemu słuchowego człowieka — (Diagnostics of auditory system usingauditorybrain-stemresponse). In: Podstawy Inżynierii Biomedycznej: [podręcznik akademicki], T. 1/pod red. nauk. Ryszarda Tadeusiewicza, Piotra Augustyniaka, pp. S.205–S.214. Bibliogr. s.213–s.214. Wydawnictwa AGH, Kraków (2009) (Wydawnictwa Naukowe/Akademia Górniczo-Hutnicza im. Stanisława Staszica w Krakowie; KU 0352). — ISBN 978-83-7464-263-7

[3] Izworski, A., Wochlik, I., Bułka, J., Pracuch, P., Bruski, Ł., Skarżyński, P.: Systemy teleinformatyczne w platformie badań zmysłów — (Telemedical system for screening examination platform). In: Biocybernetyka i Inżynieria Biomedyczna: XVI Krajowa Konferencja: Warszawa, 26–29 kwietnia 2010/IBIB PAN, PTIB, PAN. [IBIB PAN], Warszawa: Opis częśc. wg okł, — S. [1] (2010)

[4] Strzelczyk, P., Wochlik, I., Izworski, A., Bułka, J., Orzechowski, T.: Inteligentny system telemedyczny do wsparcia diagnostyki badań ABR — Intelligenttelemedical system supporting the diagnosis of ABR examinations. Bio-Algorithms and Med-Systems (Print)/Jagiellonian University. Medical College; 6(12(suppl. 1)), s.189–s.190 (2010) ISSN 1895-9091; Tekst pol.-ang. MCSB 2010: Modelowanie Cybernetyczne Systemów Biologicznych/Zakład Bioinformatyki i Telemedycyny. Uniwersytet Jagielloński. Collegium Medicum, Laboratorium Biocybernetyki. Katedra Automatyki. Akademia Górniczo-Hutnicza w Krakowie. CMUJ, Kraków (2010)

[5] Strzelczyk, P., Wochlik, I., Tadeusiewicz, R., Izworski, A., Bułka, J.: Telemedical system in evaluation of auditory brainsteam responses and support of diagnosis. In: Nguyen, N.T., Le, M.T., Świątek, J. (eds.) ACIIDS 2010, Part II. LNCS (LNAI), vol. 5991, pp. 21–28. Springer, Heidelberg (2010)

[6] Czekierda, Ł., Masternak, T., Skałkowski, K., Wyszkowski, P.: Applying SOA and virtual organization concepts in an interactive telemedical consultation system. In: MCCSIS 2012 [Dokumentelektroniczny]: IADIS Multi Conference on Computer Science and Information Systems: Lisbon, Portugal, July 17–23, 2012: Proceedings. Wersja do Windows. Dane tekstowe/International Association for Development of the Information Society. [S.\,l.]. IADIS Press, 1 dyskoptyczny. ISBN 978-989-8533-08-1. S.13– S.20. Wymaganiasystemowe: Adobe Acrobat Reader; napęd CD-ROM. Tytułprzejęto z ekranutyt. (powybraniuopcji: e-Health 2012). Bibliogr. s. 20, Abstr. Macedo, M. (ed.) Publikacjazamieszczona w części: e-Health 2012: Proceedings of the IADIS International Conference (2012) ISBN 978-972-8939-70-0

[7] Augustyniak, P.: Data queuing for non-uniform telemedical reporting. In: Burduk, R., Kurzyński, M., Woźniak, M., Żołnierek, A. (eds.) Computer Recognition Systems 4. AISC, vol. 95, pp. 539–547. Springer, Heidelberg (2011); Bibliogr. s. 547, Abstr.

[8] Augustyniak, P.: Discrete nonlinear control of the diagnostic quality in distributed telemedical systems. In: Piętka, E., Kawa, J. (eds.) Information Technologies in Biomedicine. AISC, vol. 69, pp. 355–366. Springer, Heidelberg (2010); Bibliogr. s.366, Abstr.

[9] Jaworek, J., Kańtoch, E.: Elektrofizjologiczny system monitorowaniapacjenta w ruchu — Electrophysiology-based monitoring system for human in motion. Bio-Algorithms and Med-Systems (Print)/Jagiellonian University. Medical College 6(12(suppl. 1)), 77–78 (2010), Tekst pol.-ang. MCSB 2010: Modelowanie Cybernetyczne Systemów Biologicznych/Zakład Bioinformatyki i Telemedycyny. Uniwersytet Jagielloński. Collegium Medicum, Laboratorium Biocybernetyki. Katedra Automatyki. Akademia Górniczo-Hutnicza w Krakowie. CMUJ, Kraków (2010)

[10] Duplaga, M., Zieliński, K.: Evolution of IT-enhanced healthcare: from telemedicine to e-health. In: Zieliński, K., Duplaga, M., Ingram, D. (eds.) Information Technology Solutions for Healthcare: With 98 Figures (3 in Color). Springer-Verlag London Limited, London (2006), (HI Health Informatics. Tyt. poprz.: Computers in Health Care/Hannah, K.J., Ball, M.J. (series eds.)) Opis częśc. wg okł. S.1– S.21. Bibliogr. s. 19– s. 21, Summ.

[11] Tadeusiewicz, R.: Lewatywa przez Internet, czyli blaski i cienie Telemedycyny — [Enema via Internet oradvantages and disadvantages of telemedicine]. In: Klikasz i Znikasz?: Psychologia Internetu, Kraków, Marca 17-19, konferencja/PSSiAP Polskie Stowarzyszenie Studentów i Absolwentów Psychologii [S. l. : s. n., 2006]. — S. [1–2] (2006)

[12] Augustyniak, P.: Modeling the adaptive telemedical system with continuous data-dependent quality control. In: SMC 2007 [Dokumentelektroniczny]: XII International Conference – System Modelling and Control, Zakopane, Poland. Wersja do Windows. Dane tekstowe/Institute of Computer Science, Technical University of Łódź, Poland (et al.), October 17-19, S.[1–8]. TU, Łódź (2007); 1 dyskoptyczny. ISBN 978-83-909816-2-8. Wymaganiasystemowe: Adobe Acrobat Reader; napęd CD-ROM. Bibliogr. s. [8], Abstr.

[13] Augustyniak, P.: Personal wearable monitor of the heart rate variability. Bio-Algorithms and Med-Systems (Print)/Jagiellonian University. Medical College 6(12(suppl. 1)), s. 39 (2010) ISSN 1895-9091; MCSB 2010: Modelowanie Cybernetyczne Systemów Biologicznych/Zakład Bioinformatyki i Telemedycyny. Uniwersytet Jagielloński. Collegium Medicum, Laboratorium Biocybernetyki. Katedra Automatyki. Akademia Górniczo-Hutnicza w Krakowie. CMUJ, Kraków (2010)

[14] Słowikowski, P., Zieliński, K.: Security and safety of telemedical systems. In: Zieliński, K., Duplaga, M., Ingram, D. (eds.) Information Technology Solutions for Healthcare: With 98 Figures (3 in Color). Springer-Verlag London Limited, London (2006) (HI Health Informatics. — Tyt. poprz.: Computers in Health Care/ Hannah, K.J., Ball, M.J. (series eds.)). Opis częśc. wg okł. S.63–S.84. Bibliogr. s.82–s.84, Summ.

[15] Soja, J., Duplaga, M., Dańda, J., Wąsowski, D., Skucha, W., Bestyńska-Krypel, A., Sładek, K., Niżankowska-Mogilnicka, E.: Specjalistyczne konsultacje telemedyczne w chorobach układu oddechowego realizowane w oparciu o połączenia satelitarne — Second-opinion telemedical interactions in pulmonarymedicine based on satellite data transmission. Nowiny Lekarskie = Medical News: dwumiesięcznik naukowy Uniwersytetu Medycznego im. Karola Marcinkowskiego w Poznaniu 76(suppl. 1), s. 69, 194 W (2007) ISSN 0860-7397. 63. Zjazd Towarzystwa Chirurgów Polskich = 63rd Congress of the Association of PolishSurgeons: Poznań, 12–15 września 2007 r.: streszczenia wykładów i zgłoszonych prac naukowych = abstracts of lectures and submitted scientific papers. Katedra i Klinika Chirurgii Ogólnej, Gastroenterologicznej i Endokrynologicznej UM, Poznań (2007)

[16] Dańda, J.: Zobacz pełny wykaz publikacji Autora/Autorów

[17] Tadeusiewicz, R.: Systemowe podejście do wybranych zagadnień telemedycyny [System approach to some telemedicine problems] pod red. In: Kulczyckiego, P., Hryniewicza, O., Kacprzyka, J., Autorzy, Bubnicki, Z., et al.: Techniki Informacyjne w Badaniach Systemowych, pp. S.341–S.359, Bibliogr. s.355–s.359, Streszcz. Wydawnictwa Naukowo-Techniczne, Warszawa (2007)

[18] Izworski, A., Wochlik, I., Bułka, J., Pracuch, P., Bruski, Ł., Skarżyński, P.: Systemy teleinformatyczne w platformie badań zmysłów [Telemedical system for screening examination platform]. In: Biocybernetyka i Inżynieria Biomedyczna: XVI Krajowa Konferencja: Warszawa, 26–29 kwietnia 2010/IBIB PAN, PTIB, PAN. [IBIB PAN], Warszawa, Opis częśc. wg okł. S. [1] (2010)

[19] Jaworek, J., Kańtoch, E.: Telemedycyna – nauka przydatna dziś, niezbędna jutro [Telemedicine – science usefultoday, necessarytomorrow]. In: Tadeusiewicza, R., Augustyniaka, P. (pod red. nauk)/ Podstawy Inżynierii Biomedycznej : [Podręcznik Akademicki], T. 2. Wydawnictwa AGH, Kraków (Wydawnictwa Naukowe/Akademia Górniczo-Hutnicza im. Stanisława Staszica w Krakowie; KU 0352), pp. S.393–S.398. Bibliogr. s. 398 (2009) ISBN 978-83-7464-263-7

[20] Tadeusiewicz, R.: Telemedycyna dla Społeczeństwa Informacyjnego [Telemedicine for Information Society]. Nauka i Biznes/Wyższa Szkoła Biznesu Dąbrowa Górnicza (18), s.8–s.9. Bibliogr. s.9 (2008) ISSN 1505-0653

[21] Tadeusiewicz, R.: Telemedycyna jako ważny, ale trudny składnik społeczeństwa informacyjnego [Telemedicine as important but difficult element of the informationsociety]. In: Bliźniuk, G., Nowak, J.S. (red. nauk.) Społeczeństwo Informacyjne: Doświadczenie i Przyszłość. PTI Polskie Towarzystwo Informatyczne, Katowice, OddziałGórnośląski, pp. S.99–S.123. Bibliogr. s.111–s.118 (2006)

Part III
Construction and Investigation of the Dynamic Characteristics of Tracked Vehicles

Nowadays the unmanned vehicles are becoming more and more common. Therefore there is a need for innovative mechanical constructions capable of adapting to various applications regardless the ground, air or water/underwater environment. Military operations in order to complete successfully require a vast amount of information and high quality equipment and vehicles capable of movement both in a city and terrain. An example of such are tracked vehicles. As well as in today's and tomorrow's army, tracked vehicles will be one of the most important equipment. As it is known this equipment is expensive, and extremely stressful experiments on this equipment should be conducted carefully. An answer for this problem is to conduct experiments on a computer model. Design, construction of such models require research inter alia in the field of kinematic and dynamic relationships inside entire suspension system. Elements of the vehicle are grouped in few subsystems, like suspension, wheels or tracks. During prototyping phase of the object a high fidelity model is constructed and later verified using the prototype. For instance the damping value estimation can be estimated based on tolerance area of characteristics of real damper.

One of the main threats to tracked vehicles are mines and IEDs. Therefore it is extremely important to research the impact of explosive on the tracked vehicles and especially the crew inside. It is known that the impact of explosion on the crust (hull) of a vehicle creates a secondary wave phenomenon. The phenomenon rests on the interaction of the bottom and sides of the vehicle as a membrane. The membrane moves the pressure pulse acting on the armour inside the vehicle. The increase of pressure inside the vehicle causes the damage to internal organs ultimately leading to the death of soldiers. In the case of armour penetration, even a more rapid increase of pressure is possible. One of the basic ways to protect crews of vehicles is opening the hatches of vehicles. Nevertheless, this method is not applicable when the vehicle moves in a contaminated area. High mobility in all terrain and weather conditions is one of the most significant features of the light armoured vehicles. As well as that, such vehicles have good independence, setting them as a primary military equipment to be used by stabilization forces of many countries. However, they are exposed to many threats among which are improvised explosive device (IED). As a consequence, contemporary technical and tactical requirements put strong emphasis on proper layout of the hull in order to provide a good cover for the crew. Therefore it is required to analyse the impact and effects in both ways: numerical and experimental. The chapter presents also experimental results of impact of IED on vehicles and special manipulators. It is essential that maximum stress during the explosion do not exceed the material parameters.

High Fidelity Model Construction and Its Verification Based on 2S1 Tracked Vehicle

Tomasz Nabagło[1], Janusz Kowal[2], and Andrzej Jurkiewicz[3]

[1] Faculty of Mechanical Engineering, Cracow University of Technology, al. Jana Pawła II 37, 31-864 Krakow, Poland
[2,3] Faculty of Mechanical Engineering and Robotics, AGH University of Science and Technology, Al. A. Mickiewicza 30, 30-059 Krakow, Poland
pmnabagl@cyf-kr.edu.pl, {jkowal,jurkand}@agh.edu.pl

Abstract. We present information about construction tasks of tracked vehicle model in MSC.ADAMS computer program. Kinematic and dynamic relationships inside entire suspension system are described here. Elements of the vehicle are grouped in few subsystems, like suspension, wheels or tracks. The significant accent is put on the tracks with its force connections. Those connections are between all track links as also between wheels, links and ground surface. Also significant accent is put on damping value estimation. It is estimated based on tolerance area of characteristics of real damper element, which is applied in this type of vehicles. In next part of the article, process of appropriate damping characteristics selection is described. This process is applied for high fidelity model building.

Keywords: tracked vehicle, dynamic model, simulation, ADAMS.

1 Introduction

As well in today's as tomorrow's army, tracked vehicles will be one of the most important equipment. As it is known this equipment is expensive, and extremely hard experiments on this equipment should be conducted carefully. An answer for this problem is to conduct experiments on a computer model. Dynamics of this model should be as similar to the real object dynamics as possible. During researches a new tracked vehicle model was constructed. It was made in MSC.ADAMS computer program. To realize this task, the authors had construct the model with a dynamic track model subsystem, which was considered by Nabagło [1]. Such kind of track model, with high fidelity of single track segment shape, is very important in two conditions. First one results from Assanis et al. [2] article and appears when the vehicle moves over hard and rough surface of road. Second one is described by Ryu et al. [3] and appears when the vehicle moves with high velocity. In both of them, high accuracy of the model improves results accuracy in comparison to measurements taken from the real object. In many articles about tracked vehicles, a problem of joints and contacts between parts is taken under consideration. One of the most important

A.M. Nawrat (ed.), *Innovative Control Systems for Tracked Vehicle Platforms*, Studies in Systems, Decision and Control 2,
DOI: 10.1007/978-3-319-04624-2_12,

problem in this area is connected with joints between track segments. Ryu et al. [3] taking under consideration single pin track link. He describes here so-called rubber bushing model, which connects two track links. Lee [4] in his article describes contact forces between wheels, track links and ground surface. There is explained a numerical method of force contacts calculation. Details of this connection forces, as also contact forces between wheels, track links and road surface are described in ADAMS Documentation [5]. Selected parts with high fidelity of real parts shapes may be designed in other CAD program. Łukaszewicz [6] in his article, explains multi-body CAD designing problem for simulation applications. In continuation of simulation topic, Kubela et al. [7] describes multi-body mechanism simulation in ADAMS computer program, in this specific case, in co-simulation with Matlab-Simulink program.

2 Tracked Vehicle Model Construction

In this paragraph, a conception of tracked vehicle construction in ADAMS program is described. It is divided between few aspects such as parts, connections and specific connections associated with track links.

2.1 Elements Groups in Tracked Vehicle Construction

In figure 1 the full tracked vehicle model of Platform 2S1 was shown. In the whole tracked vehicle model few groups of elements could be indicated. First of them is a vehicle hull and all elements, which could be inside of it. Also turret could be treated as element of this group.

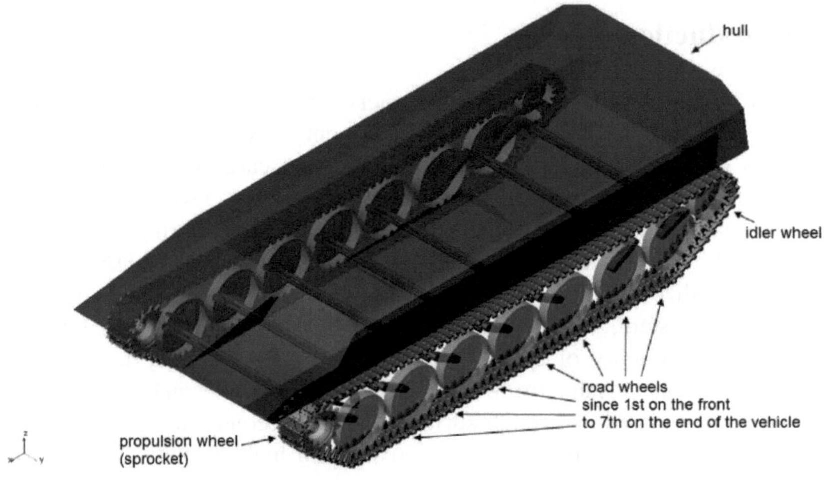

Fig. 1. Full tracked vehicle model

The hull is connected with elements of another groups with rotational, translational or fixed joints as also with forces elements. Another group is suspension system, which connect hull with another group of elements. This group collects wheels in meaning road wheels as also idlers and propulsion wheels known as sprockets. A next group encloses tracks elements. Contact forces act between these elements and wheels, as also between these elements and ground.

2.2 Kinematic and Dynamic Relationships between Parts

Suspension system is based on torsion bars (Fig. 2). A single repeatable suspension of a road wheel is constructed with a torsion bar element (Fig. 3).

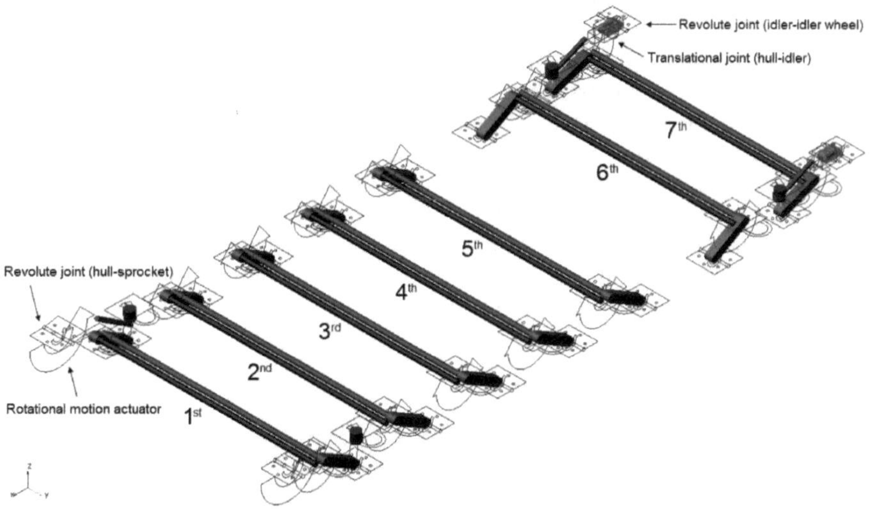

Fig. 2. Suspension system with numbered axels

Forces acting on the torsion bars are modeled with torsion springs elements. Torsion bar and suspension arm are connected together with fixed joint. In the vehicle model, dampers are mounted on the first and the last axle. Also bumpstops are mounted on the same axles. The bumpstops act on lower end of suspension arm with MSC.ADAMS sforce, modeled with impact function. Bumpstops are connected with the hull part with fixed joints. Before first axle, sprockets are mounted. They are connected with the hull element with rotational joints. To accelerate the vehicle to desired velocity and then to hold the same velocity level, a velocity actuators are mounted in sprockets. Idler arm is mounted to the hull with translational joints. Between the idler arm and the hull a spring force acts. This spring have a very large stiffness. It simulates situation where idler wheel is mounted in desired position but with very small susceptibility. Idler arm is connected with idler wheel similar like rest of wheels with rotational joints. In the model, torsion springs have linear stiffness characteristics. Those elements generate pure tensional spring force without tensional damping factor. Assumed stiffness coefficient equals 1.11E+002 Nm/deg.

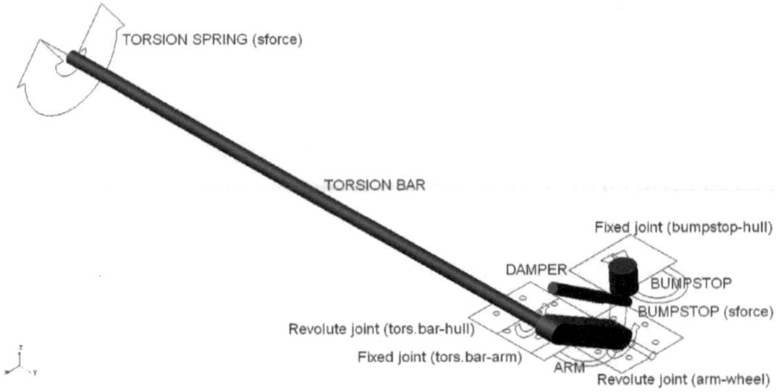

Fig. 3. Suspension repeatable unit

In the model, sprocket geometry reflect shape of the real vehicle sprocket with high fidelity. Shapes of others wheels are not exactly the same like real elements shapes but they comply important details of the real elements, and shape accuracy is very high. In dampers, nonlinear damping characteristics is assumed. This characteristic is calculated from characteristics of damping tolerance area (Fig. 4). Initially it was curve constructed from set of average maximum and minimum values of tolerance area. During simulations also others characteristics curves was analyzed. Mainly, curves enclosed in tolerance area was taken into consideration, but for comparison, a model with no damping was tested.

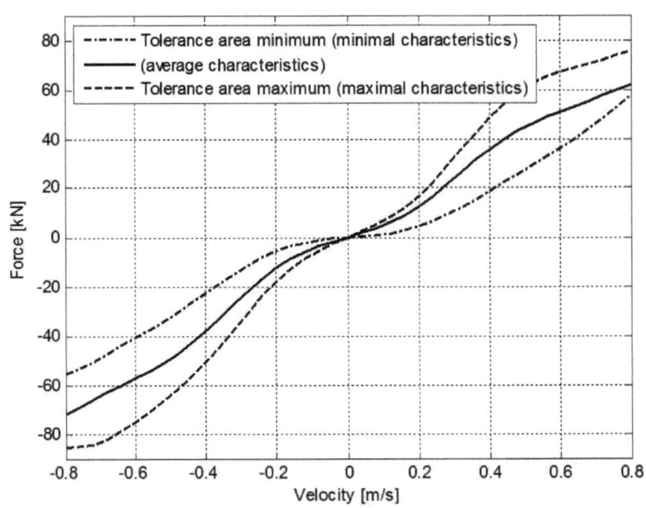

Fig. 4. Damping characteristics tolerance area

Average characteristics is compared according to equation 1. In this equation $\overline{F}$ is weighted mean values vector for calculated characteristics. F_{max} is vector of values of maximal characteristics, but F_{min} is vector of values of minimal characteristics. Weight w can be in the closed range from 0 to 1.

$$\bar{F} = w \cdot F_{max} + (1 - w) \cdot F_{min}, when\ w\epsilon[0,1] \quad (1)$$

In this specific situation $w = 0.5$ was chosen, but according this equation also others characteristics from closed range $[F_{min}, F_{max}]$ may be generated.

2.3 Track Model Construction Problem

Next group collects track elements with kinematic and force dependencies between them. This group also collects dependencies between track elements and elements, which belong to another groups, like wheels group and group, which collects ground elements. Shape of single track link is identical with shape of the real track segment (Fig. 5). To represent real relation between two track links, a model of rubber bushings are inserted between a connecting pin and the links. In this model, the pin is used only as conventional element.

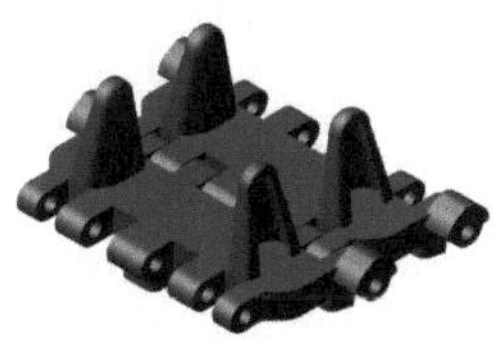

Fig. 5. Two segments of track model

In figure 6, details of connection between links, pin and bushings are shown. To define the connection between two links, a continuous model is used. This is a model of force acting between this two elements, represented as nonlinear function of the two links coordinates. As it is applied in MSC.ADAMS program, under marker object conception a coordinate system is understood. In this specific situation, two body coordinate systems are created. $X_b^i Y_b^i Z_b^i$ and $X_b^j Y_b^j Z_b^j$ for the track links and $X^i Y^i Z^i$ for joint [Ryu]. Relative displacement of this system elements is a function of the bushing stiffness. Deviation, which is presented in figure 6 is represented in vector $\delta_R = [\delta_x, \delta_y, \delta_z]$ and may be used in calculation of the generalized forces acting between these two links.

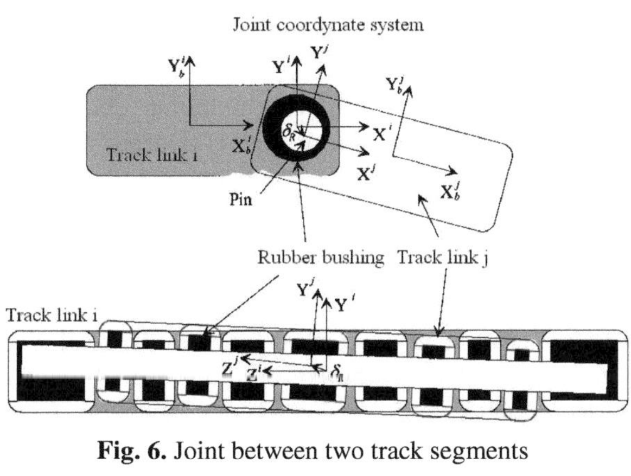

Fig. 6. Joint between two track segments

The bushing force and torque in the system of two track links is given in equation 2.

$$\begin{bmatrix} Q_R^j \\ Q_\theta^j \end{bmatrix} = -\begin{bmatrix} K_R & 0 \\ 0 & K_\theta \end{bmatrix}\begin{bmatrix} \delta_R \\ \delta_\theta \end{bmatrix} - \begin{bmatrix} C_R & 0 \\ 0 & C_\theta \end{bmatrix}\begin{bmatrix} \dot{\delta}_R \\ \dot{\delta}_\theta \end{bmatrix} \tag{2}$$

In this equation K_R, K_θ, C_R and C_θ are 3×3 diagonal matrices. These matrices contain respectively, stiffness parameters in K_R and K_θ and damping parameters in C_R and C_θ of the bushing for translational and rotational deformations.

3 Simulation of the Vehicle Model Dynamics

During simulation process, vehicle model starts its movement from equilibrium state. It accelerates from 0 km/h velocity and after 8 seconds it gains 3.8 km/h and holds it to the end of simulation. With this constant velocity, wheels move over single obstacle as shown in figure 7. This is triangular obstacle with base length 142 mm and height 71 mm (Fig. 7). The authors assumed that during the obstacle crossing all movements of the vehicle body caused by acceleration process are suppressed. Only right track moves over the obstacle. Such conditions are forced by conditions of real object experiment. During experiment on a proving ground the real 2S1 tracked vehicle was tested. In tests, measurements of vertical forces on road wheels was collected. These measurements are compared with vertical forces on the wheels, calculated during simulations.

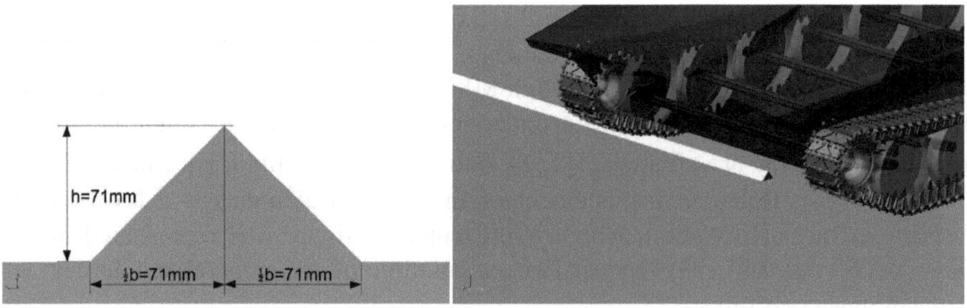

Fig. 7. Single obstacle

Simulations tend to indication a correct damping characteristics enclosed in damping characteristics tolerance area (see Fig. 4). During this empiric process, the authors tend to find higher accuracy of constructed model to its real 2S1 original.

4 Result Analysis

During the model simulation, data of vertical force on the road wheels are collected. Values of these forces are calculated based on torque reduction on torsion bars. This torque is reduced to a point of axle-carrier wheel connection. In the real object, the values of these forces was got by measurement of suspension arm tension connected with carrier wheel. The tension was measured with tensometric sensors. In this point, the authors present a simulation results comparison for vertical force time traces, which act on road wheels. These time traces was taken from the model and from the real 2S1 tracked vehicle. The comparison allows estimation of identity level between a tracked vehicle model and the real object. Simulation aim was estimation of correct damping characteristic. There was tested four cases of damping characteristics. First one was connected with a model without suspension dampers. Remaining three cases are connected with damping characteristics enclosed in the tolerance area (Fig. 4). The second case of the damping characteristics is equal to lower border of this area. Third case of characteristics is equal to mean characteristics between upper and lower border of the tolerance area. And finally fourth case is connected with damper model with damping characteristics identical with the tolerance area upper border.

4.1 Time Domain Analysis

During analysis, time-traces of the vertical forces that was measured on the real vehicle wheels and the model wheels was compared. Because of big amount of time traces, only two are presented in figure 8. In this figure, comparisons of forces on second right model's and object's wheels are presented. In the left side of this figure, results of model with minimal damping characteristics is shown. In the right side of the same figure, results of model with maximal damping characteristic is shown. For all of time-traces comparisons, a coefficient of similarity was calculated. This coefficient bases on Mean Squared Error (MSE), as shown in equation 3.

$$MSE = \frac{1}{n}\sum_{i=1}^{n}\left(X_i - \overline{X}\right)^2 \tag{3}$$

$$RMSE = MSE/MSE_{MAX} \tag{4}$$

For more clear form, a Relative Mean Squared Error (RMSE) was used (Eq. 4). It was calculated through division MSE by maximal value of MSE in all time-traces comparisons.

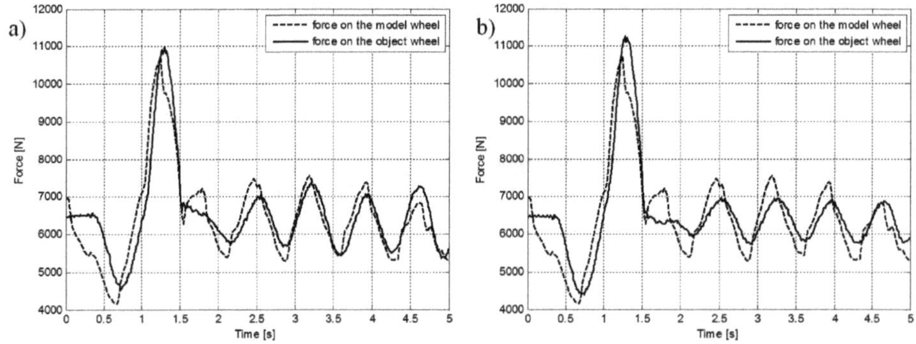

Fig. 8. Wheels vertical forces time-traces: a - force data for 2nd wheel in model with minimal damping characteristics, b - force data for 2nd wheel in model with maximal damping characteristics

The highest similarity is when RMSE coefficient has the lowest value. All RMSEs are collected in table 1. As there could be noticed the lowest mean values are in third row of the table for model with average damping characteristics. To make this conclusion stronger, in the last right column a mean coefficient is presented. This coefficient is an arithmetical mean from coefficients for all of seven right wheels.

Table 1. Relative Mean Squared Error (RMSE) for wheels vertical forces time-traces

Coef. for / Model with	1st wheel	2nd wheel	3rd wheel	4th wheel	5th wheel	6th wheel	7th wheel	Mean coefficient
no damping	0.711	0.793	0.611	0.461	0.416	1.000	0.550	0.649
minimal damping characteristics	0.317	0.414	0.463	0.426	0.351	0.795	0.496	0.408
average damping characteristics	0.309	0.414	0.530	0.462	0.322	0.662	0.480	0.397
maximal damping characteristics	0.349	0.474	0.601	0.514	0.350	0.666	0.462	0.427

4.2 Frequency Domain Analysis

For complete picture of the results, a frequency analysis was conducted. Periodogram of power spectral density estimate is a good tool for analyzing a signal in frequency domain. The integral of the Power Spectral Density (PSD) over a given frequency band computes the average power in the signal over that frequency band. The RMSE coefficient can be calculated also for periodograms of PSD estimate of the vertical forces data. In continuation of previous subsection in figure 9, the periodograms of PSD estimate are shown. Periodograms in the figure are connected also with right second wheels of the object and the model. Left part of the figure is connected with case of model with minimal damping characteristics, but right part is connected with case of model with maximal damping characteristics. As it is possible to notice directly on the picture, similarity level is higher for the second case of model.

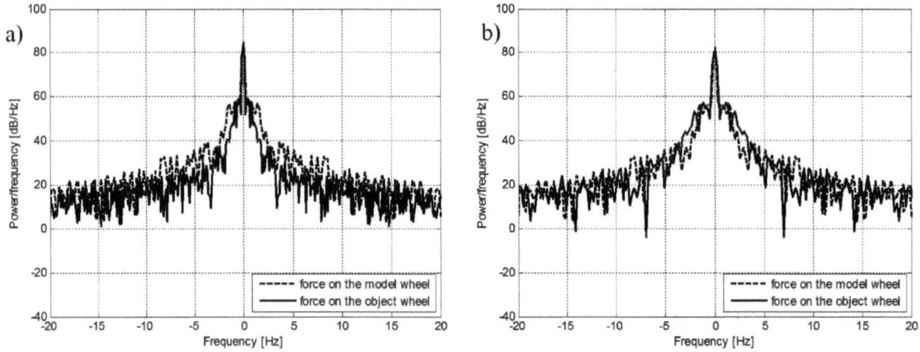

Fig. 9. Periodogram of Power Spectral Density Estimate: a - force data for 2nd wheel in model with minimal damping characteristics, b - force data for 2nd wheel in model with maximal damping characteristics

This is verified also in table 2, where one can see, that the lowest RMSE coefficient appears in the last row, especially for mean coefficient. It suggests that in measurements on the real object some measurement deviation could appear and it could be corrected trough multiply by appropriate coefficient. Finally the model with average or maximal damping characteristics appears to be appropriate to description of damping in real 2S1 vehicle object.

Table 2. Relative Mean Squared Error (RMSE) for periodograms of PSD estimates

Coef. for / Model with	1st wheel	2nd wheel	3rd wheel	4th wheel	5th wheel	6th wheel	7th wheel	Mean coefficient
no damping	0.509	0.520	0.414	0.495	0.459	0.391	0.425	0.459
minimal damping characteristics	0.898	0.672	0.903	0.786	0.697	0.636	0.929	0.690
average damping characteristics	0.888	0.660	0.694	0.581	0.683	0.720	1.000	0.653
maximal damping characteristics	0.441	0.502	0.367	0.361	0.411	0.464	0.595	0.393

4.3 Analyses Result

In next step, results of Time domain analysis and Frequency domain analysis is taken into consideration. The mean coefficient for RMSE appears to be the most representative indicator of similarity between model and real object. As it is known, minimal damping characteristics, average damping characteristics and maximal damping characteristics can be represented by appropriate weight value (see equation 1). In respect to these information, characteristics of similarity was constructed (Fig. 10).

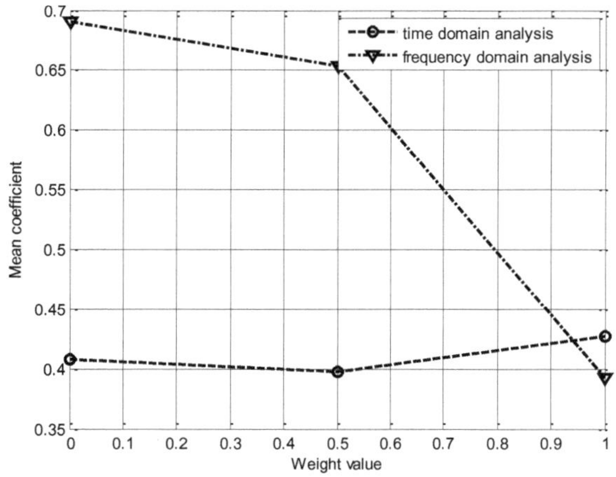

Fig. 10. Characteristics of similarity for different damping characteristics

This are mean RMSE coefficient values in function of weight values from equation 1. There are two characteristics, first one for time domain analysis but second one for frequency domain analysis. Influence of both characteristic may be considered with different weights. In this specific case, the authors assumed time domain analysis as more important, therefore its weight will be higher despite of lower values of its diagram in comparison to frequency domain analysis diagram.

5 Summary and Conclusion

In result of comparison analysis vertical forces acting on road wheels of the real vehicle and the vehicle model, one can see high level of their similarity. This result suggests high fidelity level of the full model and the object dynamics. In table 2, the highest RMSE indicator appears for 7th wheel. It is result of discrepancy between vertical force time traces for wheel number 7, where contact force value between wheel and road is low, in the model case as also in the object case. This situation could cause serious difference between vertical force values acting on the 7th road wheel for model and object. Eventually, differences between time traces for wheels 7 and 1 could be connected with hydraulic dampers in the real object and its models in the vehicle model. There should be noticed that dampers models is made based on characteristic involved by technical documentation but not by measurements taken from dampers mounted on the real object. As it is known, dampers are wear and tear during exploitation and in connection their physical characteristics are being changed. Through time domain analysis combined with frequency domain analysis, the authors constructed two characteristics, which indicate model and object similarity level (Fig. 10). According to these two characteristics, to improve model fidelity, characteristics constructed with weight values between 0.5 and 1 should be tested more precisely. Nevertheless, through simulations, the authors states that model fidelity for weight 0.5 is good enough for continuation simulation researches.

References

[1] Nabagło, T.: Tracked vehicle modeling and simulation in MSC.ADAMS program, submitted to Technical Transactions (2011)

[2] Assanis, D.N., Bryzik, W., Castanier, M.P., Darnel, I.M., Filipi, Z.S., Hulbert, G.M., Jung, D., Ma, Z., Perkins, N.C., Pierre, C., Scholar, C.M., Wang, Y., Zhang, G.: Modeling and Simulation of an M1 Abrams Tank with Advanced Track Dynamics and Integrated Virtual Diesel Engine. Mech. Struct. & Mach. 27(4), 453–505 (1999)

[3] Ryu, H.S., Bae, D.S., Choi, J.H., Shabana, A.A.: A compliant track link model for high-speed,high-mobility tracked vehicles. Int. J. Numer. Meth. Engng. 48, 1481–1502 (2000)

[4] Lee, K.: A Numerical Method for Dynamic Analysis of Tracked Vehicles of High Mobility. KSME international Journal 14(10), 1028–1040 (2000)

[5] ADAMS ver. 2010, Help Documentation (ADAMS/View, Solver). MSC.Software (2010)

[6] Łukaszewicz, A.: Modelling of Solid Part Using Multibody Techniques in Parametric CAD Systems. Solid State Phenomena, 147-149, 924 (2009)

[7] Kubela, T., Pochyly, A., Singule, V.: Advanced Tools for Multi-Body Simulation and Design of Control Structures Applied in Robotic System Development. Solid State Phenomena 164, 387 (2010)
[8] Kowal, J., Nabagło, T., Jurkiewicz, A.: R&D Project No. 0 R00 0048 05 -Final Rapport. University of Science and Technology AGH, Krakow (2011)

Modification of 2S1 Tracked Vehicle Suspension System for Improving Its Performance and Safety

Tomasz Nabagło[1], Andrzej Jurkiewicz[2], Marcin Apostoł[3], and Piotr Micek[4]

[1] Faculty of Mechanical Engineering, Cracow University of Technology, al. Jana Pawła II 37, 31-864 Krakow, Poland

[2,3,4] Faculty of Mechanical Engineering and Robotics, AGH University of Science and Technology, Al. A. Mickiewicza 30, 30-059 Krakow, Poland

pmnabagl@cyf-kr.edu.pl, {jurkand,apostol,micek_pt}@agh.edu.pl

Abstract. In the article, three alternative models of 2S1 platform suspension system are presented. First model is based on existing construction of 2S1 platform suspension. Two next were modernized by usage of new solutions in tracked vehicles suspension technology. The solutions are especially associated with torsion spring element and idler mechanism. The authors have assumed simulation conditions for straight line driving of the vehicle models, while they overcome a single obstacle with one track. Results of all models simulations are compared and analyzed to improve stability of the vehicle while driving. There are also used a human response filter to determine less harmful driving conditions for vehicle crew.

Keywords: tracked vehicle, dynamic model, simulation, ADAMS.

1 Introduction

During all modernization processes, model with high fidelity of details is as useful as the real object. The greatest advantage of the model is one of its features, which lets researcher not to care about consequences of tests in hard conditions. The real object might be seriously damaged or even destroyed, in the similar conditions. In the article, the authors describe three versions of tracked vehicle model suspension. The first one is basic model, based on torsion bars system. The construction process of this model as also the simulation of the model was described by Nabagło et al. [1]. The second one is so called first modernized model, where torsion bars in second axle was replaced with spiral springs. And finally, the last one model is presented. In this model, all of torsion bars are replaced with spiral springs. In next sections it is called a second modernized model. As it was said, during researches, suspension system of 2S1 tracked vehicle was modernized by usage of spiral springs. But not only by this, also others features were changed in these two above modernized models. The second one element was an idler mechanism. This mechanism was replaced with a constant force actuator, which acts on the idler wheel. This solution lets to hold a constant value of track tension. In the basic model, the idler mechanism was blocked in fixed

A.M. Nawrat (ed.), *Innovative Control Systems for Tracked Vehicle Platforms*, Studies in Systems, Decision and Control 2,
DOI: 10.1007/978-3-319-04624-2_13, © Springer International Publishing Switzerland 2015

position. In above mentioned article [1], the vehicle model was divided between few elements groups, such like hull, suspension, wheels and tracks. Track links are connected together by so called rubber-bushings, where stiffness and damping parameters of a single bushing is adequate to real parameters of track links connection. Ruy et al. [2] describe more detailed rules and parameters of this connection in case of two links of a track. From mathematical point of view, the same solution was used in all of three mentioned suspension models. Contact forces act between a single track link and a single wheel as also between a single track link and a ground. A contact forces, which are applied between wheel and ground, were described by Lee [3]. Specific parameters of these connections as also parameters of forces are described in ADAMS program documentation [4]. In simulation process with contact forces usage, geometries fidelity of parts being in contact is very important. For this reason, during the model construction, designers have to solve multi-body CAD designing problem, what was put forward by Łukaszewicz [5]. In continuation of simulation topic, Kubela et al. [6] describes multi-body mechanism simulation in ADAMS computer program. They shows co-simulation environment with usage of Matlab-Simulink program. Such solution could be applied for calculation of more complicated control tasks or nonlinear elements reactions in the Simulink program. Huh et al. [7] take under consideration a specific idler mechanism for track tension control. Simulations as also measurements on a real tracked vehicle was conducted in conditions when vehicle turns on a flat road. In contrast to Huh article, in above article, in case of the modernized tacked vehicle, the authors also describe track tension mechanism, but it is modeled with usage of single force acting in longitudinal axle of the vehicle. Also driving conditions are different, because the authors assume the straight line driving on a flat road with a single obstacle. For estimation all above mentioned vehicle solutions the authors use an indicator, which shows absorbed power by human body. This indicator bases on human body power absorption in vertical direction, which was described by Lee and Lins [8]. This work is associated with model, developed by U.S. Army Tank-Automotive Research and Development Center (TARDEC). This model treat a human body as a filter of vibrations power, which is absorbed by it. Because the human body reacts with different sensitivity on a different vibration frequencies. According to this work, to hold a comfort feeling, the power absorption of the human body must not exceeds 6 watts, what was described by Donahue and Hedrick [9].

2 Tracked Vehicle Model Modernization

As it was said before, in the article three solutions of tracked vehicle suspension model are taken under consideration. First one is so called basic model. Its suspension system is based on torsion bars. Construction of this suspension model is described in an article titled *Construction and simulation of a 2S1 tracked vehicle model and its verification using vertical forces on the road wheels while overcoming a single obstacle* [1]. Second model is modernized with usage of two spiral torsion springs models, which were mounted in second axle. Third suspension model is based on the spiral springs in all axles. In all of these three models, dampers are mounted at the first and last road wheel axles (Fig. 3 and 4). Damping characteristics, which was assumed for the dampers, were more precisely described in the article [1] and shown in figure 1.

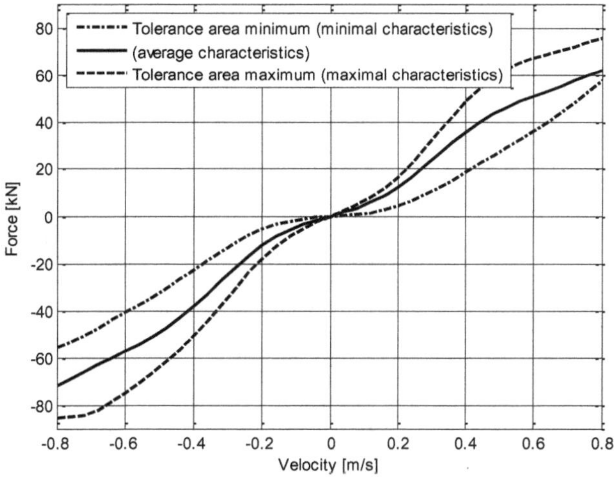

Fig. 1. Damping characteristics tolerance area

Repeatable units of suspension, based on torsion bars or spiral springs, are presented in figure 2 in part a and b. In figure 2b, packet of spiral springs consists of 15 parallel connected spiral springs.

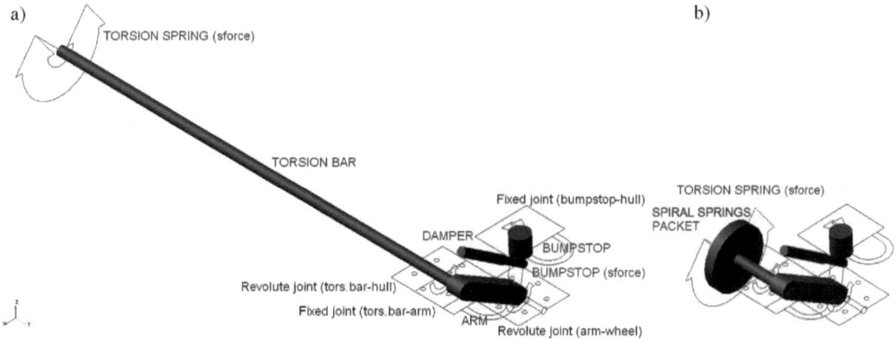

Fig. 2. Units of suspension: a - with torsion bar, b - with spiral springs packet

2.1 Basic Model

The suspension model of 2S1 vehicle in original version is equipped with torsion bars in axles of all road wheels. Original hull mass in the model is assumed as 6.7E+003 kg. In the model, torsion stiffness of torsion bars was assumed as linear, with coefficient 1.11E+002 Nm/deg. All wheels in the models are numbered since front of the vehicle to its end, what is presented on the basic model in figure 3. This is the basic model of 2S1 vehicle in its initial equilibrium state, from which the dynamics simulation is started.

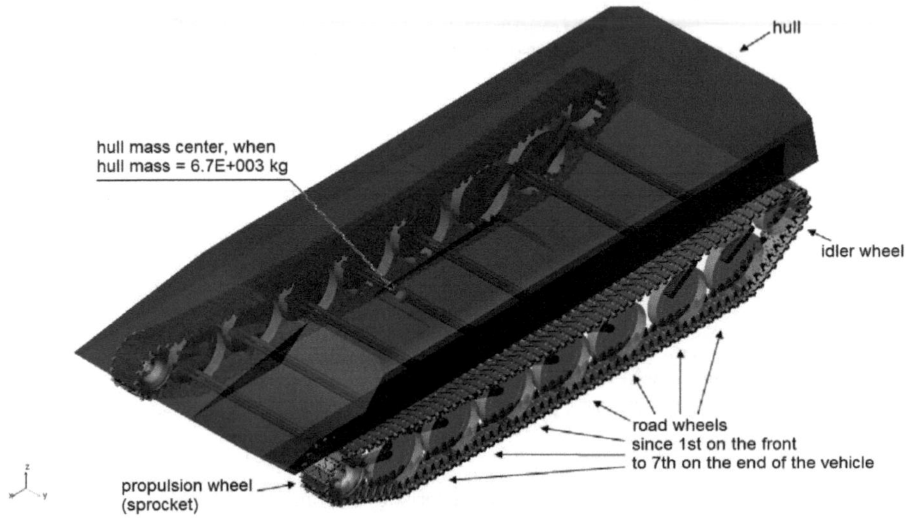

Fig. 3. Basic model of the full tracked vehicle

2.2 First Modernized Model

In this model two torsion bars are removed and in their place the spiral torsion springs are inserted (Fig. 4). These torsion springs models are based on stiffness characteristics of single real spiral spring. Abscissa of this characteristics is multiplied by 15, because 15 parallel spiral springs are assembled in one spring module in the real suspension system. The torsion stiffness characteristics for packet of 15 springs is shown in figure 5b. This characteristics is result of measurement series on the real spiral springs pocket in laboratory conditions. In the real vehicle these springs pockets works in various ride conditions. For this reason the authors had to take simplification assumptions, such like constant temperature range and linear characteristics in whole range of bump rebound movement. A single spiral spring is shown in figure 5a. In the springs packet model, a potential reaction between adjacent single springs as also between parts of the same spring are omitted. In the modernized vehicle model, mass of the hull is increased about 4.87E+003 kg to 11.57E+003 kg. Also hull mass center is moved closer to geometrical hull center (Fig. 4). The last modernization is associated with idler mechanism. This mechanism acts on the idler wheel in longitudinal axis of the vehicle with constant force value, which equals 39 kN. This solution is used for maintaining a constant tension of the tracks. It is simplified version of real idler wheel mechanism which is based on eccentric mechanism. However, this simplification realizes the main aim of the constant track tension idea.

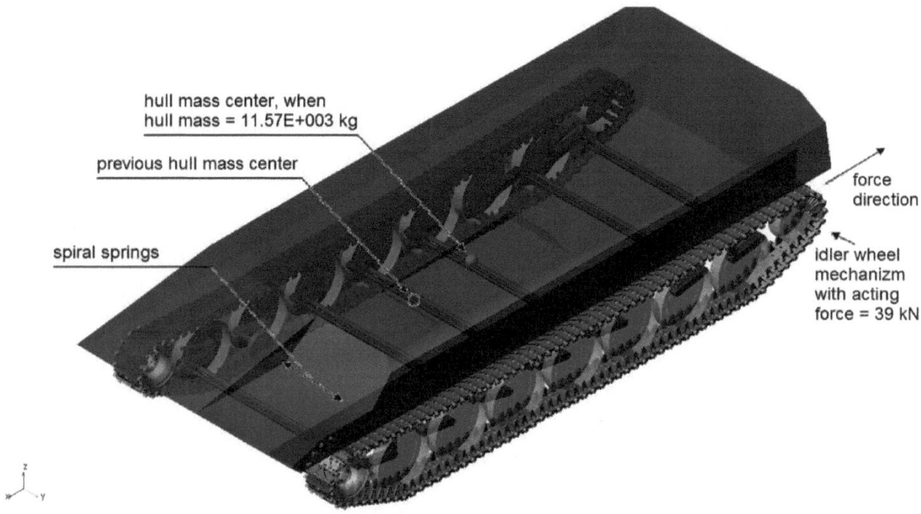

Fig. 4. Modernized model of the full tracked vehicle

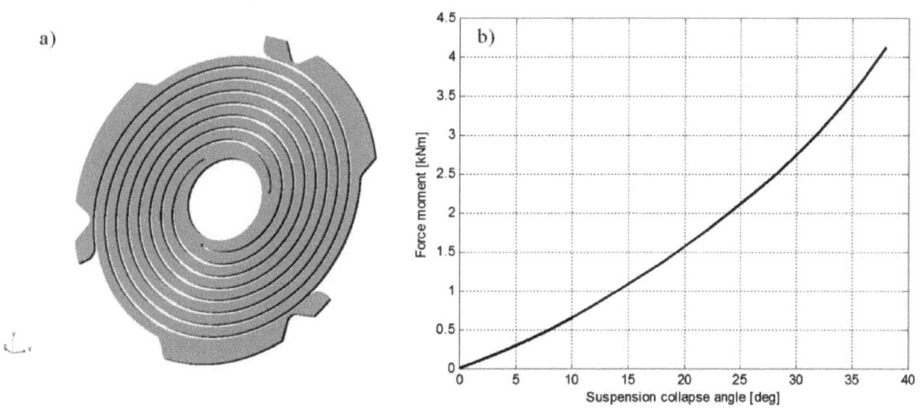

Fig. 5. Single spiral spring (a) and stiffness characteristics of 15 springs packet (b)

2.3 Second Modernized Model

This model has the same parameters like the first modernized model, but the springs modules, consisting of spiral springs, are mounted on all road wheels axles. It is used for estimation of spiral springs effectiveness through comparison of the basic model with the second modernized model. Therefore, stiffness in all axles increases, but rest of parameters are the same like in previous model. Such construction solution gives more free space inside the hull, which could be used for installing better floor-armor or other equipment. This solution improves also safety, because all reactive elements of the suspension system are removed from floor plate to both sides, where they may be better covered.

3 Simulation of the Vehicle Models Dynamics

During simulation, all mentioned models are simulated in conditions of the straight line driving on a flat road with a single obstacle. The obstacle has triangular shape with length of the base 142 mm and height 71 mm (Fig. 6a). This obstacle is overcome only with right side track (Fig. 6b).

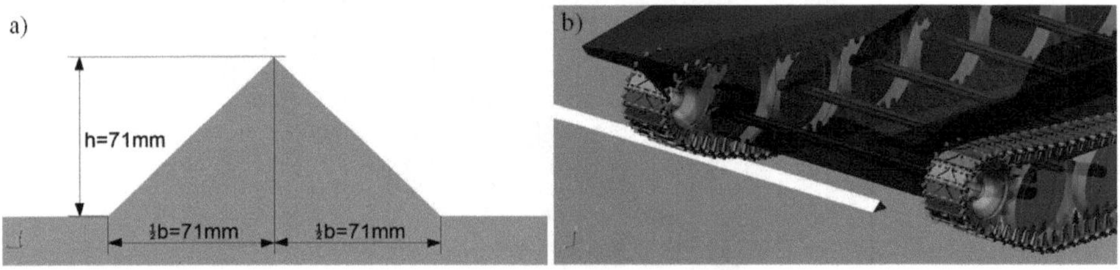

Fig. 6. Single obstacle: a - dimensions, b - view during simulation

When the obstacle is crossed, model moves with constant velocity, equal 3.8 km/h. The same velocity was used in proving ground tests on a real 2S1 vehicle with parameters of the basic model [1]. In a first phase of movement, the vehicle accelerates. It takes 8 seconds, but after acceleration time the vehicle starts to move with constant velocity. In this quasi steady state, the vehicle crosses the obstacles. The first road wheel reaches the obstacle after 20 seconds from simulation start moment, than there may be assumed, that whole vehicle is in above mentioned state because all acceleration effects should be suppressed.

4 Result Analysis

In this section, two types of results are presented. First one is associated with time domain analysis, second one with frequency domain analysis. Only analysis in these both domain can gives full view on the model. It shows not only its behavior in time but also this, which frequencies may be transmitted from road wheels to crew bodies.

4.1 Time Domain Analysis

Results of three above described models are compared during their analysis. In this subsection, the authors present time traces of vertical acceleration of the vehicle hull mass center (Fig. 7a). The smallest amplitudes of acceleration are observed in the case of first modified model. Such situation is associated with bigger mass of the vehicle hull. In the case of second modified model, where all axles are equipped with spiral springs, amplitudes of acceleration are higher, but not such high as in the basic model case. This higher amplitudes of the vertical acceleration could be caused by higher stiffness of the spiral springs pockets, in comparison to stiffness of the torsion bars. In case of the hull mass vertical acceleration, the difference between basic model and the second modified model is not so big. In this situation another indicator is more helpful. As it is known, for efficiency of vehicle crew, a comfort is very important. When crew is exposed to activity of too high vibration energy, their efficiency is getting lower and their reaction time increases radically. To estimate influence energy for human body, the authors used Human Response Filter (HRF) developed by U.S. Army TACOM [6]. Evaluation of level of absorbed power by the human body is possible with usage of this filter. Based on it, the authors calculated time traces of Relative Cumulative Absorbed Power (RCAP), which are shown in figure 7b. All values of these time traces are relative to power absorbed by human body placed in center of the hull mass in the basic model.

$$CAP = \frac{1}{T}\int_0^T P\,dt \tag{1}$$

$$RCAP = CAP/CAP_{max} \tag{2}$$

where T is time of power absorption, P is absorbed power in vertical direction, calculated with usage of vehicle hull vertical acceleration filtered by HRF. RCAP for above three models are shown in figure 7b.

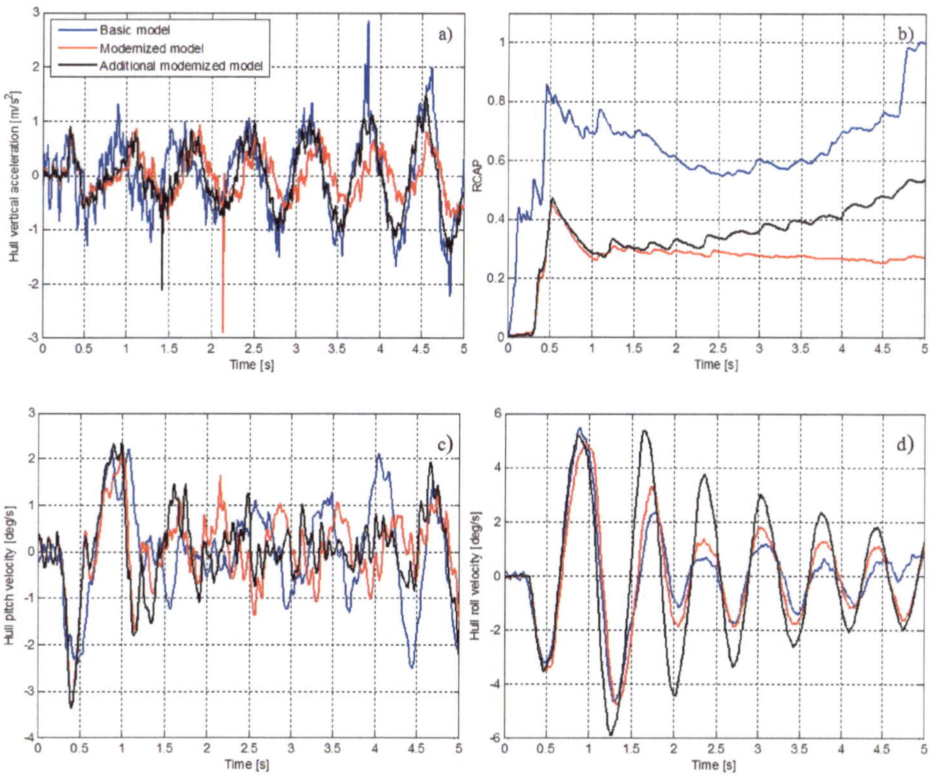

Fig. 7. Time traces of vertical acceleration of the vehicle hull (a), human body cumulative absorbed power (b), hull pitch angular velocity (c), hull roll angular velocity (d)

These time traces clearly show that absorbed power in case of both modified vehicle models are much lower than in basic model case. Therefore in this situation, solutions of the modified models give noticeably better results than basic model solution. In the figure 7c, pitch velocity is presented. This velocity time traces behaves similarly for all three models, and this effect could be neglected because maximum level of this velocity is about 3 deg/s. The last part of this figure (7d) contains a comparison of roll velocity of the vehicle hull. In this case, situation is different. The most stable model is the basic model, and the worst is the second modified model. Above mentioned figure 7 contains results for models with so called average damping characteristics (see fig. 1). To create a full picture of the damping influence on the vehicle dynamics, also minimal and maximal damping characteristics are analyzed. Results of time domain analysis are presented in table 1, and results of frequency domain analysis in table 2. For three variables: hull vertical acceleration, hull pitch velocity and hull roll velocity, three coefficients are calculated. These are namely, a coefficient of relative acceleration for the first variable and relative angular velocity for next two variables.

$$MSE = \frac{1}{n}\sum_{i=1}^{n}\left(X_i - \overline{X}\right)^2 \tag{3}$$

$$RMSE = MSE/MSE_{MAX} \tag{4}$$

where X_i is momentary value of a considered variable and $\overline{X}$ is desired value. In this specific case $\overline{X} = 0$. An ideal situation might appear, when vertical acceleration of the hull equals 0 m/s^2 and pitch and roll velocities equal 0 deg/s. In such situation, the hull is completely isolated from road roughness. This situation is adequate to so called "Sky-hook" system. In table 1, three coefficients based on RMSE indicator, for time domain analysis, is presented: C_{ta} - coefficient for vehicle hull vertical acceleration, $C_{tp\omega}$ - coefficient for pitch velocity of the hull, $C_{tr\omega}$ - coefficient for roll velocity of the hull. Those coefficients equals RMSE values for above variables. Because the relative values were used, all obtained coefficients are dimensionless. They are more convenient in a comparative analysis. Based on three above mentioned coefficients, a mean coefficient is calculated and placed in the next column. This is arithmetical mean of these coefficients.

Table 1. Relative Mean Squared Error (RMSE) for wheels vertical forces time-traces

	1st model					2nd model				
Model with	C_{ta}	$C_{tp\omega}$	$C_{tr\omega}$	**Mean coeff.**	**RCAP**	C_{ta}	$C_{tp\omega}$	$C_{tr\omega}$	**Mean coeff.**	**RCAP**
minimal damping characteristics	0.285	0.718	0.518	**0.507**	**0.328**	1.000	1.000	1.000	**1.000**	**1.001**
average damping characteristics	0.204	0.587	0.425	**0.405**	**0.272**	0.451	0.677	0.729	**0.619**	**0.535**
maximal damping characteristics	0.180	0.609	0.362	**0.384**	**0.259**	0.351	0.695	0.652	**0.566**	**0.451**

4.2 Frequency Domain Analysis

For complete picture of the simulations, results of frequency domain analysis are shown. In figure 6a, one can see, that in the basic model case, main transmitted low frequency of the hull vertical acceleration is about 1 Hz, while in the modified models case, it encloses in range from 2 Hz to 2.5 Hz. Models transmit the similar low frequencies in case of hull pitch angular velocities and hull roll angular velocities.

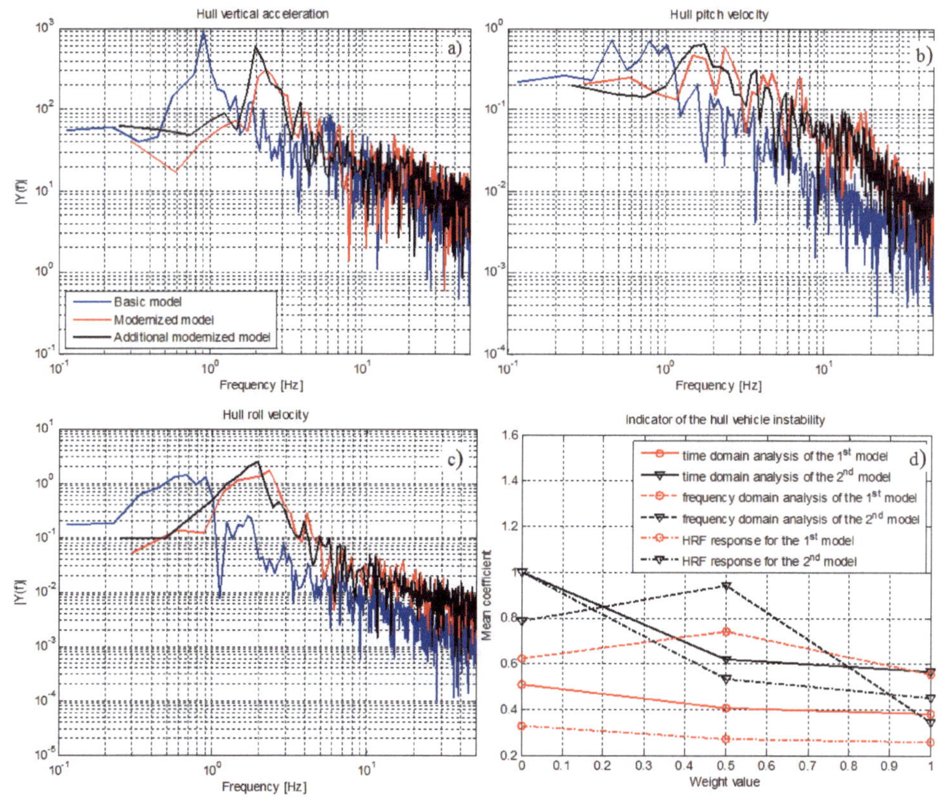

Fig. 8. Frequency characteristics of vertical acceleration of the vehicle hull (a), hull pitch angular velocity (b), hull roll angular velocity (c)

Table 2. Relative Mean Squared Error (RMSE) for wheels vertical forces time-traces

	1st model				2nd model			
Model with	C_{fa}	$C_{fp\omega}$	$C_{fr\omega}$	**Mean coeff.**	C_{fa}	$C_{fp\omega}$	$C_{fr\omega}$	**Mean coeff.**
minimal damping characteristics	0.501	0.804	0.565	**0.623**	1.000	0.677	0.680	**0.786**
average damping characteristics	0.513	1.000	0.717	**0.743**	0.884	0.950	1.000	**0.945**
maximal damping characteristics	0.377	0.821	0.468	**0.555**	0.283	0.398	0.360	**0.347**

The difference between transmitted frequencies is mainly associated with hull mass difference in the basic model and the modernized models. In table 2, three coefficients based on RMSE indicator are presented: C_{fa} - coefficient for vehicle hull vertical acceleration, $C_{fp\omega}$ - coefficient for pitch velocity of the hull, $C_{fr\omega}$ - coefficient for roll velocity of the hull.

5 Summary and Conclusion

In conclusion of results analysis, one can notice advantages of usage of the spiral spring. Because of higher stiffness level of spiral springs packets than in traditional torsion bars case, constructors can increase the vehicle hull mass. This lets also to resign from axial offsets of wheels, so the right and left wheels may be mounted in the same axles. Usage of the spiral springs packets gives more space between wheels. This space may be used for additional equipment or even additional armor. Usage of spiral springs increases also safety level, because torsion bars solutions is associated with reactive elements of suspension under the floor of crew cabin. During explosion of a charge placed under the vehicle these reactive elements could be harmful for the crew. The spiral springs packets are placed on both side of the vehicle and could be covered. During results analysis, a positive effect of higher damping level is observed. By higher damping, a higher border of the mentioned damping characteristics is understood. The higher damping improves vehicle stability as well as comfort. By comfort, a reduced level of energy absorption by the human body is understood.

References

1. Nabagło, T., Jurkiewicz, A., Apostoł, M., Micek, P.: Construction and simulation of a 2S1 tracked vehicle model and its verification using vertical forces on the road wheels while overcoming a single obstacle. Solid State Phenomena 177, 168–176 (2011)
2. Ryu, H.S., Bae, D.S., Choi, J.H., Shabana, A.A.: A compliant track link model for high-speed,high-mobility tracked vehicles. Int. J. Numer. Meth. Engng. 48, 1481–1502 (2000)
3. Lee, K.: A Numerical Method for Dynamic Analysis of Tracked Vehicles of High Mobility. KSME international Journal 14(10), 1028–1040 (2000)
4. ADAMS ver. 2010, Help Documentation (ADAMS/View, Solver). MSC.Software (2010)
5. Łukaszewicz, A.: Modelling of Solid Part Using Multibody Techniques in Parametric CAD Systems Solid State Phenomena, vol. 147-149, pp. 924–929 (2009)
6. Kubela, T., Pochyly, A., Singule, V.: Advanced Tools for Multi-Body Simulation and Design of Control Structures Applied in Robotic System Development. Solid State Phenomena 164, 387–391 (2010)
7. Huh, K., Cho, B.H., Choi, J.H.: Development of a track tension monitoring system in tracked vehicles on flat ground. ProcInstnMechEngrs, Part D: Journal of Automobile Engineering 215, 567–578 (2011)
8. Lee, R.A., Lins, W.F.: Human Vibration Measuring Instrument, Technical Report No. AD785648 U.S. Army Tank-Automotive Command, Warren, Michigan (1973)
9. Donahue, M.D., Hedrick, J.K.: Implementation of an Active Suspension, Preview Controller for Improved Ride Comfort. In: Nonlinear and Hybrid Systems in Automotive Control, pp. 1–21. Springer (2003) ISBN 978-1-85233-652-3

Numerical and Experimental Study of Influence of IED Blast on Crew Inside Vehicle

Wiesław Barnat[1] and Aleksander M. Nawrat[2]

[1] Department of Mechanics and Applied Computer Science, Military University of Technology, Gen. S Kaliskiego St. 2, 00-908 Warsaw, Poland
wbarnat@wat.edu.pl
[2] Institute of Automatic Control, Silesian University of Technology, Akademicka 16, 44-100 Gliwice, Poland

Abstract. The paper presents an approach to modelling of explosion on a vehicle. The studies take into account the impact of the secondary wave on a crew inside the vehicle. It is known that the impact of explosion on the crust (hull) of a vehicle creates a secondary wave phenomenon. The phenomenon rests on the interaction of the bottom and sides of the vehicle as a membrane. The membrane moves the pressure pulse acting on the armour inside the vehicle. The increase of pressure inside the vehicle causes the damage to internal organs ultimately leading to the death of soldiers. In the case of armour penetration, even a more rapid increase of pressure is possible. One of the basic ways to protect crews of vehicles is opening the hatches of vehicles. Nevertheless, this method is not applicable when the vehicle moves in a contaminated area.

Keywords: secondary wave, safety crews.

1 Introduction

The basis of requirements for vehicles ballistic protection is NATO STANAG 4569 standardization. Designers and companies producing armoured vehicles search for new high strength ballistic materials with better energy absorbing properties securing both vehicle and the personnel inside in case of mine or IED attack [1, 8]. After detonation, depending on the material size, type and potential additives such as fragments, a blast wave arises and hits the bottom or sides of a vehicle with potential fragments.

The main problem in safety of vehicles is a growing mass of an explosive. Existing Defence and NATO Standards do not take into account such big charges. Standard procedures are based on explosives that are of standard anti-tank 3 kg mines.

In the literature, there are many publications concerning the explosion of classic materials and blast wave propagation in air [2,3] and in metals or soil [4, 6, 7].

A.M. Nawrat (ed.), *Innovative Control Systems for Tracked Vehicle Platforms*,
Studies in Systems, Decision and Control 2,
DOI: 10.1007/978-3-319-04624-2_14,

In case of loading the vehicle hull with a rapid and intensive pressure impulse from detonation of the explosive charge, the main threat is vehicle elimination due to plastic deformation of vehicle's hull sides and bottom. In extreme situations the vehicle can overturn, which leads to serious crew injuries. Additionally, the overpressure inside the vehicle – internal membrane pressure wave – is another threat to the crew. In case of a high peak value it can be lethal or lead to serious internal injuries as presented in fig. 1.

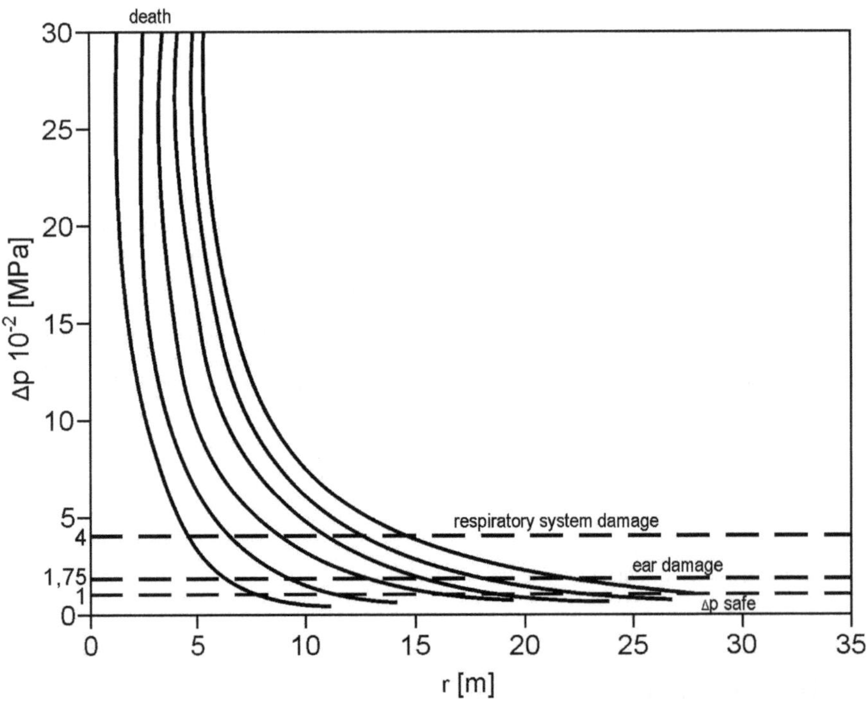

Fig. 1. Overpressure versus distance from explosive charge [4]

An example of an overpressure injury on a human body versus distance is presented in Fig. 1. Both auditory and respiratory systems damage are indicated.

An internal membrane pressure wave arises from deformation of the hull surface (an obstacle) due to the incident pressure wave impact. A general scheme of secondary membrane pressure wave initiation is presented in Fig. 2.

At the beginning of the internal membrane pressure wave (fig. 2b), the wave front (3) caused by the detonation of the explosive (2) reaches surface 1 (fig. 2a). Then, the obstacle (4) deforms, which leads to initiation of the internal membrane pressure wave (5) (fig. 2c) and its propagation (fig. 2d).

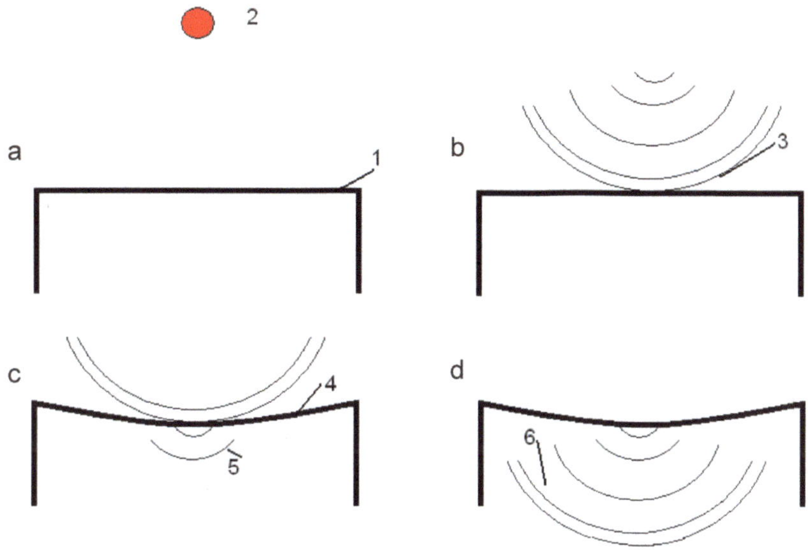

Fig. 2. Scheme of secondary membrane pressure wave initiation: 1 – surface on which the incident wave impacts, 2 – explosive charge, 3- incident pressure impulse on membrane, 4 – deformed membrane, 5 - initiation of internal pressure wave, 6 – internal pressure wave propagation

2 Experimental Tests

2.1 Test Stand

For the purpose of conditioning and amplification of signals from incident wave pressure gauge transducers, an LTT 500 amplifier was used. It was produced by Tasler GmbH and is designed to register rapid changing signals. The registration process was carried out using NI-USB 6833 measurement card with 16 bit analog-digital transducer (sampling 2 MHz on every channel) made by National Instruments. Additionally, Toshiba Satellite laptop was used to run the tests. All the equipment is shown in fig. 3.

The indicent wave pressure run was meassured using special pressure gauge 137A21 produced by PCB Piezoelectronics (fabric no. 9080) with sensitivity 143.3 mV/MPa range 6.894 MPa. The gauge was placed 400 mm from the explove axis in the middle of the test stand.

2.2 Experimental Results

During the examination of the secondary pressure wave , an unique test stand in the form of a cube with 0.5 m edge length. All walls were made of ARMOX steel plates 6 mm thick. In order to obtain the membrane, the lid was made of 2 mm plate. During the experiment, the reflected wave pressure (affecting the lid) and internal membrane pressure were measured. An internal pressure gauge was placed in the symmetry

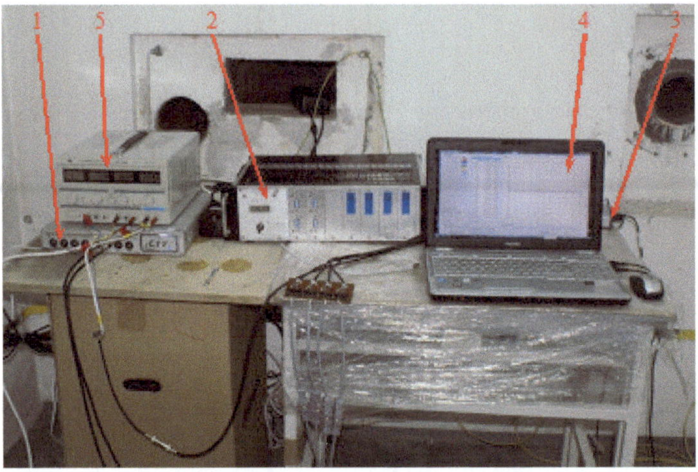

Fig. 3. Test apparatus for the purpose of experiment 1 – LTT 500 amplifier, 2 – tensometric amplifier MS1001, 3 – measurment card NI-USB 6833, 4 – computer with proper software, 5 – power supply

Fig. 4. Scheme of test stand

plane of the box. Both the pressure and strains were measured using an electrical resistance method. In order to measure the pressure inside the box, the bottom was bolted to the walls and sealed with a special tape. A general view is shown in Fig. 4-6.

Fig. 5. Mounting base of internal pressure gauge

Fig. 6. Incident pressure wave gauge and strain gauges

As a result, a reflected wave pressure plot and an internal membrane wave pressure plot were obtained. They are presented in Fig. 7. The value -5 MPa refers to incident pressure wave. Therefore, it is important to read pressure values properly.

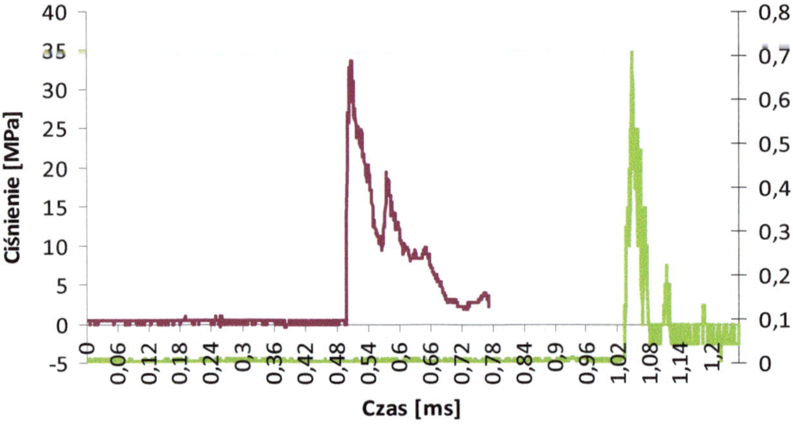

Fig. 7. Pressure values of reflected wave and internal pressure wave from detonation of 200 g of TNT

As a result of a cylindrical shape of the explosive and its mass – 200 g, large overpressure of the reflected wave was obtained with peak value 35 MPa. It is due to the shape and mass of the explosive and wave formulation. In the case of the internal (reflected) wave, the peak of overpressure reached 0,7 MPa. Based on the plot presented in Fig. 1, a serious auditory system injury of occupants occurs.

3 Numerical Analysis

3.1 Physical Model

The analysis of the pressure impulse impact on the structure was carried out using a Finite Element Method with an explicit scheme of integration implemented in DYTRAN software [5]. In the algorithm, the equilibrium equation of discrete model (1) is solved by explicit integration over time. Those equation are coupled so the computational cost is high and, therefore, the time required for computation is increased.

$$[M]\{\ddot{x}\}_n = \{F\}_n - \{F_{\text{int}}\}_n, \tag{1}$$

Where:

$[M]$ – mass matrix,

$\{F\}$ – matrix of the external forces,

$\{F_{\text{int}}\}$ – matrix of internal forces.

Supposing that mass matrix is diagonal, movement equations can be solved using an explicit Euler method. In that case, the acceleration vector $\{\ddot{x}\}$ can be obtained using equation [8]:

$$\{\ddot{x}\}_n = [M]^{-1}(\{F\}_n - \{F_{\text{int}}\})_n \tag{2}$$

In equation (2) damping can also be taken into account, however, only in the diagonal matrix. Another advantage is uncoupling of differential movement equations which allows solving them separately. Velocity $\{\dot{x}\}_{n+1/2}$ and displacement $\{x\}_{n+1}$ vectors at subsequent time steps are obtained by integrating over time using a finite difference method [8]:

$$\{\dot{x}\}_{n+1/2} = \{\dot{x}\}_{n-1/2} + \{x\}_n \Delta t_n \tag{3a}$$

$$\{x\}_{n+1} = \{x\}_n + \{\dot{x}\}_{n+1/2} \Delta t_{n+1/2} \tag{3b}$$

However, the method is only stable under certain conditions. In order to ensure stability, the time step of integration must be limited:

$$\Delta T \leq 2 / \omega_{\max}, \tag{4}$$

Where $\omega_{\max}$ is the highest undamped frequency of a natural mode of the discrete model. It means that the time step has to be shorter than wave propagation through the smallest element in the entire computational model. As a consequence, the more precise model with smaller elements the shorter time step and longer computation time. Therefore, the problem of determining a proper time step disappears.

Mechanics problems solved using modelling and numerical techniques require an appropriate description of the analyzed materials. It has to take into account specific features such as physical state, ductility, brittleness or hardness. All the necessary data were obtained during experimental research held at the Department of Mechanics and Applied Computer Science, Military University of Technology.

3.2 Description of the Numerical Model

The analysis Numerical analysis was carried out for case of the numerical model of a vehicle hull loaded by a pressure wave. The pressure wave came from the detonation of a large explosive charge placed on the side of the vehicle, according to a project of the latest standards. In order to increase the effect of the impact of explosions on the vehicle, interior thickness of sides wall of the vehicle was reduced.

The pressure wave caused by detonation, was distributed in a cube shape volume, with the assigned appropriate outflowing, boundary conditions for a proper representation of the reality. The theoretical solution of propagation of strong discontinuities with a spherical shape initiated from a point source exists in the form of analytical similarities equations formulated by Taylor. After transformation they can be written as [5]:

$$p(r) = 0{,}155\, E_o\, r^{-3} \tag{5}$$

where: E_0 - initial internal energy, r - current radius of the sphere.

This allows for the computer simulation of a shock wave propagation process by giving appropriate initial conditions (density, energy, pressure) to some selected elements of the Euler domain and solution laws of conservation of mass, momentum and energy. Typical values for the explosive substances are: density - 1600 kg/m3, and specific internal energy – 4.2 MJ/kg.

The space, in which the shock wave spread, was modelled with the use of Euler type elements Hex 8 characterized by the ideal gas properties of $\gamma = 1.4$ and density corresponding to the density of atmospheric air at standard conditions ($\rho = 1.2829$ kg/m3). The calculations take into account the changes caused by deformation of the structure, since the structure was modelled using the elements of the Lagrange type Shell Quad 4. Those elements have been given the following mechanical properties: E = 2.15·109MPa, $\nu = 0.31$. To describe the behaviour of the steel, piecewise linear plastic material model DYMAT 24 was used. A kinematical model of hardening of material, namely Johnson - Cook model, was used. The maximum strain failure criterion was adopted [5].

Developing a model of the structure was preceded by the additional laboratory testing of mechanical properties of a steel armour. The laboratory research was carried out at the Department of Mechanics and Applied Computation Science, Military University of Technology.

A general view of a numerical model of the vehicle and the cross-section of the entire system is shown in Figure 8.

In all models used in this study the structural element nodes (nodes Lagrangian domain) were not imposed with any initial conditions. It means that all speeds and displacements for the time t = 0 were set to zero.

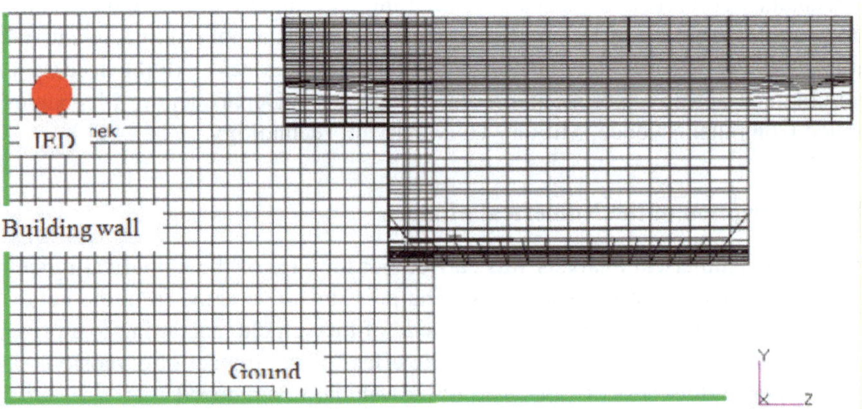

Fig. 8. Scheme of the numerical model of vehicle and ground and side placed IED load

3.3 Results of Numerical Analysis

This In the case of side detonation of the IED, the model was loaded with a charge of the weight equal to 8 kg. As a result of numerical analysis, there were obtained among others: pressure, maps and charts of deformation, acceleration and speed for characteristic points of the structure.

The analysis was based on a cubic, but suitably dense, grid of Euler elements. The influence of the shape of the grid on the pressure on the front of shock wave is relatively small. Pressure pulse propagation in a denser medium causes that the value, and the velocity are different than the propagation of explosion in the looser medium.

Figure 9a shows the pressure distribution at t = 2.38 E-6 sec for the initial stage of pressure pulse propagation. Figure 9b and 9c show the initial unfolding of a pressure wave for the next moments of time. In addition, Figure 9c presents initially developing Mach wave resulting from the imposition of the incident and reflected wave (from a rigid barriers - the land). Figure 10a for the time moment (0.00045s) presents strengthening of the pressure pulse reflected from the upper side of the vehicle. Moreover, the diagram for the reflected wave is presented in picture 11. A very interesting problem is further propagation of Mach wave, which is shown in picture 10b (for time 0.000528 s). Reflection of the peak of pressure from a lower wall of the vehicle is presented in picture 10c (for time moment in 0.0006517 s).

a)

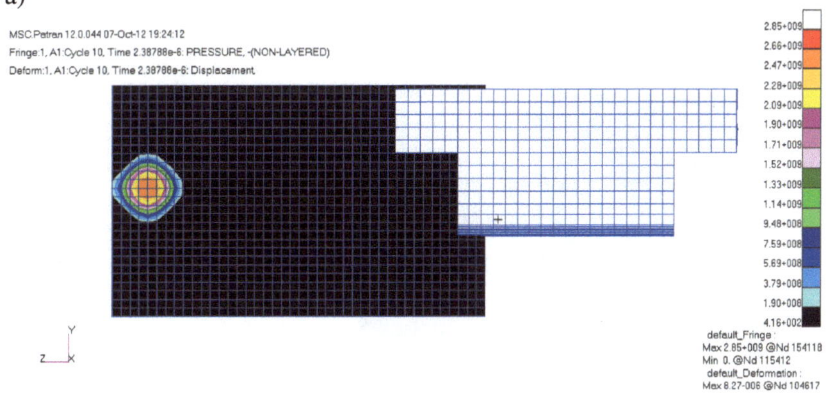

b)

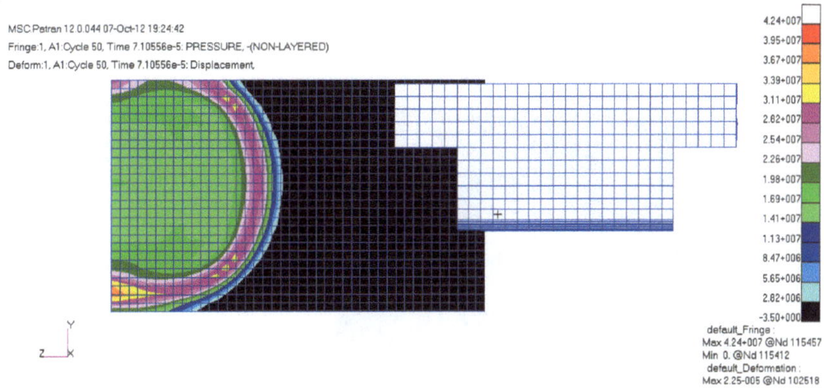

Fig. 9. The way of pressure wave propagation in various moments of time a) 2.38 E+6 s, b) 7.1E+5s, c) 0.00022195 s

c)

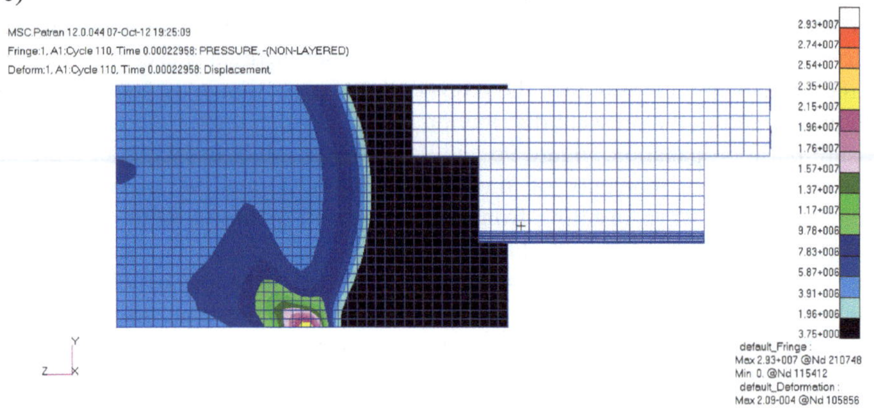

Fig. 9. (*Continued*)

a)

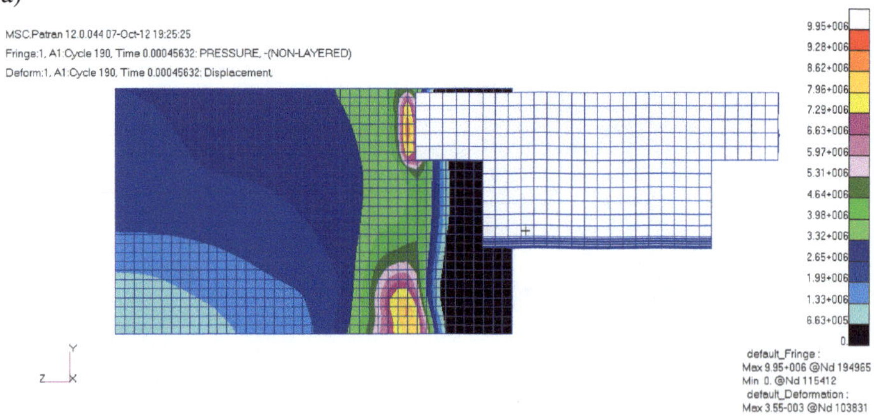

b)

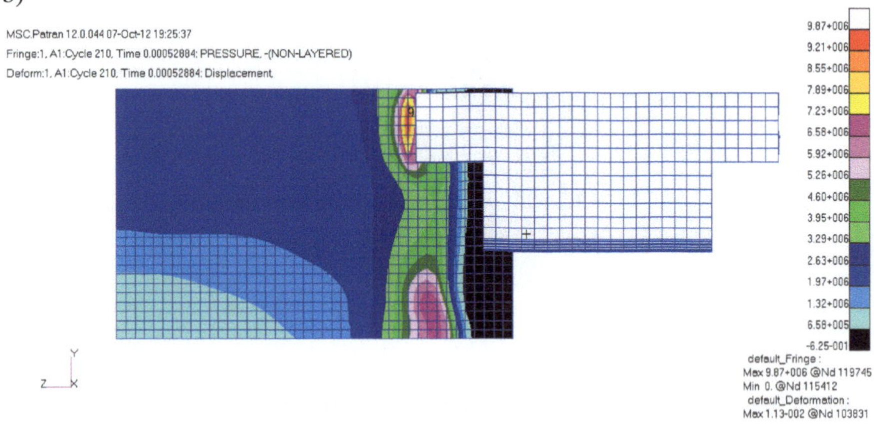

Fig. 10. The way of propagation the pressure wave in various moments of time: a) 0.00045 s, b) 0.000528 s, c) 0.0006517 s

c)

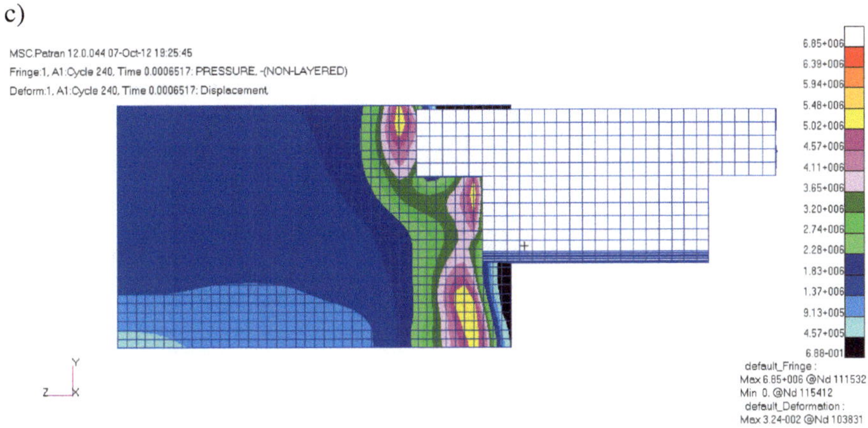

Fig. 10. (*Continued*)

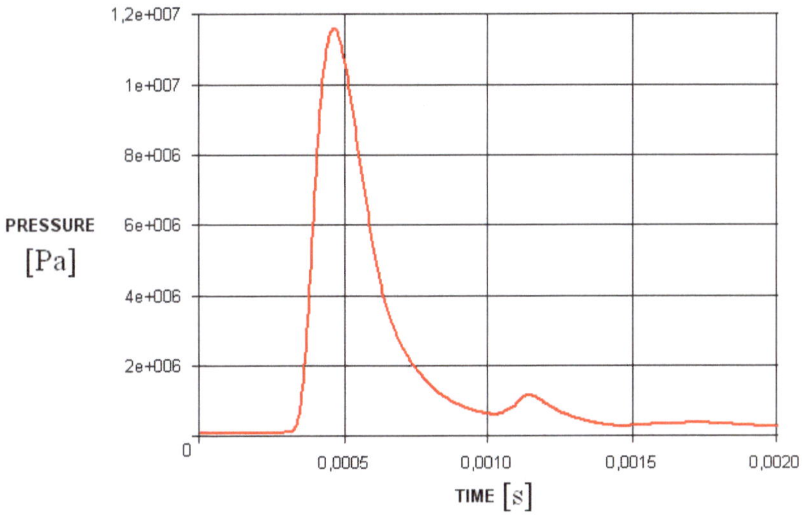

Fig. 11. The diagram of pressure impulse of the wave incident in the first plane on the vehicle side

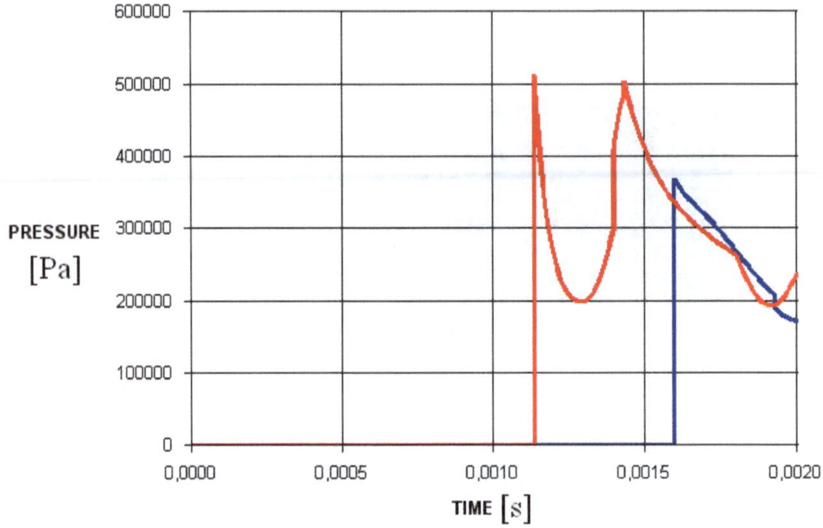

Fig. 12. The diagram of internal membrane wave pressure (secondary)

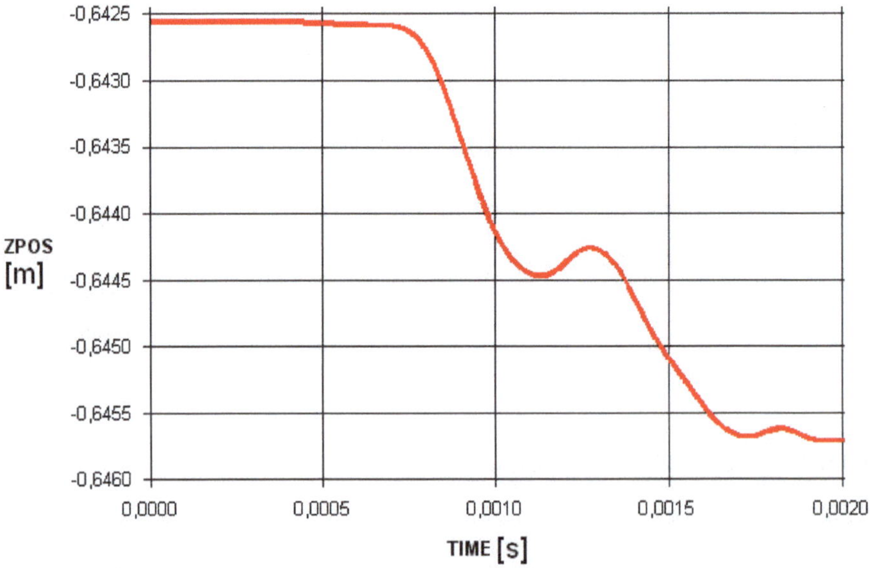

Fig. 13. The diagram of the node displacement inside ceiling of vehicle in "z" direction

a) b) c)

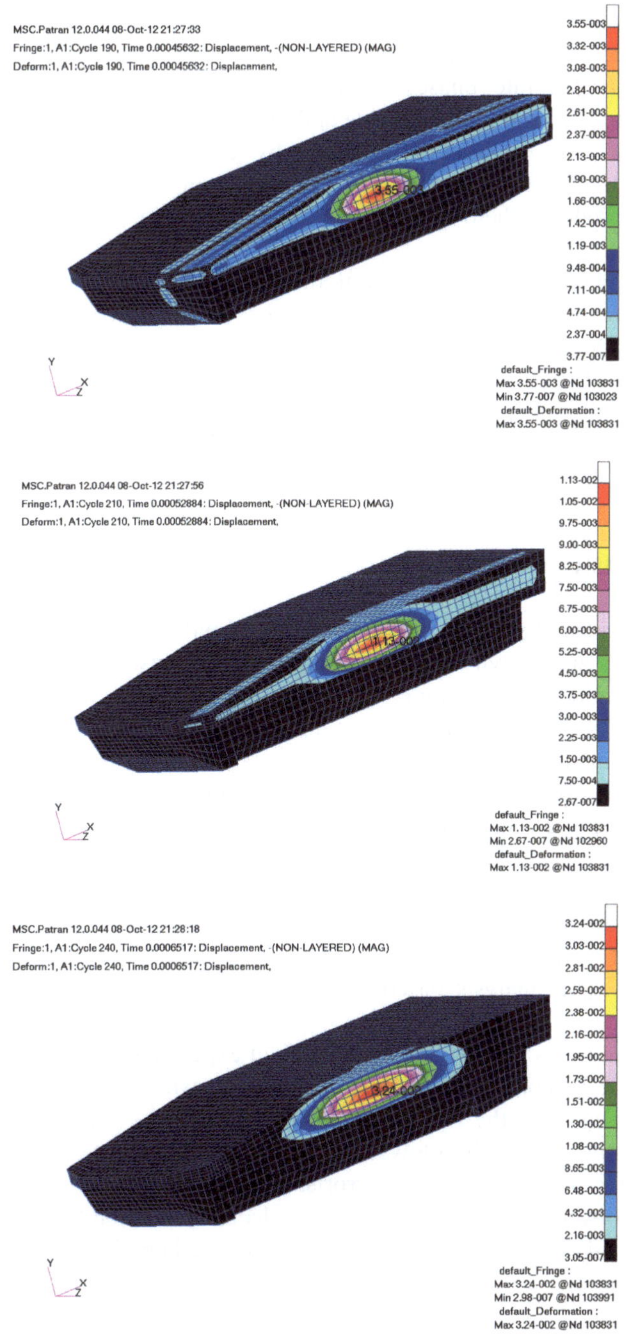

Fig. 14. Deformation of the side of the vehicle in different moments in time: a) 0.00045 s, b) 0.000528 s, c) 0.0006517 s

The diagram of increase internal pressure in the vehicle is presented in Fig. 12. The diagram of displacement of the central node (on the vehicle ceiling) in z direction is shown in Fig. 13. The graph is a polyline curve. The character of the graph shows the strengthening in separate phases of the influence of the pressure wave on the construction (impact into the upper side and impact into the lower side).

In this case, the pressure is induced by deformation of partition amounts approximately 0.5 MPa. Comparing the results from Fig. 1, it can be easily concluded that inside the vehicle there appears the disorder, which could be the reason of a hearing loss.

The use of both a ground and wall model causes the transfer of strengthened impact pressure on the side of the vehicle. The reflection contributed to the phenomenon of Mach wave.

The analysis of IED lateral influence on the vehicle hull shed light on the nature of dynamic interactions between a large load and the vehicle shell. Additionally, hull deformation for various moments in time (appropriately shown in Fig. 10) are shown in Fig. 14.

4 Conclusions

The paper presents the results of experimental studies and numerical analysis of coupling of two Euler environments (describing air - outside and inside the vehicle) and Lagrange elements (describing the vehicle). In the case of experimental studies, a very important problem was proper arrangement of sensors measuring the pressure increases.

In the case of numerical analysis, the gas is characterized with air parameters. The interaction between a vehicle and a charge is achieved by general coupling. All kinds of material data have been selected from experimental studies. For a vehicle description, a bilinear material model with dynamic strengthening was used. This approach fully describes the phenomena occurring in the system.

Since the influence of the shape of a grid on the value of the front wave pressure is relatively low, in analysis based on a cubic, but suitably thick, Euler elements mesh.

The article presents the results of a numerical experiment for the elements of main structures of the combat vehicle loaded by impact, which was generated by a large charge blast on a side of the vehicle taking into consideration the wave reflected from the ground. This approach allows for a good approximation of the numerical simulation of the real conditions of terrorist attacks. An explosive device - IED - has been simulated using the power of properly selected (based on literature studies - the initial energy and density) device. The analysis of the lateral blast on the structure of the vehicle included an additional obstacle in the form of the building increasing pressure impulse.

References

[1] Babul, W.: Odkształcanie metali wybuchem. WNT, Warszawa (1980)
[2] Baker, W.E.: Explosions in Air. University of Texas Press, Austin and London (1973)
[3] Barnat, W.: Wybrane zagadnienia ochrony życia i zdrowia załóg pojazdów przed wybuchem. Wydawnictwo MilitaryRok (2011)
[4] Barnat, W., Panowicz, R., Moneta, G.: Numeryczne badanie oddziaływania fali wtórnej na załogę pojazdu. Journal of Transdisciplinary Systems Science 16(3) (2012)
[5] Dytran Theory Manual, 2004, LS-DYNA theoretical manual (1998)
[6] Prace minerskie i niszczenia, Sztab Generalny, szefostwo Wojsk Inżynieryjnych, Warszawa (1995)
[7] Włodarczyk, E.: Wstęp do mechaniki wybuchu, PWN, Warszawa (1994)
[8] Krzystsła, E., Kciuk, S., Mężyk, A.: Identyfikacja zagrożeń załogi pojazdów specjalnych pod-czas wybuchu. Politechnika Śląska, Gliwice (2013)
[9] Jedrasiak, K., Daniec, K., Nawrat, A.: The Low Cost Micro Inertial Measurement Unit. In: Industrial Electronics and Applications (ICIEA), pp. 403–408 (2013)
[10] Bereska, D., Jędrasiak, K., Daniec, K., Nawrat, A.: Gyro-stabilized platform for multispectral image acquisition. In: Nawrat, A., Kuś, Z. (eds.) Vision Based Systems for UAV Applications. SCI, vol. 481, pp. 115–121. Springer, Heidelberg (2013)
[11] Daniec, K., Iwaneczko, P., Jędrasiak, K., Nawrat, A.: Prototyping the Autonomous Flight Algo-rithms Using the Prepar3D Simulator. In: Nawrat, A., Kuś, Z. (eds.) Vision Based Systems for UAV Applications. SCI, vol. 481, pp. 219–232. Springer, Heidelberg (2013)
[12] Babiarz, A., Bieda, R., Jędrasiak, K., Nawrat, A.: Machine vision in autonomous systems of detection and location of objects in digital images. In: Nawrat, A., Kuś, Z. (eds.) Vision Based Systems for UAV Applications. SCI, vol. 481, pp. 3–25. Springer, Heidelberg (2013)
[13] Bereska, D., Daniec, K., Fraś, S., Jędrasiak, K., Malinowski, M., Nawrat, A.: System for multi-axial mechanical stabilization of digital camera. In: Nawrat, A., Kuś, Z. (eds.) Vision Based Systems for UAV Applications. SCI, vol. 481, pp. 177–189. Springer, Heidelberg (2013)

[illegible]

References

[illegible]

[illegible] Vision Based Systems for UAV Applications, SCI 481, pp. 173–189. Springer, Heidelberg (2013)

Tuning the Dynamic Characteristics of Tracked Vehicles Suspension Using Controllable Fluid Dampers

Sławomir Kciuk, Sławomir Duda, Arkadiusz Mężyk, Eugeniusz Świtoński, and Klaudiusz Klarecki

Silesian University of Technology, The Faculty of Mechanical Engineering, Department of Theoretical and Applied Mechanics, Konarskiego 18a Street, 44-100 Gliwice, Poland
skciuk@polsl.pl

Abstract. The paper addresses analytical, numerical and experimental aspects of the design of controlled fluid damper. The mathematical model served the development of a numerical model in MATLAB/Simulink environment. The selected important aspects of the tracked vehicles dynamic modelling process related to dynamics of the drive system and the suspension of selected tracked vehicles are presented. The main aim of the article was to present the investigation results of the experimental studies of a prototype rotary shock absorber at various magnitudes of control current was presented in this paper. Simulation and preliminary experimental results comprising resistance force values as a function of angular displacement and angular velocity are presented.

Keywords: rotary shock absorber, damper, tracked vehicle, vehicle suspension, dynamics, hydraulic system.

1 Introduction

The long-standing technical research has proven that the main factor reducing the traveling speed of high-speed tracked vehicles is the design of the suspension [1]. Presently, the research concerning this problem is aimed at finding damping solutions allowing for active vibration control methods. Due to the above, many scientific works presenting models or prototypes of dampers based e.g. on the magnetorheological fluid properties have been published [2, 3, 5]. Following this tendency of research for the solution which may be practically applied in the stabilization of high-speed track vehicles, it has been attempted to provide an innovative technical solution consisting in the application of a controllable system in a traditional fluid damper allowing for the shaping of the damper operation characteristics.

The work presents results of conceptual research concerning the modification of the fluid dampers in existence in the context of applying them in a new, active suspension system for high-speed tracked vehicles. The second part of the article presents experiment results of the fluid damper.

A.M. Nawrat (ed.), *Innovative Control Systems for Tracked Vehicle Platforms*,
Studies in Systems, Decision and Control 2,
DOI: 10.1007/978-3-319-04624-2_15,

Increasing the speed of combat vehicles is limited by the efficiency of suspension systems. In case of military vehicles, the variability of the suspension characteristics must be ensured (roads, unbeaten tracks, testing ground conditions, etc.). As to the above, modern, semi-active suspension systems allow for the further development of the new generation of high-speed military vehicles and the increase in the mobility of the modernized vehicles.

The suspensions of wheel and tracked vehicles differ significantly. In both cases, however, the independent suspension is predominant. The independent suspension is increasingly often designed to be semi-active – dampers with regulated absorbing characteristics or active – hydropneumatic dampers.

The set of the suspension system of a tracked vehicle with an independent suspension usually includes: drive wheels, stretching wheels, rollers of the tracks, torsion bars, elastomer bumpers, rockers with friction or hydraulic dampers, the stretching mechanism and tracks with rubber pads or grousers. The design solutions of the modern vehicles' suspension systems most often incorporate torsion bars and hydraulic or friction dampers (Fig. 1).

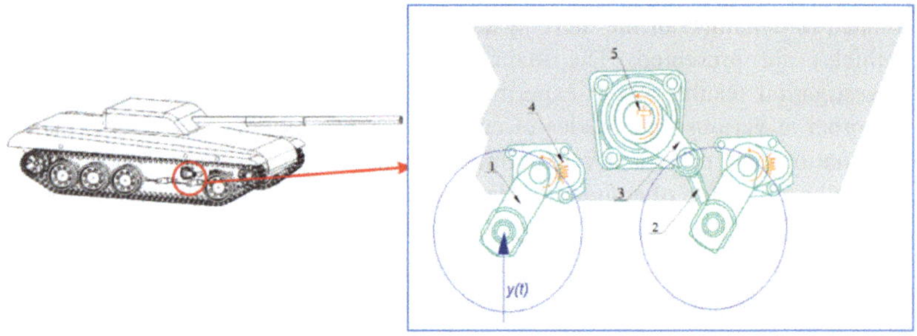

Fig. 1. T-72 Tank suspension diagram: 1 – rocker, 2 – link bar, 3 – damper arm, 4 – elastic element – the torsion bar, 5 – vibration damping element

2 Design of a Damper Controlled by a Pressure Proportional Valve

The operation of a fluid damper consists in forcing the oil through the damper chambers as a result of exciting movement at its arm, which conditions the absorption of the shock, vibration and impact energy (Fig. 2). The arm of the damper is connected to two vanes using splines. The generation of movement at this arm causes the compression of fluid in one of the chambers of the damper while opening the throttle-non-return valve placed in one of the two permanent barriers of the damper. The barriers are seated in the external body. The damper conditions the movement of the damper in one direction only – the return is an idle movement.

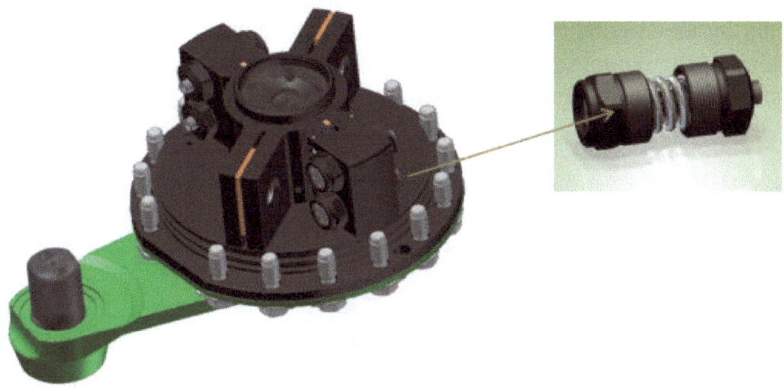

Fig. 2. The design of the damper without its body. Straight-through valve has been marked.

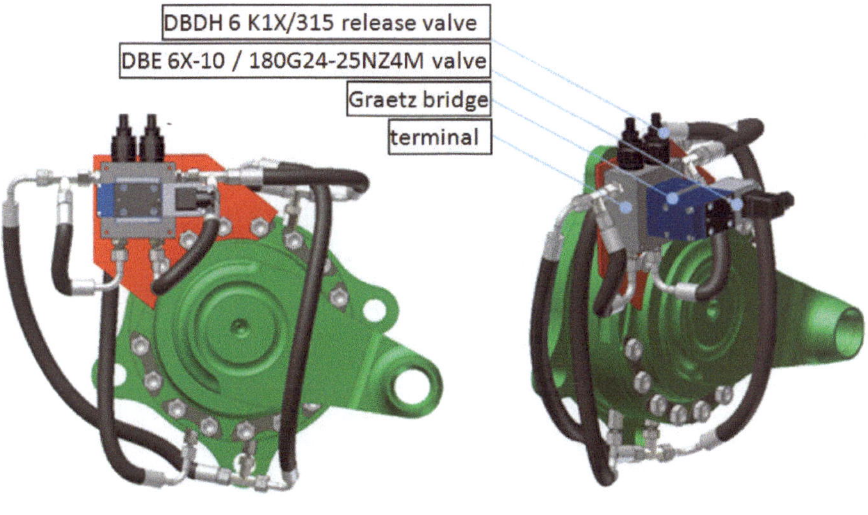

Fig. 3. The structure of a prototype controlled fluid damper

The construction of the controlled damper consisted in a modification of an existing passive damper presented in Fig 2. The element constituting the difference between a standard damper and the controlled damper is the external valve, allowing for the fluent control of working fluid pressure (Fig. 3). The electrically controlled proportional valve serves the fluid pressure regulation that is independent from the flow. The pressure is regulated by current set by a control system. The principle of operation of the proposed damper consists in the movement of the arm (vane) of the damper causing fluid compression in the two opposite chambers. In a traditional dampers, the through valves are opened due to such state, which is determined by the deflection of the damper's spring. In the proposed solution the fluid pressure is regulated by an external valve, the operational characteristics of which are regulated. The value of the damping force variation is thus – in this case – dependent upon the method of control of the proportional valve.

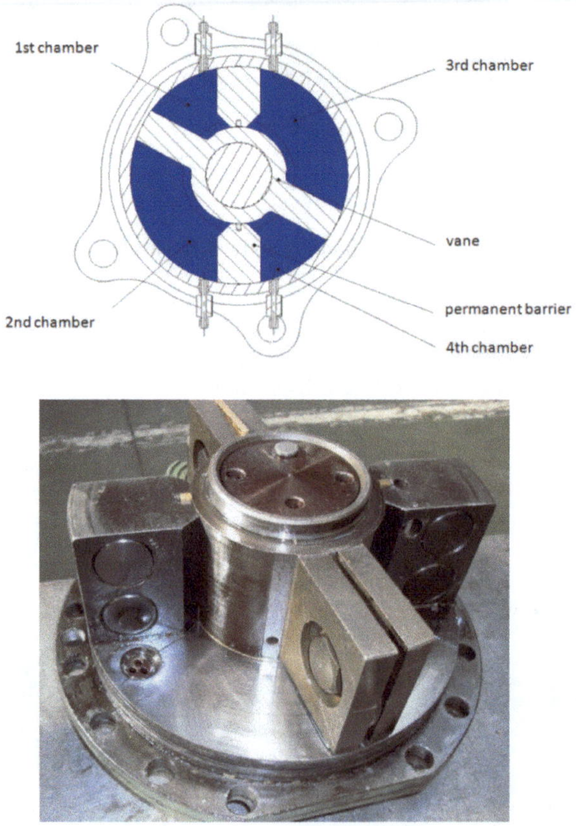

Fig. 4. The internal structure of the damper including the division of the chambers

For the testing of the pressure proportional valve, a solution consisting in the connection of the opposite chambers of the damper has been applied (chambers 1 and 4; chamber 2 and 3). The chambers have been connected using high-pressure hydraulic ducts. The ducts have been coupled with the Graetz bridge using a T-connection. The Graetz bridge conditions the flow of the fluid in one direction, thus ensuring the correct operation of the pressure proportional valve. The release valve protects the connection system of the damper against overpressure, while the terminal including the Graetz bridge and the proportional valve constitutes a base for the fixing of the individual hydraulic control components.

The tests of the damper have been performed using a test stand at the Zakłady Mechaniczne BUMAR plant in Zabrze Mikulczyce and concerned the measurement of the damping force for two ranges of the damper operation, the angle of the damper crank in the tests, the pressure in the chambers and the control current of the solenoid valve [4]. Due to an older construction of the test stand in the plant, the newly manufactured dampers are only tested as far as thermal loads are concerned. The test stand had to be modified to determine the damping forces. Bolts with extensometers have been placed in sockets, constituting an extensometric bridge which plays the role of force sensors. For the measurement of the force in the two stages of the damper operation – the upward and downward movement of the crank respectively – two such sensors had to be installed at the bridge. It should be noted that the primary stand has once allowed for the force measurement, yet due to the age of the equipment, the force measurement using the original sensors was no longer possible.

The new force sensors have been calibrated at the MTS strength equipment, thus obtaining a precise force-displacement characteristics in the range between 0 and 10 kN.

To perform continuous pressure variation measurement depending on the arm angle variation, a prototype goniometer has been installed (Fig 5.). The equipment has been constructed based on an electronic resistance setting device – potentiometer. For the measurement of resistance force at the damper arm, prototype sensors made of WCL steel with rosette-type extensometers at their circuit have been applied. The pressure measurement in the hydraulic system (independently for the upward and downward arm movement) has been performed using electric manometers connected to measurement sockets, the so-called minimes couplings.

The measurement and the processing of the signals obtained from the sensors described above and the signal registration was possible due to the use of the TRAVELER system.

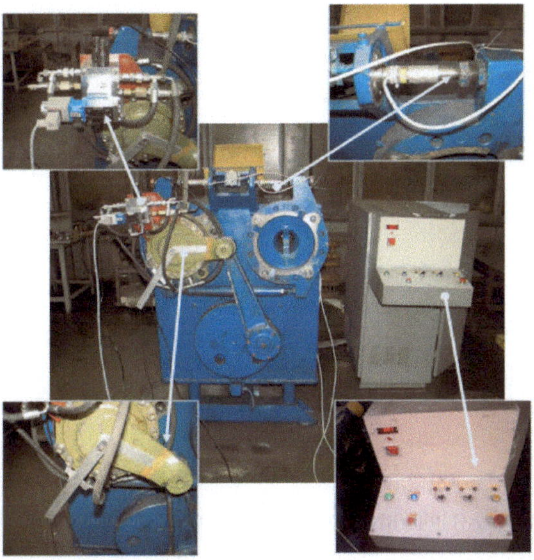

Fig. 5. The design of the test stand for fluid dampers [4]

The test stand for the fluid dampers is comprised of:

- The machine at which the tested damper is placed,
- The control panel, the signal of which (on/ off, left/ right rotation) was processed by the STARVERT "is5" Variable Frequency Drive controller into a rotational speed of the electric engine correct for the testing purposes,
- Gear and coupling transmitting the torque from the engine onto the working element (connecting rod),
- Three-phase inductive 55kW ZELTECH motor of the nominal speed of 1475 rpm.

3 Test Results

The drawings below present the waves of the force variation measured by the sensors in the function of the damper crank displacement (angles from 0 - 88 degrees) and the angular speed of the crank as well as the changes of the pressure in the damper chambers in the same domains as the forces.

Due to the controllable operation of the damper, in each drawing four waves have been plotted: the red color (___) 0 V control current, the blue color (___) 3 V control current, the green color (___) 6 V control current, the magenta color (___) 10 V control current.

a) for 0.65 [rad/s]

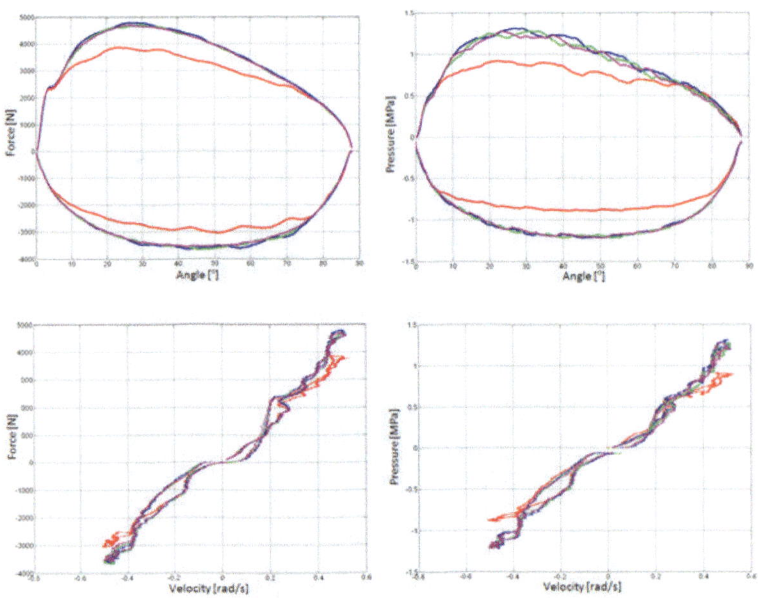

Fig. 6. The waves of the damping force and pressure variation in the function of the crank displacement and the angular speed

b) for 3.8 [rad/s]

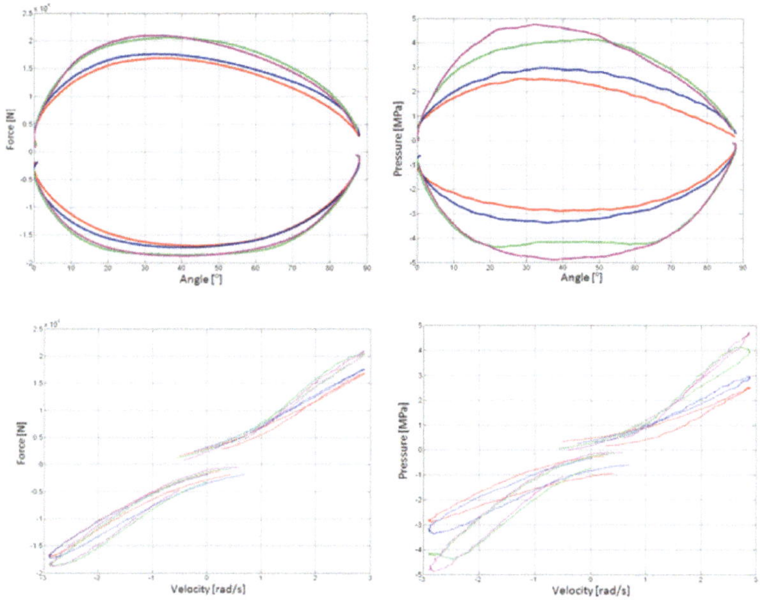

Fig. 6. (*Continued*)

The test stand was driven by an asynchronous motor supplied from an inverter, due to which the step-less regulation of the engine operation speed was possible. After an earlier analysis of the damper operation, the following angular speeds of the damper operation cycle were selected: 0.65 [rad/s], 1.15 [rad/], 1.5 [rad/s], 2.3 [rad/s], 2.8 [rad/s] and 3.8 [rad/s]. Due to the number of waves, the work only includes the waves of the boundary angular seeds of the damper operations cycle.

4 Model Research of the Movement of a Quarter of a Car with the Application of the Proposed Vibration Damper

The vehicle suspension system equipped with a hydraulic damper with controlled characteristics is an example of a semi-active mechatronic suspension. Such a system may be applied in both wheeled and tracked vehicles[2]. The dynamic model characterized by two degrees of freedom, describing the vehicle dynamics (the so-called model of a vehicle quarter), giving consideration to dumping control with an implemented dynamic model of the prototype fluid damper has been presented in Fig. 7.

The research in the scope of dampers with controlled characteristics for (among others) special vehicles, have been conducted since few years at the Department of Theoretical and Applied Mechanics of The Silesian University of Technology's Faculty of Mechanical Engineering. As a result of the performed computer simulations and experiments with the use of the test stand (Fig. 5.), a prototype hydraulic damper has been designed and its dynamic characteristics have been determined.

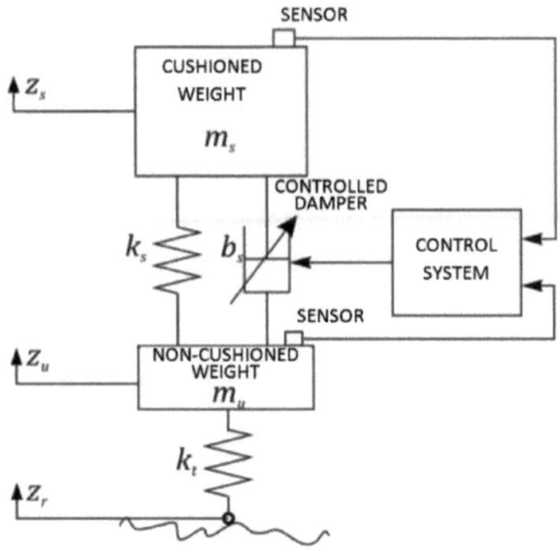

Fig. 7. The model of a semi-active suspension system with a hydraulic damper and a control system

Fig. 8 presents the diagram of the active hydraulic vibration-damping system. The damper is equipped with two rotational vanes coupled with an external arm and two permanent barriers, in the standard versions of which – throttle non-return valves have been placed. As a result two pairs of chambers have been created in the damper, between which the hydraulic fluid is transferred (HL-46 mineral oil).

In the serial version, the oil is transferred through valves in the permanent barriers. Such a construction ensures that the hydraulic resistance in the standard damper is constant and the resistance torque depends upon the angular speed of the damper arm, the medium viscosity and the direction of rotation (throttle non-return valves have been placed in the barriers).

The active damper requires the application of resistance torque control at its arm. This may be conducted by introducing an external flow between the chambers by means of an additional system with a controllable hydraulic resistance. The following valves may be applied as hydraulic resistance control valves:

- Throttle controllable valves,
- Flow controller,
- Overflow valves.

The application of external controllable throttle valves allows for the setting of the resistance torque of the damper. It does not, however, make it independent from the damper arm's angular speed.

The use of the flow controller will cause the damper to be a 2-state element. At low angular speeds of the arm, for which the flow between the chambers will be lesser than the controller setting, it would act as a normal throttle valve. At higher angular speeds of the arm, the controller limits the flow to the pre-set value, while the remaining stream is directed to a suction chamber by the overflow valve (which is pre-set at the highest allowable pressure value and protects the damper against damage). This means an increase of the pressure at the pressurized side of the damper and a rapid increase of the resistance torque.

The solution allowing the direct setting of the resistance torque is the application of the overflow valve as a regulated hydraulic resistance.

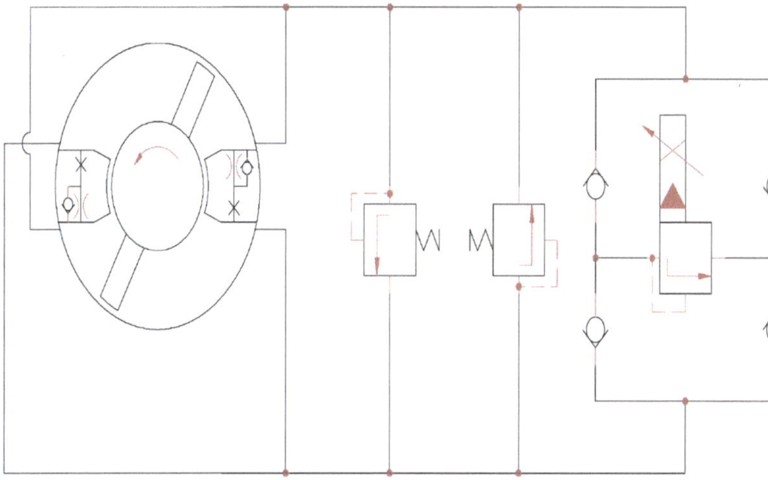

Fig. 8. The diagram of the controllable damper hydraulic system

Fig. 9 presents the physical model of a simplified damper in the form of a hydraulic diagram.

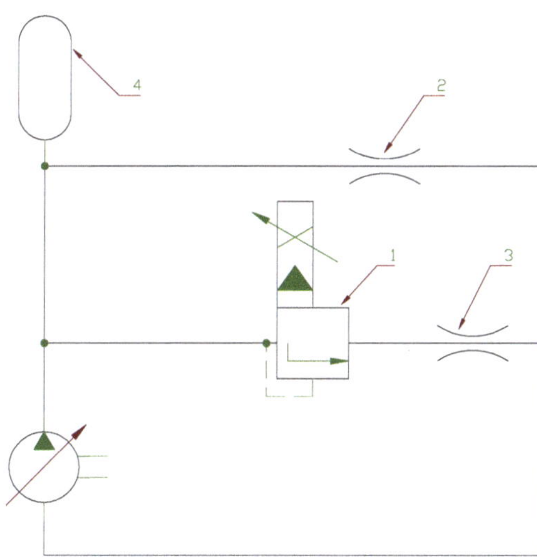

Fig. 9. The physical model of the active damper: 1 – proportional overflow valve, 2 – throttle modeling the influence of damper clearance, 3 – throttle modeling hydraulic resistance of the valve block and the connecting ducts, 4 – the volume of the chambers in which the oil is compressed

The damper clearance may cause a formation of a stream of leakages of a QL density provided by the following dependence:

$$Q_L = \frac{(D + B - d) \cdot g \cdot e^3}{6 \cdot \nu \cdot \rho \cdot L} \cdot p \tag{1}$$

Where:

D – the external diameter of the damper chambers,
d – the internal diameter of the damper chambers,
B – the height of the damper chambers,
L – the width of the damper vanes,
e – the clearance between the cooperating surfaces of the damper,
ν – the kinematic viscosity of the oil,
ρ – the density of the oil,
p – the pressure in the pressurized chambers of the damper.

The resistances of the flow through the valve block and the ducts have been determined experimentally. At a 25 dm3/min flow, a pressure of 5 MPa has been observed, thus the hydraulic resistance constant amounts to:

$$Q_{op} = C \cdot \sqrt{p_{op}}$$

$$C = \frac{Q_{op}}{\sqrt{p_{op}}} = \frac{25/60000}{\sqrt{5 \cdot 10^6}} = 1{,}8634 \cdot 10^{-7} \tag{2}$$

The stream generated by the pressurized chambers of the damper QT amounts to:

$$Q_T = \frac{(D^2 - d^2) \cdot B}{4} \cdot \omega \tag{3}$$

Where:

ω - the angular speed of the damper arm.

A flow balance equation has been laid out for the system as presented in Fig. 2:

$$Q_T - Q_{sc} - Q_{ZP} - Q_L = 0 \tag{4}$$

Where:

QZP – The flow through the proportional overflow valve (this, at the same time is the flow through the valve block),

Qsc – flow caused by the compressibility of the medium at the pressurized side.

The flow in the overflow valve is provided by the dependence:

$$Q_{ZP} = \sqrt{\frac{2p_{ZP}}{\rho}} \cdot k_V \cdot \pi \cdot d_{ZP} \cdot x \tag{5}$$

Where:
pzp – the drop of the pressure at the overflow valve,
kv – the flow coefficient,
dZP – the diameter of the closing element,
x – the opening of the valve.

At the same time – in line with the Kirchhoff's laws – the following equations occur:

$$\begin{aligned} p &= p_{ZP} + p_{op} \\ Q_{ZP} &= Q_{op} \end{aligned} \tag{6}$$

The flow caused by the compressibility of the medium is described by the dependence:

$$Q_{sc} = \frac{E}{V_{sc}} \cdot \frac{dp}{dt} \tag{7}$$

while the volume of the pressurized chambers is expressed as:

$$V_{sc} = \frac{\pi\left(D^2 - d^2\right) \cdot B}{2} \cdot \left(\alpha_0 - \omega t\right) \tag{8}$$

where:
α0 – the initial angle of the pressurized chamber.

After the introduction of the equilibrium equation of the overflow valve, a system of equations has been obtained, which constitute a mathematical model of the active damper.

The mathematical model served the development of a numerical model in MATLAB/SIMULINK environment.

The aim of the simulation testing was to obtain waves of pressure in the pressurized chambers of the damper during the movement of the arm upwards (corresponding to the movement of the vehicle through an obstacle).

The simulations have been performed for conditions corresponding to the conditions of the test stand for experiments, that is, a pendulous movement of the damper arm has been assumed (at the test stand this was constituted by an element of a four-bar linkage), driven by a crank characterized by the rotational speeds of: 14, 23, 45 and 76 rpm.

Subsequently, the following settings of the proportional overflow valve have been assumed:

0 . 30% iZN , 60% iZN , 100% iZN.

The timing of the obtained values has been presented in Fig. 10 – Fig. 15.

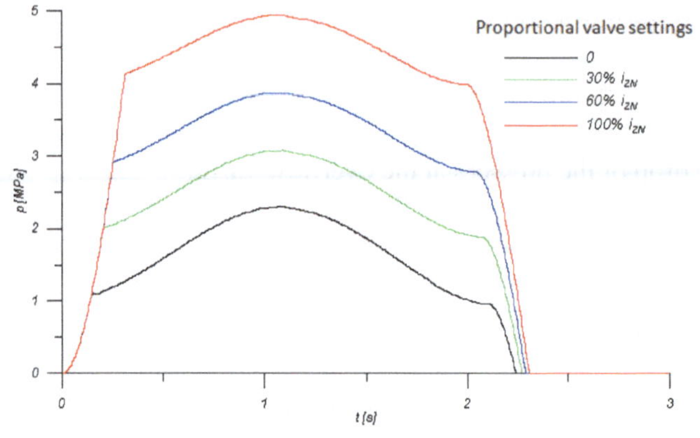

Fig. 10. Pressure waves for 14 cycles/min

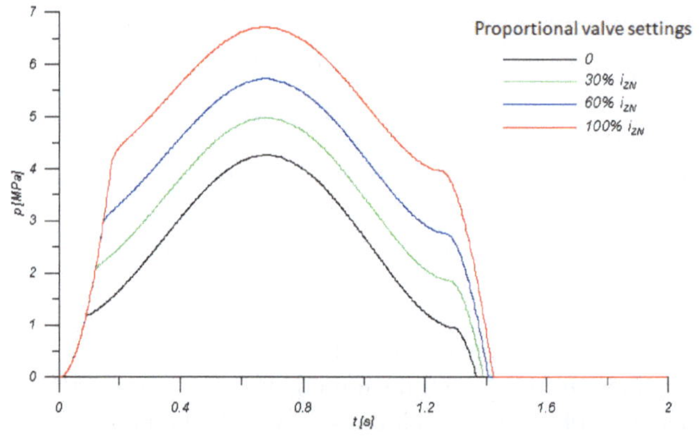

Fig. 11. Pressure waves for 23 cycles/min

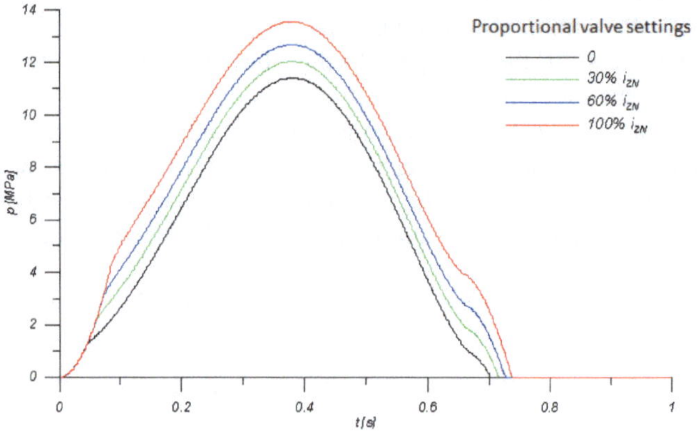

Fig. 12. Pressure waves for 45 cycles/min

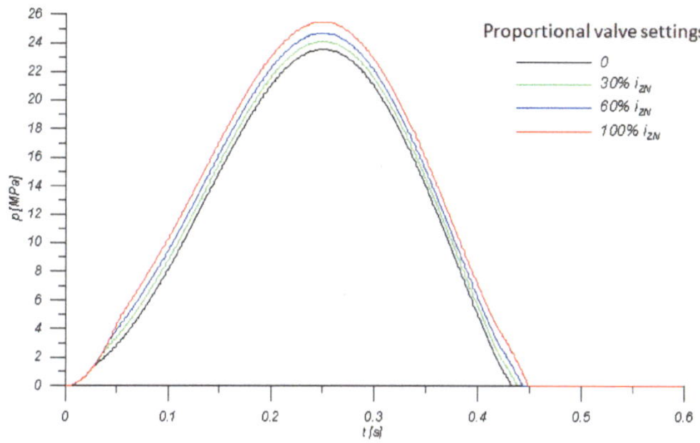

Fig. 13. Pressure waves for 76 cycles/min

The obtained pressure waves are evidence of a low sensitivity of the active damper to the signal controlling the proportional overflow valve. This is especially evident in higher speeds of the damper arm.

The initial analysis of the obtained results leads to a question: why, in the pressure waves, no phase of constant pressure, corresponding to the stabilized work of the overflow valve, is observed?

To answer this question, another series of simulations has been performed. The model, however, has been modified: a proportional valve of nominal size 16 has been assumed and resistance caused by the flow through the valve block has been omitted. The obtained results have been presented in Fig. 14.

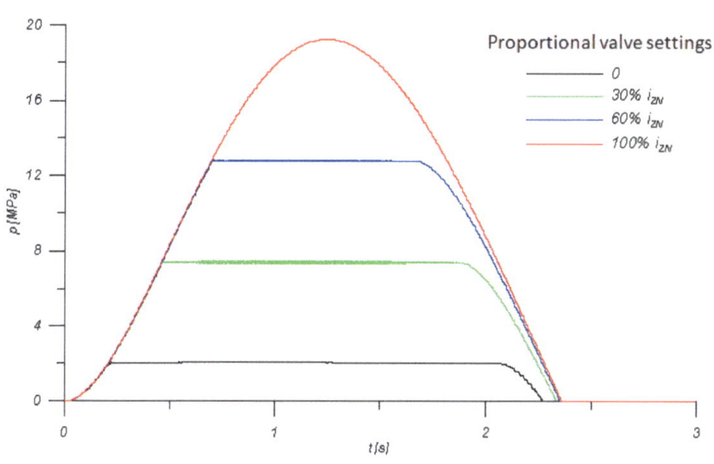

Fig. 14. Pressure waves for 14 cycles/min and the nominal valve size 16

The above waves are clearly influenced by the impact of the valve setting on the maximal pressure in the pressurized chambers. At the same time, it may be noted that at low number of cycles, due to leakages in the damper, the required pressure value may not be obtained. This is clearly visible in Fig. 15, in which the waves of the modified system for 7 cycles a minute have been plotted.

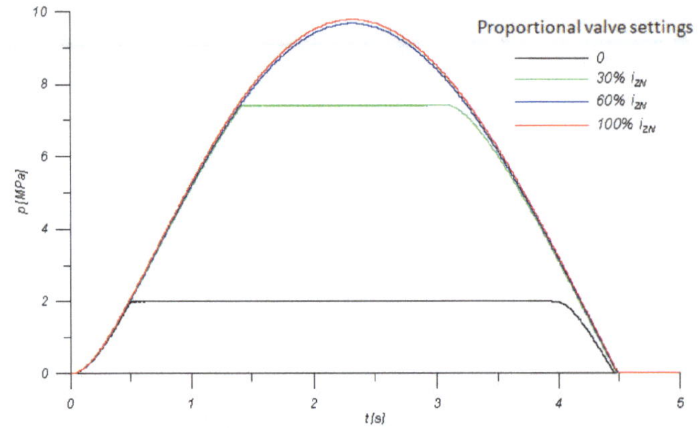

Fig. 15. Pressure waves for 7 cycles/min and the nominal valve size 16

Based upon the registered characteristics obtained through experiment, describing the dynamics of the prototype hydraulic damper, while considering the geometric form (the working stroke of the piston), a dynamic model of the damper has been developed in Simulink software.

As a result of the conducted numerical simulations, the waves of displacement, speed, relative and absolute acceleration of the springed and non-springed models of vehicle quarters have been obtained (Fig. 7). Excitation is assumed to be a harmonic function. The selected results of the simulation have been presented in the drawings below.

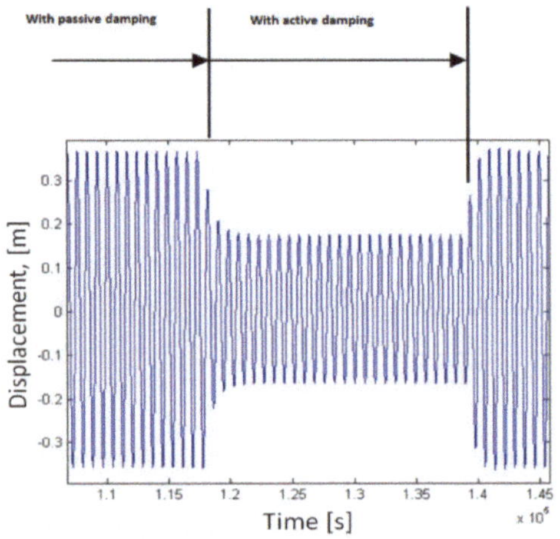

Fig. 16. The wave of absolute displacements of the vehicle body

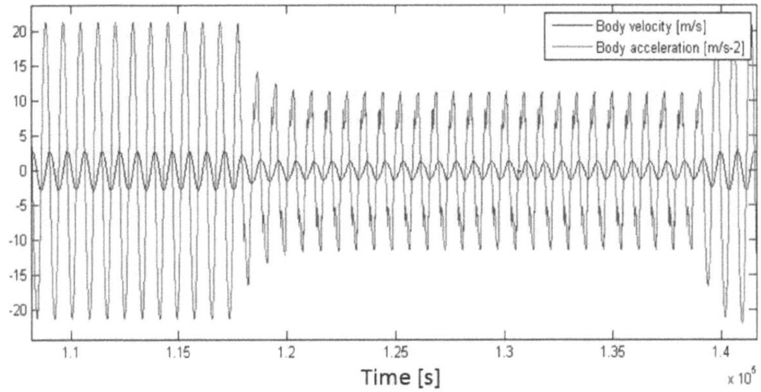

Fig. 17. The speed and absolute acceleration waves of the vehicle body

The above chard shows the differences in the displacement of the mass with shock absorption and the magnetorheological damper control turned on and off for the resonance frequency of 1.25 Hz.

The simulations of the tracked vehicle movement require the development of considerably more complex models, which take the lack of suspension symmetry to the lateral axis of the vehicle and in some conditions – no longitudinal symmetry of the vehicle (such as turret rotation, equipment exchange, etc.)

5 Conclusions

The concept of the active vibration damper using a proportional overflow valve as a controllable hydraulic resistance – in the light of the obtained results – appears to be wrong at first glance. The unfavorable qualities of the analyzed active damper, especially its low sensitivity to the dumping settings – does not however result from the essence of the operation of the proposed active damper. It is the result of excessive flow resistance in the valve block and the ducts. Increasing the nominal value of the hydraulic elements and the controlled overflow valve will allow to obtain the required qualities including the dependence between the proportional valve setting and the resistance torque. Additional improvement of functional properties may be obtained by decreasing the radial and axial clearances between the walls of the chambers and the damper vanes.

References

[1] Czapla, T., Wrona, J.: Technology development of military applications of unmanned ground vehicles. In: Nawrat, A., Kuś, Z. (eds.) Vision Based Systems for UAV Applications. SCI, vol. 481, pp. 293–309. Springer, Heidelberg (2013)

[2] Chooi, W.W., Oyadiji, S.O.: Design, modelling and testing of magnetorheological (MR) dampers using analytical flow solutions. Computers and Structures 86, 473–482 (2008)

[3] Kciuk, S., Martynowicz, P.: Special application magnetorheological valve numerical and experimental analysis, Diffusion and Defect Data – Solid State Data. Pt. B. Solid State Phenomena, vol. 177: Control engineering in materials processing s. 102–115. — Bibliogr. p. 115, Abstr. (2011) ISSN 1012-0394

[4] Kciuk, S., Kciuk, M., Turczyn, R., Martynowicz, P.: An Investigation of The Behaviour of Magnetorheological Fluids in the Rotary Shock-Absorber. Advanced Materials Research 628, 512–517 (2013)

[5] Użycki D., Begier T., Sobala S.: Współczesne Gąsienicowe Wozy Bojowe" Wydawnictwo Lampart (1996)

[6] Daniec, K., Jedrasiak, K., Koteras, R., Nawrat, A.: Embedded Micro Inertial Navigation System. Applied Mechanics and Materials 249, 1234–1246 (2013)

[7] Jędrasiak, K., Bereska, D., Nawrat, A.: The Prototype of Gyro-Stabilized UAV Gimbal for Day-Night Surveillance. In: Nawrat, A., Simek, K., Świerniak, A. (eds.) Advanced Technologies for Intelligent Systems. SCI, vol. 440, pp. 107–116. Springer, Heidelberg (2013)

[8] Jędrasiak, K., Nawrat, A., Wydmańska, K.: SETh-link The Distributed Management System for Unmanned Mobile Vehicles. In: Nawrat, A., Simek, K., Świerniak, A. (eds.) Advanced Technologies for Intelligent Systems. SCI, vol. 440, pp. 247–256. Springer, Heidelberg (2013)

[9] Jędrasiak, K., Nawrat, A.: Image Recognition Technique for Unmanned Aerial Vehicles. In: Bolc, L., Kulikowski, J.L., Wojciechowski, K. (eds.) ICCVG 2008. LNCS, vol. 5337, pp. 391–399. Springer, Heidelberg (2009)

[10] Nawrat, A., Jedrasiak, K.: SEThSystem Spatio-Temporal Object Tracking Using Combined Color and Motion Features. In: WSEAS International Conference Proceedings, Mathematics and Computers in Science and Engineering, vol. 9 (2009)

[11] Jedrasiak, K., Daniec, K., Nawrat, A.: The Low Cost Micro Inertial Measurement Unit. In: Industrial Electronics and Applications (ICIEA), pp. 403–408 (2013)

[12] Bereska, D., Jędrasiak, K., Daniec, K., Nawrat, A.: Gyro-stabilized platform for multispectral image acquisition. In: Nawrat, A., Kuś, Z. (eds.) Vision Based Systems for UAV Applications. SCI, vol. 481, pp. 115–121. Springer, Heidelberg (2013)

[13] Daniec, K., Iwaneczko, P., Jędrasiak, K., Nawrat, A.: Prototyping the Autonomous Flight Algorithms Using the Prepar3D Simulator. In: Nawrat, A., Kuś, Z. (eds.) Vision Based Systems for UAV Applications. SCI, vol. 481, pp. 219–232. Springer, Heidelberg (2013)

[14] Babiarz, A., Bieda, R., Jędrasiak, K., Nawrat, A.: Machine Vision in Autonomous Systems of Detection and Location of Objects in Digital Images. In: Nawrat, A., Kuś, Z. (eds.) Vision Based Systems for UAV Applications. SCI, vol. 481, pp. 3–25. Springer, Heidelberg (2013)

[15] Bereska, D., Daniec, K., Fraś, S., Jędrasiak, K., Malinowski, M., Nawrat, A.: System for Multi-Axial Mechanical Stabilization of Digital Camera. In: Nawrat, A., Kuś, Z. (eds.) Vision Based Systems for UAV Applications. SCI, vol. 481, pp. 177–189. Springer, Heidelberg (2013)

The Prototype of an Unmanned Underwater Vehicle – Mechanical Construction, the Operator Panel

Krzysztof Jaskot, Marcin Sroka, and Aleksander M. Nawrat

Silesian University of Technology, Institute of Automatic Control,
Akademicka 16, 44-100 Gliwice, Poland
{krzysztof.jaskot,aleksander.nawrat}@polsl.pl

Abstract. The paper presents design of Remotely Operated Vehicle - ROV prototype. To date, the operator has the ability to seamlessly move and maneuver both horizontal (geographic change in direction, moving from point to point) and vertical (change of depth of submerged position). Solid construction provides resistance to mechanical damage, and the ergonomically designed operator panel, offers a convenient and intuitive use of all modules.

Keywords: unmanned underwater vehicle, propellers, ballast tank, a central controller.

1 Introduction

Unmanned underwater vehicles are the most different types of underwater robots or robotic platform controlled by an operator (also called "a pilot") located on the surface, usually on the ship. They are usually equipped with a camera (or camera system) and lighting, to ensure adequate visibility, allowing the operator to perform a task under water. Depending on the application, on board can be mounted various additional sensors and devices, such as, for example: supporting sonar to find buried objects or robotic arm, allowing the lifting of objects from the bottom and perform other underwater works [1,2,3]. There are also highly specialized underwater robots, with an additional drive crawler or wheeled [4,5,6], armed dedicated to one specific application equipment, such as: vehicles stacking cables or pipelines on the ocean floor [7,8,9].

Remotely Operated Vehicle (ROV) panel is connected by a cable operator [10], which is beyond the transmission of control signals and video and very often also the power, allows the vehicle pulling out of the water.

In the basic version, the operator usually has a control panel (also called the operator panel), providing access to all functions of the vehicle, the TV screen displaying images from video cameras, joysticks for motor control and screen display vehicle parameters (depth, an indication of the compass onboard, diagnostic data from internal sensors, etc.) [1,3].

In a more extended ROV, operator panels, and actually have control centers, can occupy the entire wall with an area of several square meters, and to require the cooperation of several service providers. In Figure 1 shows the appearance of an underwater

A.M. Nawrat (ed.), *Innovative Control Systems for Tracked Vehicle Platforms*, Studies in Systems, Decision and Control 2,
DOI: 10.1007/978-3-319-04624-2_16,

Fig. 1. Underwater robot Victor 6000 and the control room [47]

robot Victor 6000 (weight: 4 tons) and the control room, which is controlling the robot. According to NORSOK standard [11], which was developed by a number of Norwegian companies related to the petroleum industry, unmanned underwater vehicles can be divided into five classes:

- Class 1 - little overall dimensions, lightweight vehicles equipped with propellers, camera and lighting, serving only to observe the underwater environment.
- Class 2 - larger vehicles, able to lift the load in the form of additional video cameras, sonar, or devices that enable acquisition of different types of environmental samples (biological, benthic flora).
- Class 3 - underwater robots - a larger unit, equipped with additional sensors and manipulators for performing work under water (eg removal of blockages from the propellers).
- Class 4 - underwater robots moving on the bottom of the water tank on wheels or tracks - the largest and heaviest overall dimensions vehicles designed to perform specific tasks, such as laying cables, digging, mining.
- Class 5 - prototypes and development unit - the other vehicles that do not include in any of the above categories.

The vehicle presented in this work may be classified as Class 5, but eventually you will be able to put it on the border between Class 2 and 3.

In the initial phase, the development of underwater vehicles dealt mainly with the U.S. Navy [12]. In 1966, a submersible vehicle U.S. Navy - Curve (called Cable Controlled Underwater Vehicle), retrieved from the bottom of the sea, near the fishing village of Palomares, Spain, the atomic bomb that fell into the water after a collision in the air during refueling maneuver, two aircraft [48]. The accident and bomb mining shares led the ROV began to become more popular. However, research and construction of such vehicles require huge financial resources, because only after several years, thanks to advances in technology, more and more private companies began to be interested in ROV. These were primarily related businesses to the petroleum industry, which need such devices to work in the performance of wells and offshore inspection.

Nowadays, the army still actively involved in research into new designs of unmanned underwater vehicles. U.S. Navy for almost 10 years program "Unmanned Undersea Vehicle (UUV) Master Plan", which aims to determine the future of defense tasks for unmanned systems. In that case, the work focuses on fully autonomous vehicles, which are sent on a mission, with the programmed a particular task. These units are not connected by cables to the operator panel and the operator does not exercise continuous control over them. Relying on the installed sensors and embedded algorithms, AUV

(Autonomous Underwater Vehicle called) must themselves find the way to order and assigned to perform the task.

Polish Navy is also using unmanned underwater vehicles. The primary objective of their activities is to increase the operability of protection bodies, approaches to ports, shipping lanes and hydrotechnical facilities of military and civilian. Military submersible vehicles are also used in the work of rescue and exploration and mine action. Polish Navy probably has the largest number of ROV in the country [13]. Currently, many companies specializing in the construction of unmanned underwater vehicles. These range from small units of class 1 or 2 (according to NORSOK standard), intended "for all" and specialized units built to order, according to customer requirements.

One of the most popular manufacturers, with a wide range of ROV, are Seaeye [49] and Seabotix [50]. Vehicles of both these companies can be found also in Poland, and the model Seaeye Falcon (weight: 50 kg), presented in Figure 2 was in a sense the prototype, which was based on a project carried out within this work. Some academic

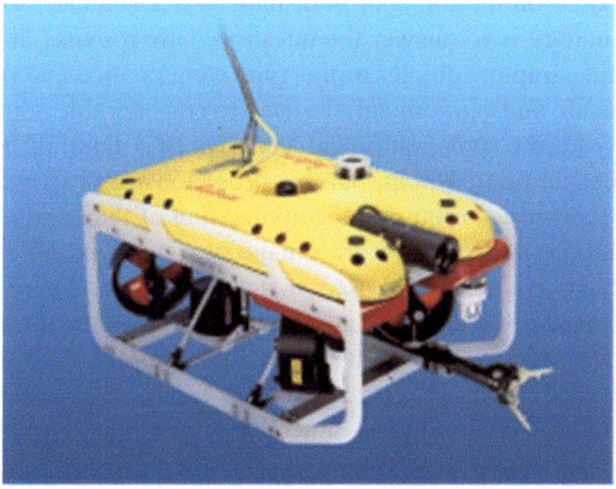

Fig. 2. Underwater vehicle - Seaeye Falcon [49]

centers have in their lab prototype unit unmanned underwater vehicles. The various scientific works dealing with them, above all, developing newer and newer, more efficient and reliable control algorithms. Due to the specific environment in which the vehicle moves underwater and on the range of tasks that should be able to implement, very frequently used genetic algorithms and neural networks. Control then it is computationally complex, but allows the vehicle to adapt to the changing physical hydro.

The solutions developed by researchers are often used by commercial or military. Vehicles located at universities, devices are usually built by students [14,15]. These are fairly simple structures, made of soft materials (eg PVC pipe [16]), and those reasons are not suitable for commercial applications. Figure 3 shows an example of ROV was built at the Faculty of Marine Technology Technical University of Szczecin. Currently the vehicle is supplied with Scientific Circle ABYSAL and serves students interested

Fig. 3. Underwater vehicle - KRAB [17]

in conducting research on motion control of underwater vehicles. The first step to understand any technology is to answer the question, why it exists at all. In the case of ROV answer is quite simple: vehicles of this type were created and formed, since there is no other option (safe and economical) descend to considerable depth below the water surface. The man with the right equipment, can dive to a depth of 50 m. It is possible that it can be immersed up to 450 m, but this project is very expensive, time consuming and of course dangerous, requiring the application of the breathing gas with lower specific gravity and the participation percentage of oxygen in the air other than [18]. Underwater vehicles are also widely used at shallow depths, where their main advantages are: the possibility of a contaminated area, high mobility, long residence time in the immersion. They increase safety standards, replacing the divers while performing dangerous tasks.

Here are some of the very numerous, practical unlimited applications in which work well ROV:

- inspection and monitoring of the quays, underwater parts of hydrotechnical structures, submerged parts of ships and other floating objects,
- search and examination den wreckage and bodies of water,
- monitoring the aquatic environment,
- assisting with the work of the dive,
- installation and recovery of equipment design,
- maintenance of underwater objects,
- to support drilling for oil and gas exploration in the seabed,
- military applications.

Many institutions around the world are involved in various projects, whose main research are unmanned underwater vehicles, both controlled by the operators as well as partially or fully autonomous. Here are some examples [13]:

- exploration and hydrographic data collection on large coastal areas (project Columbus and the Advanced Unmanned Search System)

- identify and search for mines on the small and medium depths (draft Long-Term Mine Reconnaissance)
- recognition in the coastal zone hydrographic using semi-autonomous units collaborating deep surface of the unit prior to landing troops (Project Semi-Autonomous Hydrographic Reconnaissance Vehicle),
- identification of oceanographic over large areas by means of drifting in the sea current,
- autonomous units deep dropped from the plane (Autonomous Profiling project Lagrangin Circulation Explorers).

2 General Characteristics of the Vehicle

Unmanned, a prototype remote-controlled underwater vehicle, is a functional unit, allowing the operator to observe the underwater environment. In the present state, allows for smooth movement and maneuvering in both the horizontal (geographic change in direction, moving from point to point) and vertical (changing the depth of immersion). Solid construction provides resistance to mechanical damage, and the ergonomically designed operator panel, provides a convenient and intuitive operation of all modules, and the ROV. Figure 4 is a picture of the vehicle taken during one of the practical tests. Access to the stainless steel, and the opportunity to receive professional help for her

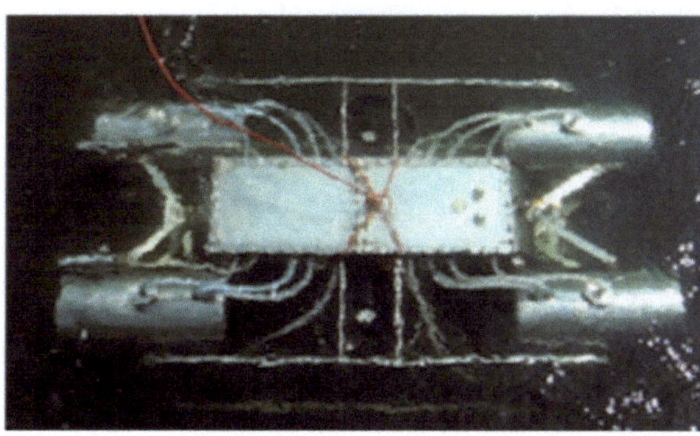

Fig. 4. Vehicle ROV in the water

treatment was the main motivator for selecting material for the construction of the hull of the vehicle. The next step was to determine the basic functionality of the vehicle. By design, the vehicle had to have the opportunity to observe the underwater environment with rotatable camera and halogen, the ability to determine important for some reason or hazardous for shipping locations in the tank water using buoys and additionally discarded, making manipulation of objects under water through mounted robotic arm [19,20,21]. In addition, the vehicle was to be powered with internal battery and be able to adjust their buoyancy through ballast water, as in submarines. These assumptions led to determine the dimensions of a vehicle large enough to accommodate the ballast tanks, launchers buoys (buoy marker), cylinder of compressed air and the hull, inside of which had to find a place for: electronics needed to control the vehicle, the battery and the solenoid valves, mechanism to move the camera / light and the mechanism of the robot arm.

Provide adequate mobility of the vehicle in the water was to be achieved through appropriate location in the vehicle design engines with propellers that in the rest of the work will be called propulsors [22]. The whole was to be surrounded by a frame, representing the vehicle additional protection against mechanical damage and facilitate its transport. The frame should be able to mount the wheels to facilitate movement of the vehicle on land, which with its large mass would be a great deal. The vehicle was

1) ballast tank	7) communication cable
2) propeller horizontal motion	8) propeller vertical movement
3) camera body drive	9) tube with air
4) launcher buoys	10) main hull
5) wheel for transport	
6) frame	

Fig. 5. Vehicle ROV - Description of the structure

made entirely by hand. Sheet steel, which is the plating of the vehicle, has a thickness of 1.5 mm for the hull of the vehicle and 2 mm for water ballast tanks and frames. Finally, the vehicle was built as planned, its general appearance is shown in Figure 5. In this figure, describes the main components of the structure. In Figure 6 are shown the dimensions of the vehicle. Inside the hull are:

- the main driver for motors, solenoids and lighting,
- arm band controller camera-lighting,
- power module, batteries,
- position sensors, and static and dynamic acceleration and pressure sensors and depth,
- solenoid valves,
- pressure reducer,
- air collector with air ducts,

- mechanism and motor assembly arm-camera lighting,
- video Converter for RS-485 standard.

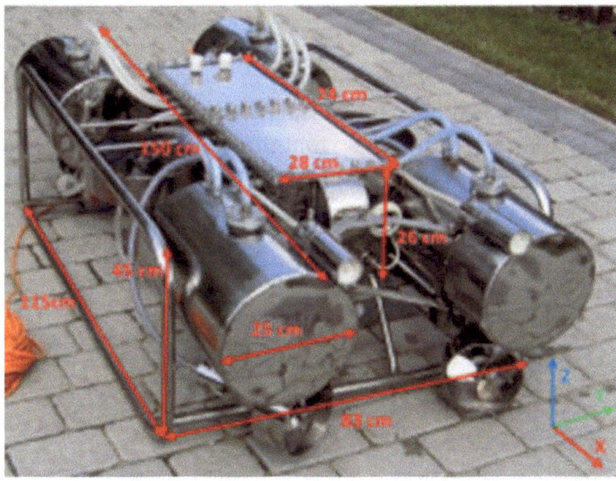

Fig. 6. Vehicle ROV - Dimensions

In front of the hull, on a movable pivot (which allows movement up and down), were fixed camera and lighting. Their location is shown in Figure 7. On one side of the movable pivot was attached heat sink on which are mounted LED Power LED serving as halogen. On the other side of the movable pivot is located in a waterproof camera housing. For both devices are brought cables that go from the front, bottom of the hull

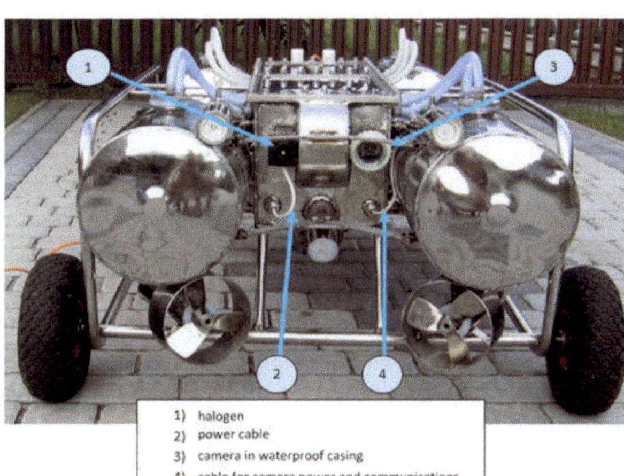

Fig. 7. Vehicle ROV - Location camera and halogen

through a watertight gland. These cables, power is supplied to the diode and the power to the video camera and a video signal is transmitted from the camera to the interior of the vehicle, which then is converted to a standard RS-485 and sent on kablolin (communication cable, Figure 5 mark 7) for the TV monitor supported by the operator [23]. The vehicle has four ballast tanks, positioned in the corners of the vehicle. Each of them, independently of the others, can be filled with air or water. They are an important element of the structure, which plays an important role in the mobility of the vehicle, used for buoyancy control device. It is thanks to the operator can freely immerse and emerge the vehicle. Their dimensions were chosen so that the displacement of the vehicle provided him with the assumptions under water capacity 30 kg more than its weight. These tanks have a total length of 44 cm and a diameter of 25 cm. Schematic cross-section switched in a tank in Figure 8. ROV is equipped with six engines, propulsion (propellers), provid-

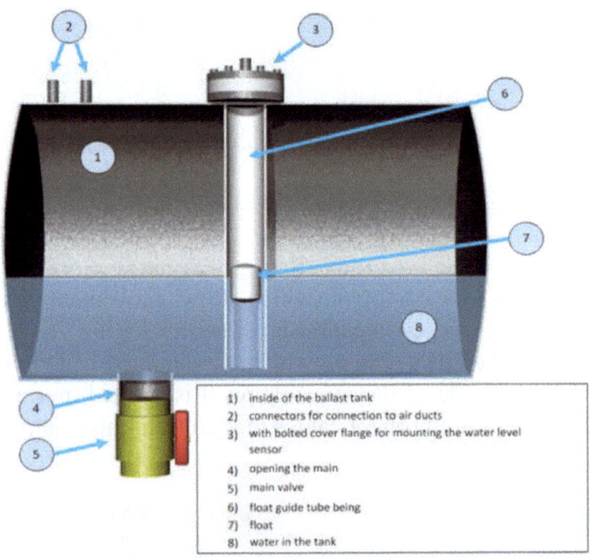

Fig. 8. Vehicle ROV - Cross ballast tank

ing him maneuverability - 4 of them, providing horizontal movement, was placed under the ballast tank and the other two, providing vertical movement, placed on both sides of the hull, in the middle of the vehicle. The position of these propulsors is presented in Figure 9. Drives horizontal movement differ from those providing vertical movement, and its propeller guard. Visual comparison of vertical and horizontal propellers are presented in Figure 10. Propellers have horizontal motion, hand-made stainless steel propellers with a diameter of 14 cm. Each lobe of the screws by hand cut from sheet metal, shaped, treated and welded to the shaft mounting. Then all the propellers were subjected to manual treatment to balance and harmonize them in all the jumps, which was to align the curvature of the airfoil profile. Vertical propulsors propellers are purchased, the factory bolts to MotorGuide's outboard boats used for sailing. They have

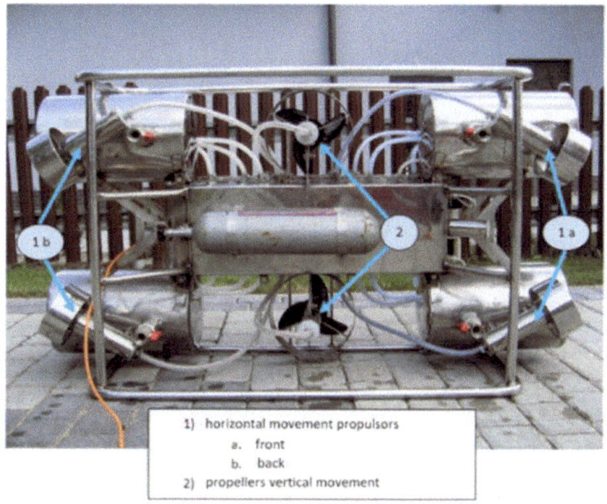

Fig. 9. Vehicle ROV - Location of propellers

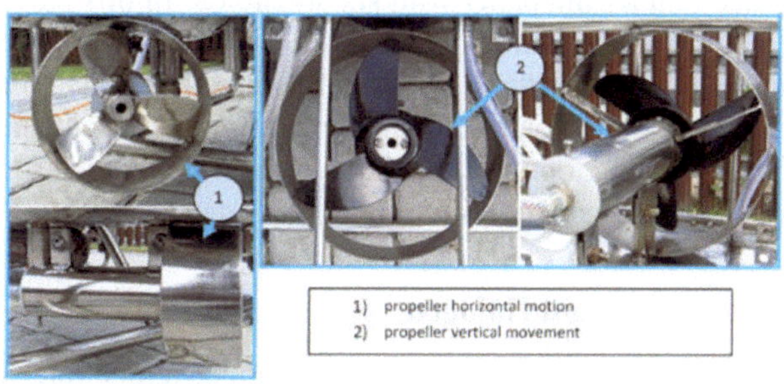

Fig. 10. Vehicle ROV - Comparison of propellers

a larger diameter (24 cm) with more thrust. All elements of the corps propellers were performed manually. Cross section of the propeller shows in Figure 11. All the power cables, conducted from inside the hull of the vehicle to the propellers, are carried in the pocket with rubber hoses, ensuring tightness. Similar tubes are used for transportation of air masses and from the ballast tanks and rocket buoys. They will work later called in air ducts or air ducts. Location of hose to the vehicle structure is shown in Figure 12. The interior of the hull is shown in Figure 13 Inside the front of the hull is such articulation drive assembly to rotate the camera-light. In this part of the hull is also a box of trim and course sensors, water detection system (alarm for leaks), the video signal converter from the camera to the standard RS-485 and the panels on which are mounted equipment. Top panel contains 4 solenoid and actuator pivot mount to rotate the camera and lights. In the bottom panel, located directly below the top, there is air collector, which are distributed with the air ducts to the solenoid. Inside the rear of the fuselage panel is the main driver of the vehicle, the power module and solenoid driver and

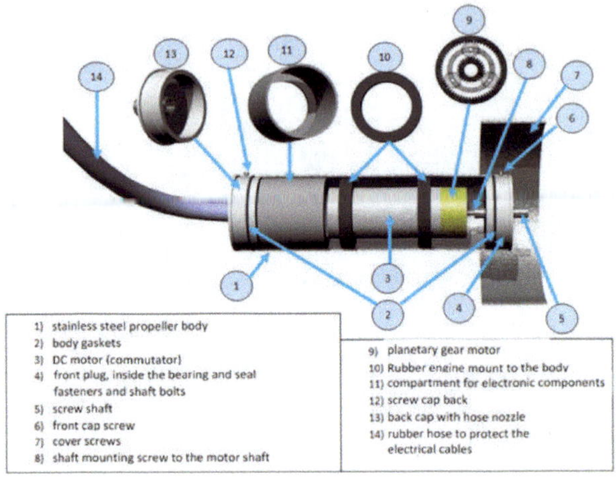

Fig. 11. Vehicle ROV - rig section

lighting. It was placed under the panel with six solenoids. In this part of the hull there is also a control module and a manually adjustable propulsors pressure reducer, which is served at the inlet air from a cylinder of compressed air, and its output is connected to the aforementioned air collector. In the middle, between the front and rear part of the hull, there are 2 gel batteries, one to power the entire vehicle electronics, one for the non-electronic parts (motors, solenoids, etc.).

2.1 Camera Specification and Halogen

In order to increase the range of possible uses of unmanned underwater vehicle equipped him with additional external components. They include camera and halogen, which allow the operator to observe the underwater environment around the vehicle. Both of these elements are placed in front of the hull, the rotational axis. The camera lens is set in the direction of travel of the vehicle to the front (position 90°) and the rotating mechanism can be tilted up or down in the range 50°to 70°(Figure 14).

The underwater vehicle was used in black and white camera with very high sensitivity. Along with lighting, it provides sufficient visibility, allowing the operator to move collision-free vehicle under water. In addition to the primary use of the camera, as described above, another very important is to use it to watch/inspection of hydraulic structures underwater and submerged parts of the hulls of vessels, as well as the search for the bottoms of bodies of water submerged objects [24,25,26]. The camera allows the identification of small defects studied the structure and use it to the operator is able to distinguish between minor elements in the water. The limit of visibility, the distance from the lens to the point where the image becomes blurred, depends mainly on the transparency of water. The more contaminated water (eg, small floating plants), the distance at which the operator sees is much smaller [27,28,29]. Parameters of the camera: Sensor: 1/3 Sharp CCD, Resolution: 420 TVL, Sensitivity: 0.3 lux, Lens: 3.6 mm, Supply voltage: 12VDC, Dimensions: 32x32x27 mm. An integral part of an underwater

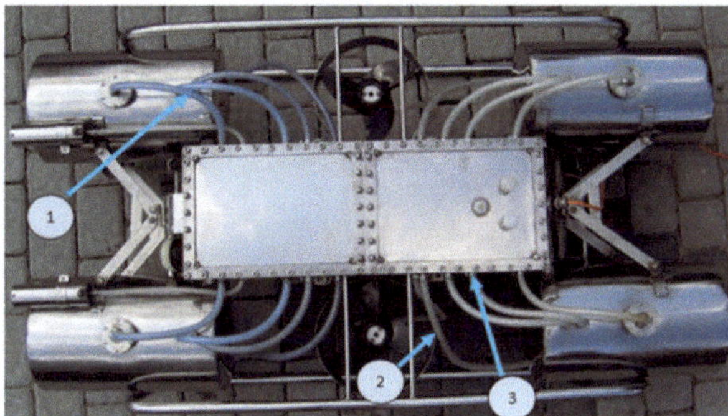

Fig. 12. Vehicle ROV - The location of the air pipes

I front part of the hull interior	II back part of the hull interior
1) sensor trim and course	8) battery in non-electronic components
2) joint drive camera	9) electronic control module propulsors
3) water sensor	10) control panel of the vehicle; below the second panel with electrovalves
4) panel of the electrovalves, below is a air collector	11) air pressure reducer
5) video signal converter	12) electrovalves control electronics module
6) wires from the air	
7) battery power electronics	

Fig. 13. Vehicle ROV - Interior of the hull

vehicle vision system, in addition to a video camera, is lighting. This largely depends on it the picture quality when viewed by the operator. Already, even in the day, at shallow depths is needed an additional source of light directed in the direction in which the camera is facing. The underwater vehicle lighting power must be adjusted to suit your needs. In highly polluted water (eg, which rises to a minor vegetation - the water green) should be used with low brightness, otherwise the light will be reflected from contamination, blinding the camera. The underwater vehicle, a light emitting diode is used for high-brightness, so called. "Power LED". This choice was dictated by the very good functional lighting with low power consumption. Additional advantages are: small size

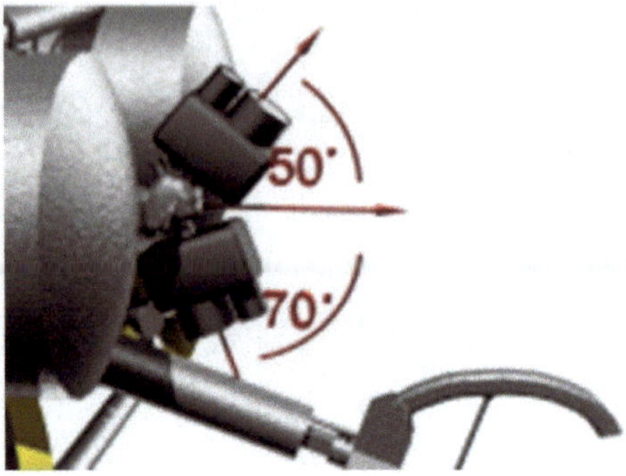

Fig. 14. Arm Camera

and ease of installation. Lighting fitted parameters CREE-MCE4WT-A2-0000-000M02 [30]:

- 4 structure of the MC-E M bin with a capacity of 10 W,
- Light: 430 lumens at 350 mA for the structure
- Color temperature: 6,500 K (cool white)
- Supply current: up to 700 mA for each structure with a serial connection, or 2.8A for the parallel connection
- The voltage drop on a single structure: about 3,2-3,4 V,
- 110 angle.

Brightness of the lighting is controlled by the main driver, as directed by the operator, with a rectangular voltage on the course of filling pulse width adjustable (called pulse-width modulation PWM). In the case of LED used was also necessary to build additional electronic circuit - current regulator to provide LED current equal to 1.4A at each value of the supply voltage.

Basic parameters of the vehicle:

- Total weight of the vehicle: 98 kg.
- Vehicle Operating weight: 130 kg.
- Maximum load capacity of the vehicle: 32 kg.
- Maximum working time under water: 10 h (depending on battery capacity).
- Maximum draft: 25 m.
- The maximum movement speed: 0.32 m / s.
- Maximum speed twist: 0.41 rad / s = 23.9 m / s.
- Maximum speed of descent / ascent: 0.24 m / s.

2.2 Buoyancy of the Vehicle

Underwater Vehicles must be able to move under water. Maintenance on the water surface or at any immersion depth depends mainly on the ratio of gravity forces to buoyancy forces acting on the object in the water. If the buoyancy force exceeds the force of gravity on an object will remain at the surface. If the force of gravity exceeds the force of buoyancy is an object sink. If these forces are balanced, the object reaches zero static buoyancy. Maintaining a zero buoyancy facility allows you to keep him in any depth. The present work in the subject vehicle has been designed to control their own buoyancy. It has four ballast tanks arranged in the corners of the hull structure. Tanks are equipped with electronically controlled solenoid valves by regulators, may be freely filled with water or emptied, so the vehicle can remain on the surface or any depth of the water.

2.3 Mobility Vehicle

One of the most important features of underwater vehicles is their mobility in the aquatic environment. Since the mobility of the vehicle depends on the effectiveness of their work in difficult conditions, such as the underwater currents, severe pollution, underwater obstacles (eg seaweed, shipwrecks, hydrotechnical constructions). Adequate mobility of an underwater vehicle can be provided in many ways, including through streamlined shapes, low weight and dimensions of the vehicle with a strong drive accordingly. The submarines and submersible vehicles such UUV or AUV3 most common solution is 1 or 2 turbine aft to give the longitudinal velocity, and depth rudders for vertical movement and stern rudders for turning. Unfortunately, vertical movement, and then turn it is possible only when the object is moving straight ahead. Type of ROV underwater vehicles perform maneuvers usually within a certain small area, so the solution does not satisfy these requirements. In this case, the simplest, but effective method is to use a number of propellers, which is sealed with screw propulsion engines, arranged in different planes in the structure of the vehicle. When planning a drive module for the construction of this vehicle are considered different technical solutions having the weight of the vehicle, its shape and location of the engines in the construction. For their selection guided by the receipt of possible high mobility and simplicity, at least cost of their own. Here are some solutions that have been taken into account:

- Two movable propellers under the hull of the vehicle.
- Two movable propellers on the sides of the vehicle.
- Drive streaming.
- 5 propulsors - 4 horizontal, one vertical.
- 6 propellers - 4 horizontal, 2 vertical - applied to the solution.

The current vehicle is equipped with 6 fixed propellers that provide him the opportunity to move in five degrees of freedom. four thrusters (Fig. 17) have a maximum strength of the f = 8.5 N, the other two (Fig. 16) f = 10.9 N (results of measurements and calculations are presented in the next section). These thrusters are arranged in two planes, as shown in Figure 15. Figure 16 shows a graph illustrating the changes during one of the propellers of vertical and horizontal one of the propellers, depending on the set of engine power. Measurements were made using a digital dynamometer FG-20KG company Lutron. Measured, the maximum string for vertical propeller was: 10.9 N, while for the horizontal: 8.5 N. The chart shows that the entire power range, vertical engine, with a screw of larger diameter, has a power greater by about 25% from the horizontal

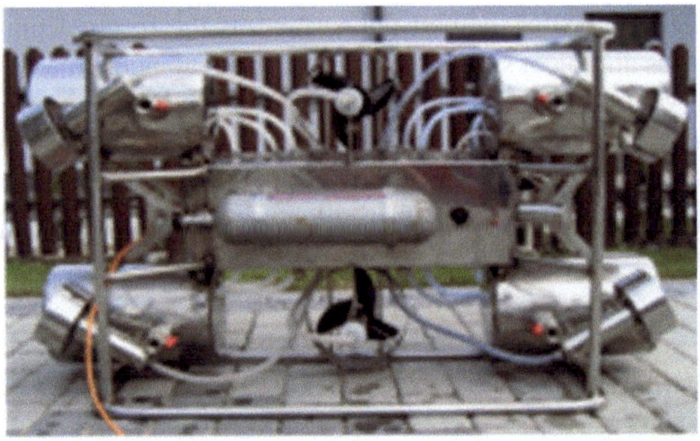

Fig. 15. The system of underwater vehicle propellers

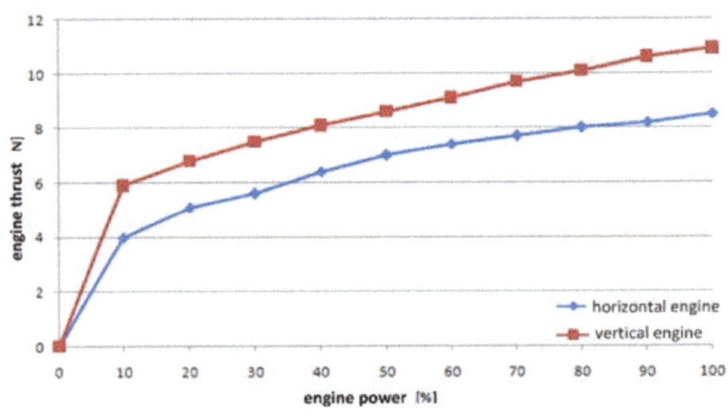

Fig. 16. Parameters propulsors - A string of vertical and horizontal engines

engine with a smaller screw. Motors in the vehicle were so arranged as to ensure maximum maneuverability of the vehicle. Horizontal axis of the vehicle's propulsors is set at an angle of 30°to the longitudinal axis of the vehicle, while the axis of propulsors used for ascent and descent is set vertically, perpendicular to the longitudinal and transverse axis of the vehicle. The paper assumes that the front engine is from the propeller. Construction of the screws used to make its engine "pulling", which moves forward in the direction of the bolts when the engine revs are right, and "pushing", which moves backward when the rotations are left. The terms "string to the front" and "reverse sequence" refers to the entire vehicle. Regardless of the propeller rotation, the use of the term "thrust forward" corresponds to the moving vehicle in front, by analogy with the use of the term "reverse sequence" corresponds to moving the vehicle backwards.

2.4 Vehicle Power

Underwater vehicle power supply is quite complicated. This vehicle must be able to work under water for several hours, and the internal battery power substantially reduces operating time. On the other hand, external power poses many problems, including voltage drop when powered by a long cable. Increasing the voltage to reduce power losses associated with expensive voltage converters, and more likely to leakage current or even shock the operator. Transferring the AC cable to minimize the cost of converters, and causes disruption of communication and still threatens to shock in case of breakdown at high voltage. The choice of how to power the vehicle so influenced by several factors. Primarily guided by the operator's safety and cost-effective solution possible. Currently, the prototype uses internal power mainly because of safety and low cost compared to other solutions. In order to avoid interference generated by the commutators of engines, power electronics completely isolated by using a second battery, a separate power source. Currently used in the vehicle 12 volt gel battery with a capacity of 45 Ah to power all components except electronics and 12 V lead battery 7 Ah to power electronics. This allows for several hours under water, depending on the intensity of the use of propellers.

3 Control and Communication with the Vehicle

Underwater vehicle control exercised by the operator located on the surface (on land or on the vessel). To control the ROV-I is used with a PC running dedicated software (a set of a computer and dedicated software, will be hereinafter referred to as "operator panel"). The operator panel of the underwater vehicle is connected by a flexible rubber cable with a length of 40 m. This cable is transmitted control signals (2 wires) and the signal from a video camera (2 cores). In both cases, the information is transmitted in the standard RS-485. This is a standard for serial data transmission, more resistant to interference than the RS-232 interface formerly used in personal computers, but still often used by electronic engineers. A very important integral part of an unmanned underwater vehicle is the operator panel. It provides an interface between the user and the ROV-I, which allows mutual transfer of information. Operator for an underwater vehicle:

- Launch the individual components of the vehicle,
- configuration and calibration of sensors and controllers,
- launch of regulators,
- maneuvering vehicle,
- support for additional devices (eg camera).

From an underwater vehicle to the operator:

- Send confirmation execute commands operator,
- send the indications of sensors and state-board equipment,
- alerting of faults,
- sending an image from the camera.

Application used to control ROV-I was prepared so that the use of it was like the most intuitive. The main application window is shown in Figure 17 It consists of two parts: the left side of the mats are different types of indicators (including the strength and direction of work on each engine, compass, przechyomierz, depth indicator), while the right side contains a number of switching cards, which are buttons , sliders, input fields from the keyboard, used to control the vehicle. Application developed in C + + using Qt 4.5.1 [51] (This library is used to create graphical user interface software).

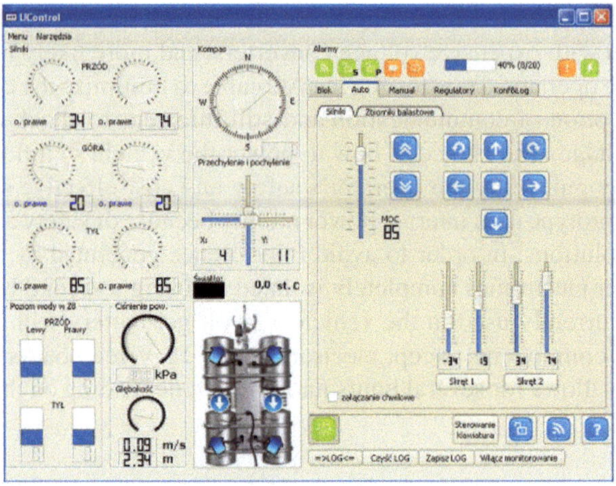

Fig. 17. Dedicated PC application

3.1 Elements of the Control System

The central element of the control system is the main driver. To him get the information from all sensors mounted on the vehicle and there are algorithms for controlling the operation of actuators. Figure 18 shows graphically the way the signals between the individual elements of the control system. Figure 19 includes only elements important for the regulation system, it is not placed on it such as: humidity sensor. There is however a device such as a video camera and TV monitor. They have been involved in the control, because in some cases, the role of feedback and control signals to develop a man has to perform - that supports the vehicle operator. The operator also exercises supreme control over the look of the entire system (through observation sensors on the panel display and video on the monitor) and in the event of an adverse situation with which they can not cope with built-in algorithms, takes appropriate action to restore the balance of the vehicle. Figure 19 shows the path control signals and regulations. Visible is the main

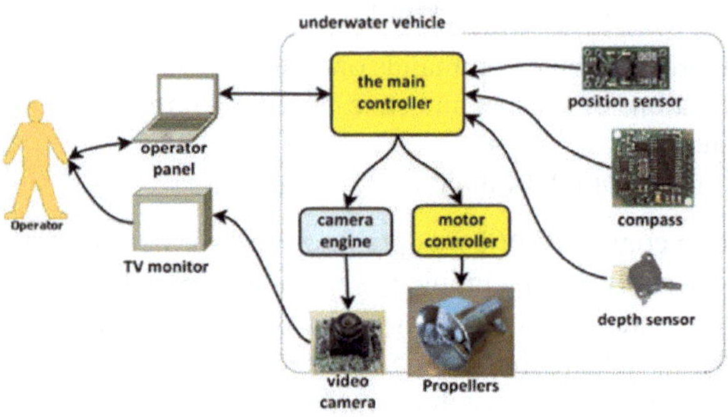

Fig. 18. Elements of the control system

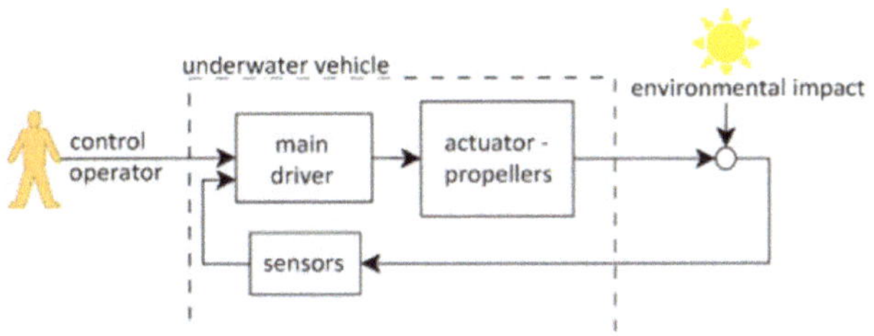

Fig. 19. The control system - block diagram

Fig. 20. The Central Unit

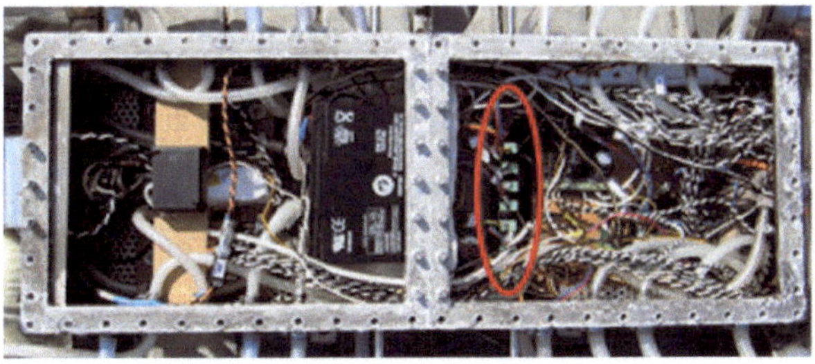

Fig. 21. Drive motor controller

feedback loop which includes the key elements of the control system. The central body will be called the main driver of the vehicle based on the ATmega128 chip. This driver is inside the rear of the hull of the vehicle with a bundle of cables and wires at the place marked on Figure 20 His job is to take any action automatic control of all vehicle peripherals, communication with all modules and sensors, and the operator panel. The main processor or microcontroller unit will be called the ATmega128 microcontroller in the central unit. Motor controller will be called Direct Control module propulsors based on ATmega168 microprocessor that accepts commands from the CPU. It is located in the rear of the hull vertically in the slot next to the main battery, its location is marked in Figure 21.

4 Summary

The aim of further research and development is a systematic improvement of the individual, existing elements of the vehicle and enrich it with additional, new capabilities, broadening the scope of its applications. In the near future, it may be, for example: mount imaging sonar to the bottom of the water and supporting the search for buried objects. Note that the structure is built and experimental prototype version. Although the current version is only a prototype, its functionality allows you to perform certain work, such as: the penetration of water reservoirs and hydrotechnical construction inspection. The solutions in the area of the drive, have been tested in the field of enabling seamless and maintenance free operation of propulsors and control electronics. One of the more useful of modernization, relating to motor control, propulsion engines will be equipped with encoders. This will allow you to control, with high accuracy speed control motors. At present, the operator only has the ability to set the engine power. Mounting encoders will implement in the software driver controls engine speed control screw market, and thus, it will be possible to maintain the preset speed and controlling the performance of propellers. Another of the improvements that could be introduced in future is the replacement of the horizontal propellers propellers on the original bolts are made of plastic. This change should improve performance, such as maximum speed and acceleration of the vehicle in a horizontal motion. This modification, however, is not necessary, because the dynamics of that vehicle has now, is quite sufficient. There are no problems with lack of power and speed and acceleration achieved enable you to perform underwater work, even with moderate strength of the waves. However, power supply, which could be achieved in this simple way, it could prove useful in the future, when the increase in weight of the vehicle when mounted arm and sonar.

References

1. Salgado-Jimenez, T., Gonzalez-Lopez, J.L., Pedraza-Ortega, J.C., García-Valdovinos, L.G., Martínez-Soto, L.F., Resendiz-Gonzalez, P.A.: Design of ROVs for the Mexican power and oil industries. In: 1st International Conference on Applied Robotics for the Power Industry, pp. 1–8 (2010)
2. Plotnik, A.M., Rock, S.M.: Visual Servoing of an ROV for Servicing of Tethered Ocean Moorings. In: OCEANS, pp. 1–6 (2006)
3. Sakagami, N., Shibata, M., Hashizume, H., Hagiwara, Y., Ishimaru, K., Ueda, T., Saitou, T., Fujita, K., Kawamura, S., Inoue, T., Onishi, H., Murakami, S.: Development of a human-sized ROV with dual-arm. In: OCEANS, pp. 1–6 (2010)
4. Inoue, T., Katsui, T., Tahara, J., Itoh, K., Yoshida, H., Ishibashi, S., Takagi, K.: Experimental research on movability characteristic of crawler driven ROV. In: OCEANS, pp. 1–6 (2008)
5. Inoue, T., Katsui, T., Murakami, H., Tahara, J., Sinbori, T.: Preliminary research on the thruster assisted crawler system for a deep sea ROV. In: OCEANS - EUROPE, pp. 1–5 (2009)
6. Shiosawa, T., Takagi, K., Inoue, T.: Experimental and theoretical study on the motion of ROV with crawler system. In: OCEANS, pp. 1–5 (2010)
7. Bird, L.E., Graves, D., Massion, G., Chaffey, M., Hamilton, A., Keaten, R.: Establishing a Benthic Cabled Observatory with ROV Based Cable Deployment. In: OCEANS, pp. 1–6 (2006)

8. Bird, L.E.: ROV based tool sled for the placement of fiber optic cable between benthic instrument nodes. In: OCEANS, vol. 4, pp. 2050–2053 (2002)
9. Salgado-Jimenez, T., Gonzalez-Lopez, J.L., Martinez-Soto, L.F., Olguin-Lopez, E., Resendiz-Gonzalez, P.A., Bandala-Sanchez, M.: Deep water ROV design for the Mexican oil industry. In: OCEANS, pp. 1–6 (2010)
10. Murashima, T., Aoki, T., Tsukioka, S., Hyakudome, T., Yoshida, H., Nakajoh, H., Ishibashi, S., Sasamoto, R.: Thin cable system for ROV and AUV in JAMSTEC. In: OCEANS, vol. 5, pp. 2695–2700 (2003)
11. NORSOK standard, Norwegian Technology Centre 2003(2)
12. Wernli, R.L.: Low cost UUV's for military applications: Is the technology ready?. Space and Naval Warfare Systems Center San Diego, California, U.S.A.
13. Kubiak, K., Olejnik, A.: Koncepcja wykorzystania bezzaogowego pojazdu podwodnego do wspomagania nurków w ochronie obiektów hydrotechnicznych". Polish Hyperbaric Research 9(1), 50–54 (2004)
14. Howse, J., The, R.O.V.: Pontus - A winning design In: Canadian Conference on Electrical and Computer Engineering, pp. 346–349 (2009)
15. Brundage, H.M., Cooney, L., Huo, E., Lichter, H., Oyebode, O., Sinha, P., Stanway, M.J., Stefanov-Wagner, T., Stiehl, K., Walker, D.: Design of an ROV to Compete in the 5th Annual MATE ROV Competition and Beyond. In: OCEANS, pp. 1–5 (2006)
16. Miller, D.P.: Design of a Small, Cheap UUV for Under-Ship Inspection and Salvage. In: Proceedings of the Symposium on Autonomous Underwater Vehicle Technology, pp. 18–20 (1996)
17. Koczko, S., Kulawiak, P., Iwankiewicz, K., Matejski, M., Graczyk, T.: Warunki pozyskiwania danych uczcych dla stworzenia neuronowego modelu ruchu pojazdu podwodnego KRAB w paszczynie pionowe, II Konferencja Studenckich K Naukowych KSKN, Szczecin (2006)
18. Olejnik, A.: Wizyjna identyfikacja zatopionych obiektw za pomoc pojazdu typu ROV na przykadzie wraku jednostki GRAF ZEPPELIN. Polish Hyperbaric Research 19(2), 17–32 (2007)
19. Solvang, B., Deng, Z., Lien, T.K.: A methodological framework for developing ROV-manipulator systems for underwater unmanned intervention. In: OCEANS, vol. 2, pp. 1085–1091 (2001)
20. Kawaguchi, K., Kaneko, S., Nishida, T., Komine, T.: Cable laying ROV for Real-time seafloor observatory network construction. In: OCEANS - EUROPE, pp. 1–4 (2009)
21. Hildebrandt, M., Christensen, L., Kerdels, J., Albiez, J., Kirchner, F.: Realtime motion compensation for ROV-based tele-operated underwater manipulators. In: OCEANS - EUROPE, pp. 1–6 (2009)
22. Gomes, R.M.F., Martins, A., Sousa, A., Sousa, J.B., Fraga, S.L., Pereira, F.L.: A new ROV design: issues on low drag and mechanical symmetry. In: OCEANS - EUROPE, vol. 2, pp. 957–962 (2005)
23. Shaheen, N., Waterworth, G.: ROV serviceable science node for cabled ocean observatories. In: OCEANS, vol. 3, pp. 2250–2256 (2005)
24. Nowak, B.M., Whitney, T., Ackley, S.F.: Analysis of ROV video imagery for krill identification and counting under Antarctic sea ice. In: Autonomous Underwater Vehicles, pp. 1–9 (2008)
25. Grzadziel, A., Olejnik, A., Szymaniuk, R.: Badania identyfikacyjne oraz inspekcja wraku"Graf Zeppelin". Polish Hyperbaric Research 21(4), 17–30 (2007)
26. Marchand, E., Chaumette, F., Spindler, F., Perrier, M.: Controlling an uninstrumented ROV manipulator by visual servoing. In: OCEANS, vol. 2, pp. 1047–1053 (2001)
27. Shahriar, N., Pezhman, F., An, R.O.V.: Stereovision System for Ship-Hull Inspection. IEEE Journal of Oceanic Engineering 31(3), 551–564 (2006)

28. Massimo, C.: Laser-Triangulation Optical-Correlation Sensor for ROV Slow Motion Estimation. IEEE Journal of Oceanic Engineering 31(3), 711–727 (2006)
29. Rife, J., Rock, S.M.: Segmentation methods for visual tracking of deep-ocean jellyfish using a conventional camera. IEEE Journal of Oceanic Engineering 28(4), 595–608 (2003)
30. "CREE XLamp MC-E LED datasheet", Cree (2008)
31. Jedrasiak, K., Daniec, K., Nawrat, A.: The Low Cost Micro Inertial Measurement Unit. In: Industrial Electronics and Applications (ICIEA), pp. 403–408 (2013)
32. Bereska, D., Jędrasiak, K., Daniec, K., Nawrat, A.: Gyro-Stabilized Platform for Multispectral Image Acquisition. In: Nawrat, A., Kuś, Z. (eds.) Vision Based Systems for UAV Applications. SCI, vol. 481, pp. 115–121. Springer, Heidelberg (2013)
33. Daniec, K., Iwaneczko, P., Jędrasiak, K., Nawrat, A.: Prototyping the Autonomous Flight Algorithms Using the Prepar3D Simulator. In: Nawrat, A., Kuś, Z. (eds.) Vision Based Systems for UAV Applications. SCI, vol. 481, pp. 219–232. Springer, Heidelberg (2013)
34. Babiarz, A., Bieda, R., Jędrasiak, K., Nawrat, A.: Machine Vision in Autonomous Systems of Detection and Location of Objects in Digital Images. In: Nawrat, A., Kuś, Z. (eds.) Vision Based Systems for UAV Applications. SCI, vol. 481, pp. 3–25. Springer, Heidelberg (2013)
35. Bereska, D., Daniec, K., Fraś, S., Jędrasiak, K., Malinowski, M., Nawrat, A.: System for Multi-Axial Mechanical Stabilization of Digital Camera. In: Nawrat, A., Kuś, Z. (eds.) Vision Based Systems for UAV Applications. SCI, vol. 481, pp. 177–189. Springer, Heidelberg (2013)
36. Babiarz, A., Bieda, R., Jędrasiak, K., Nawrat, A.: Machine Vision in Autonomous Systems of Detection and Location of Objects in Digital Images. In: Nawrat, A., Kuś, Z. (eds.) Vision Based Systems for UAV Applications. SCI, vol. 481, pp. 3–25. Springer, Heidelberg (2013)
37. Mezyk, A., Switonski, E., Kciuk, S., Klein, W.: Modelling and investigation of dynamic parameters of tracked vehicles. Mechanics and Mechanical Engineering 15(4), 115–130 (2011)
38. Martynowicz, P., Kciuk, S., Mezyk, A.: Rotary-shock-absorber with magnetorheological valves. Advanced Materials Research 628, 505–511 (2013)
39. Kciuk, S., Martynowicz, P.: Special application magnetorheological valve numerical and experimental analysis, Diffusion and Defect Data Pt. B Solid State Phenomena 177, 102–115 (2011)
40. Barnat, W., Panowicz, R., Niezgoda, T.: Numerical and Experimental Comparison of Combined Multilayer Protective Panels. Acta Mechanica et Automatica 6(1), 13–16 (2012)
41. Barnat, W., Panowicz, R., Niezgoda, T., Dybcio, P.: Numerical analysis of IED detonation effect on steel plate. Acta Mechanica et Automatica 6(1), 10–12 (2012)
42. Panowicz, R., Barnat, W., Niezgoda, T., Szymanczyk, L., Grzymkowski, J.: Numerical-Experimental Investigation of Squared-based Metal Pyramids Loaded With a Blast Wave From a Small Explosives Charge. Acta Mechanica et Automatica 6(1), 49–52 (2012)
43. Barnat, W., Niezgoda, T., Panowicz, R., Sybilski, K.: The Influence of Conical Composite Filling on Energy Absorbtion During The Progressive Fracture Process. WIT Transactions on Modelling and Simulation 51, 625–633 (2011)
44. Niezgoda, T., Panowicz, R., Sybilski, K., Barnat, W.: Numerical Analysis of Missile Impact Being Shot By Rocket Propelled Grenades with Rod Armour. WIT Transactions on Modelling and Simulation 51, 625–633 (2011)
45. Barnat, W., Dziewulski, P., Niezgoda, T., Panowicz, R.: Application of Composites to Impact Energy Absorbtion. Computational Materials Science 50(4), 1233–1237 (2011)
46. Niezgoda, T., Barnat, W.: Analysis of Protective Structures Made of Various Composite Materials Subjucted to Imapct. In: Materials Science and Engineering A, vol. 483-484(1-2 C), pp. 705–707 (2008)

47. http://www.ifremer.fr
48. http://www.rov.org/
49. http://www.seaeye.com/products.html
50. http://www.seabotix.com/products/products.html
51. Qt Online Reference Documentation, http://doc.trolltech.com/

Selected Issues of the Underwater Explosion on the Special Vehicles

Wiesław Barnat[1] and Aleksander M. Nawrat[2]

[1] Department of Mechanics and Applied Computer Science, Military University of Technology, Gen. S Kaliskiego St. 2, 00-908 Warsaw, Poland
wbarnat@wat.edu.pl
[2] Institute of Automatic Control, Silesian University of Technology, Akademicka 16, 44-100 Gliwice, Poland

Abstract. The purpose of this article is to present the influence of the environment on the propagation of shock wave and the dynamic response of the plate and the special vehicle load by shock wave. In the course of research in this field experimental study was performed. Experimental study concerns the test plate load by shock wave, formed after the detonation of an explosive device with equivalent weighing of TNT equal to 1 kg. During the numerical investigations has been tested environmental impacts on the dynamic response disc. The considered environments of explosions were air and water. Model of the phenomenon has been prepared using commercial software MSC. DYTRAN.

Keywords: under water explosion, numerical model, validation, special vehicle.

1 Introduction

All kinds of numerical studies about structures which are loaded by the blast wave, are burdened with many errors. In the case of loss of the dynamic stability protective structures distinguish global loss of stability of the structure and the local (in certain areas). This loss is due to the short pulse pressure with a high value. In the case of numerical analysis, there are a few approaches to dynamic phenomenon. The first approach is related to the vibration of the structure. This issue was developed by many scholars [13, 7]. Modern science considers the approach to the numerical analysis of type implicitly and explicitly. These approaches are associated with the duration of impulse with the frequency of the object vibrations [9, 10].

When the pulse duration is shorter than the period of natural object vibrations the problem is solved by explicit method.

The main problem in numerical analysis is the appropriate choice of material properties and load parameters. In the case, where load parameters are considered, often analytical methods are used [12]. These methods are based on many complex formulas and experiments [4].

A.M. Nawrat (ed.), *Innovative Control Systems for Tracked Vehicle Platforms*,
Studies in Systems, Decision and Control 2,
DOI: 10.1007/978-3-319-04624-2_17,

2 Influence Various Environments for the Object of Research

2.1 Object of Research

During research were carried out one experiment and two numerical analysis. Summary of the tested objects shown in Table 1.

Table 1. Summary of the tested objects

Test	Charge	Distance	Environment	Density of medium kg/m3	Experiment	Numerical analysis
1	1 kg	0.4 m	air	1,2829	yes	yes
2			water	1000	no	yes

All experimental studies were carried out during the realization of the research work, which the author was the head of [3]. The test ground consisted of a frame and a steel plate with a thickness of 5 mm. Elements of test ground were made of steel S235JR2 (according to the European standard EN 10025:1990).

Tensile test for steel S235JR2 were carried out by the Department of Applied Mechanics MUT. Exemplary test results is shown in Fig.1.

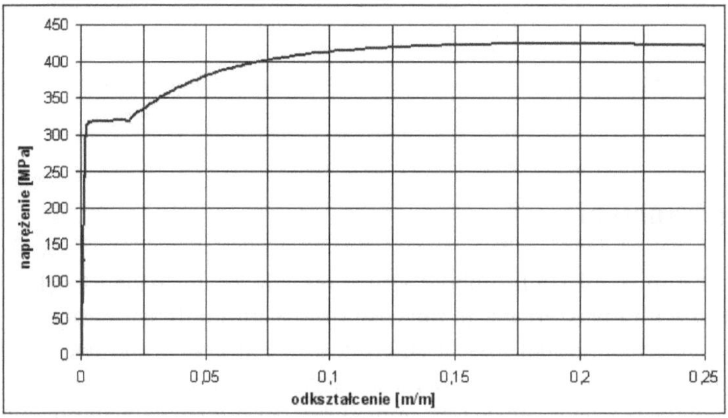

Fig. 1. The plot of the stress-strain tensile tests obtained from the plate material S235JR2 (according to the European standard EN 10025:1990)

Based on the analysis of the test results, the following material parameters are adopted for the numerical model:

- Young's modulus E = 215 GPa;
- Yield strength Re = 320 MPa;
- Tangent modulus H = 700 MPa;
- Tensile strength Rm = 430 MPa.
- Poisson ratio ν = 0.31

By welding test plate to frame allowed to avoid the local loss of stability. Beside to the strong bending, there also appeared the state of tension.

The base frame was made of steel profiles. Profiles were made of steel S235JR2 and they had square shape 120mm x 120mm and a wall thickness of 5 mm. (Fig. 2) Profiles were respectively cut and welded at the corners.

Entire test plate had dimensions 0.76 m x 0.76 m. Open test zone had dimensions 0.5 m x 0.5 m (after excluding test frame). Open test zone can be deformed freely.

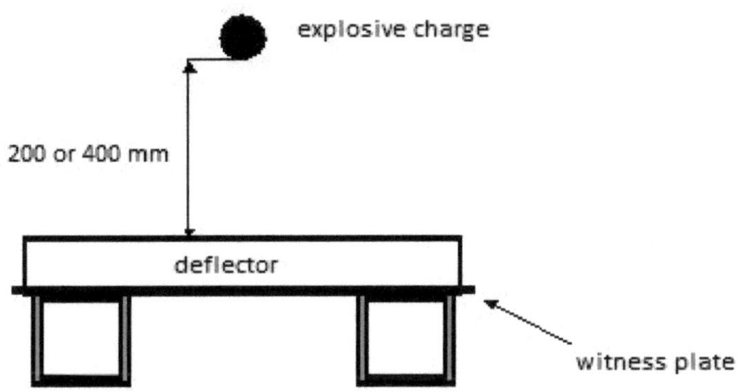

Fig. 2. The scheme of the test zone

The test-plates (Fig. 2) were loaded by pressure wave (impulse) which came from detonations of explosive charge. Charge had equivalent of 1 kg of TNT. The load was placed centrally above the plate at a distance of 0.4 m from the top surface of the plate.

2.2 The Results of Experimental Studies

Experimental studies performed in conditions testing ground. The following test zone was made based on numerical simulation and analysis of the available literature. The results for the alone plate were presented in the work which author participated [5].

During development and implementation of the test ground was taken into account the ability to change some of initial parameters of the experiment, to be able to carry out tests for different sizes of explosives. A general view of the position shown in Fig. 3. In additional studies, additional steel plate was used as a non-deformable base to leveling land effect on the permanent deformation of test plate. Additional studies have shown superfluity abovementioned plate - there was no difference in the final deformation of the tested systems.

The source of the pressure pulse was a big load of TNT placed at a distance of 0.4 m from the object. Explosive was freely placed on the test object. Induced detonation pulse load was characterized by short duration and high amplitude. The duration of this pulse pressure is an order of magnitude or even two orders shorter than the time of impact, and is a few tens of milliseconds.

Fig. 3. A general view of the position to test panels loaded by large explosives

a)

b)

Fig. 4. Appearance experimental position after the experiment - Test 1: a) the system after the explosion, b) deformation map obtained by measuring the 3D scanner

As a result of the experiment were obtained permanent deformation test plate. Size deformation was initially measured using a straight edge and caliper. Deformed base plate is shown in Figure 4. Figure 4a shows a photograph of test plate after the experimental study and 4.b shows a map of permanent deformation obtained using a 3D scanner. As a result of scanner measurements, it was found that the permanent deformation of the plate was 16.5 mm. This value was measured in the symmetry center of plate, at the site of the largest deformation.

2.3 Numerical Research

The Formulation of the Physical Model

Analysis of the impact of pulse pressure on the structure were carried out by finite element method (FEM) with the explicit scheme of integration of equations of motion implemented in the MSC DYTRAN [9]. In those algorithms the equilibrium equation of discrete model (1) is solved by direct integration of the time. These equations occur in the conjugate form, which binds with high numerical cost of their solutions and high demand for memory in calculation.

The basic equation in explicit approach, omitted damping due to the short duration of the phenomenon, is as follows:

$$[M]\{\ddot{x}\}_n = \{F\}_n - \{F_{\mathrm{int}}\}_n, \qquad (1)$$

where: [M] - mass matrix, {F} - matrix of the external forces, {Fint} - matrix of internal forces, n – step

Assuming that the mass matrix occurs in diagonal form, the equation of motion can be solved by an explicit Euler method. In this case, nodal acceleration vector $\{\ddot{x}\}$ can be determined from the formula [8]:

$$\{\ddot{x}\}_n = [M]^{-1}(\{F\}_n - \{F_{\mathrm{int}}\})_n, \qquad (2)$$

In equation (2) can also be taken into account the damping, but also just as a diagonal matrix. An additional advantage of this approach, is the uncoupling of differential equations of motion, which makes them a separate resolution.

Velocity $\{\dot{x}\}_{n+1/2}$ and displacement $\{x\}_{n+1}$ vectors in successive time steps are obtained by integrating over time with use of the central finite difference method [3]:

$$\begin{aligned}\{\dot{x}\}_{n+1/2} &= \{\dot{x}\}_{n-1/2} + \{x\}_n \Delta t_n, \\ \{x\}_{n+1} &= \{x\}_n + \{\dot{x}\}_{n+1/2} \Delta t_{n+1/2}.\end{aligned} \qquad (3)$$

Unfortunately, this method is conditionally stable. In order to ensure the stability, there should be reduced length of the integration step:

$$\Delta T \le 2/\omega_{\max}, \qquad (4)$$

where $\omega_{\max}$ is first the highest frequency of not-damped vibration discrete model. It means that the time step must be shorter than the propagation time of the wave through the smallest element occurring in the calculation model. The consequence of this is that the more accurate is the model (small elements), the integration step is shorter and computation time is longer. At the same time the problem of length selection vanishes.

Mechanical issues solved by computer modeling and simulations require the adoption of adequate description of the behavior of the materials, which were used. This description must take into account their specific characteristics, such as physical state, plasticity, brittleness, hardness.

Description of the Numerical Model

Numerical analysis was carried out for the structure which was loaded by pressure wave created by explosion of explosive charge, which was placed centrally above the test object (according to table 1). The pressure wave caused by detonation (approximately simulated detonation in point) was spread in an area of cube shape with assigned the appropriate boundary conditions. The theoretical solution of propagation of strong discontinuities with spherical shape initiated from a point source exists in the form of analytical similarities equations formulated by Taylor. After transformation can be written as [4]:

$$p(r) = 0.155E_0r^{-3}, \tag{5}$$

where: E0 - initial internal energy, r - current radius of the sphere.

This allows for the computer simulation of shock wave propagation process by giving appropriate initial conditions (density, energy, pressure) to some selected elements of the Euler domain and solution laws of conservation of mass, momentum and energy. Typical values for the explosive substances are: density - 1600 kg/m3, and a specific internal energy - 4.2 MJ/kg. The space in which the shock wave spread, was modeled by using Euler type elements Hex 8 characterized by the ideal gas properties of $\gamma = 1.4$ and density corresponding to the density of atmospheric air at standard conditions (ρ = 1.2829 kg/m3). The calculations take into account the changes caused by deformation of the structure, because of that the structure was modeled by using the elements of the Lagrange type Shell Quad 4. Those elements have been given the following mechanical properties: E = 2.15·109 MPa, ν = 0.31. To describe the behavior of the steel, piecewise linear plastic material model DYMAT 24 was used. The maximum strain failure criterion was adopted [9].

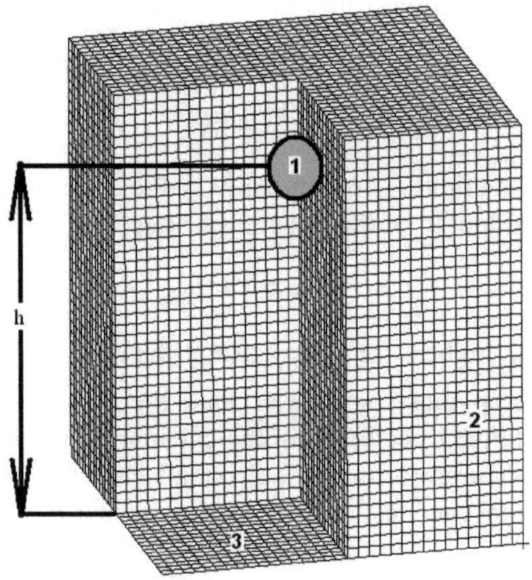

Fig. 5. Diagram of numerical model tested structure: 1 - an explosive, 2 –Eulerian domain, 3 - tested structure

Developing a model of structure was preceded by additional laboratory testing of mechanical properties of steel armor. Laboratory research was carried out in the Department of Mechanics and Applied Computation Science on Military University of Technology.

A general view of the numerical model of the test structure is shown in Figure 5. Ground was described by material model Mie-Gruneisen with the following parameters: $\gamma = 2$ and a density of 2300 kg/m3. The main parameter describing water was density equal to 1000 kg/m3.

The Results of Numerical Analysis

Numerical model was loaded In the same way as it was happened In experiment. As a result of numerical analyzes were obtained among others Pressure maps and displacements charts, accelerations and speeds for characteristic points on structure especially for center test plate. Additionally were studied values of strain energy and hourglass energy (responsible for the numerical errors). It is known that, in theory increase the medium increases the pressure incident on the barrier. This phenomenon is

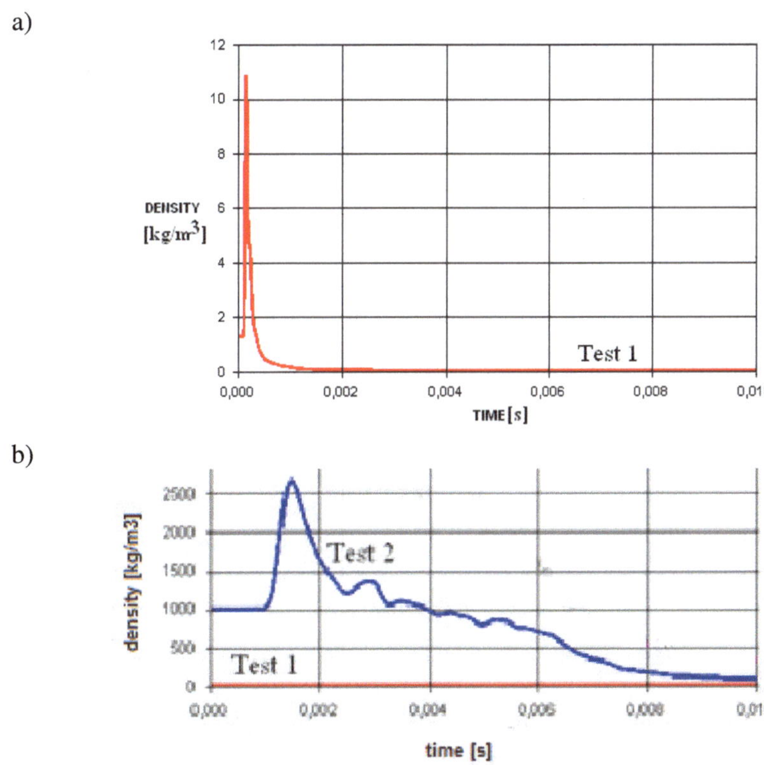

Fig. 6. Change the value of medium density at a central point of the plate a) air, b) summary air (Test1), water (Test2)

caused by the treatment medium as a continuous. This results in that the pressure pulse acting on the medium in front of him causing compaction of the medium. In Fig. 6 shows the change of medium density in the central point of the plate. In the case of air by the action of the pressure pulse density value of the medium increased nine fold. In the case of water value has increased twice. Treatment of dense medium (water) as a continuous medium resulted in behave of concentrated medium as a stamp. This phenomenon is widely known and used for explosive embossing of metal [6].

An interesting are graphs obtained in a the numeric way for individual tests. In the case of displacement the largest sustained displacement (Fig. 7) was observed for the central node for test 2 (water), and it was equal to 0.047 [m]. The smallest permanent displacements was observed for test1 (air) and it was equal to 0.017 [m]. In table presents a comparison between results obtained in the experimental method and numerical method.

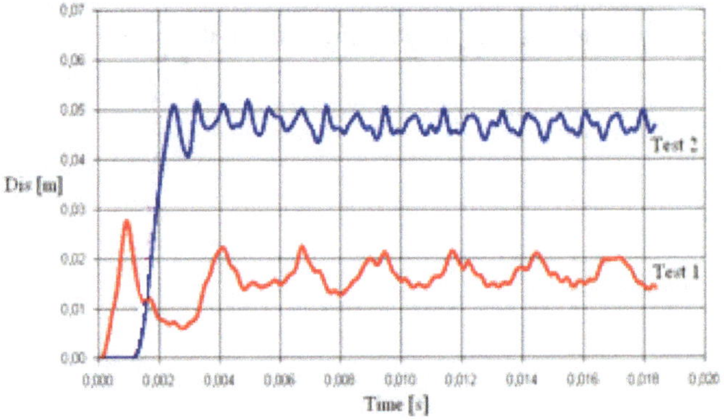

Fig. 7. The middle node displacement plots for individual tests

Table 2. Summary of displacement for each test

Test	Numerical results – maximal displacement [m]	Numerical results - permanent displacement [m]	Experimental results [m]	Difference [%]
1	0.0275	0.017	0.0165	2.9
2	0.051	0.047		

As in the case values permanent displacement of middle node, represent the results for acceleration, velocity and energy. Graphs above mentioned characteristics are shown in Figures 8-10.

The greatest value of the speed for the middle node noted for test 3 - 84 [m/s]. For tests 2 and 1 speed values were respectively 63 and 53.7 [m/s].

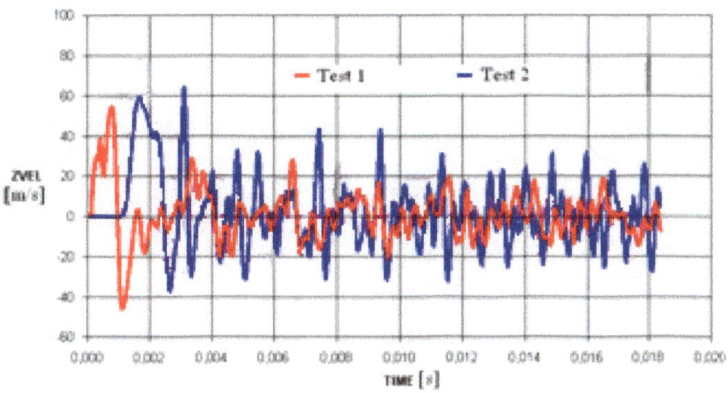

Fig. 8. Chart of speed for the node located on the upper wall of the object (in the plains of symmetry)

The greatest values of the strain energy and the hourglass were read for test 3 and amounted to 73 kJ. Comparison of all additional physical quantities are shown in Table 3.

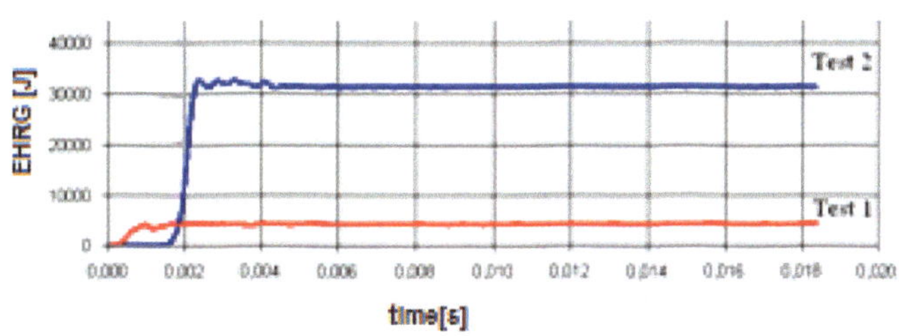

Fig. 9. The strain energy for each test

Table 3. Summary of values: speed, strain energy and hourglass energy for each test

Test	Speed [m/s]	Acceleration [m/s2]	Strain energy [J]	Hourglass energy [J]
1	53.7	347 000	5000	1
2	63	600 000	30 100	2

Similarly as in case of the graphs additional physical quantities obtained from the numerical analysis such as maps: displacement, strain and equivalent tensile stress were shown in Figure 11-13. All maps were characterized by symmetry (which proves the correctness of modeling issue).

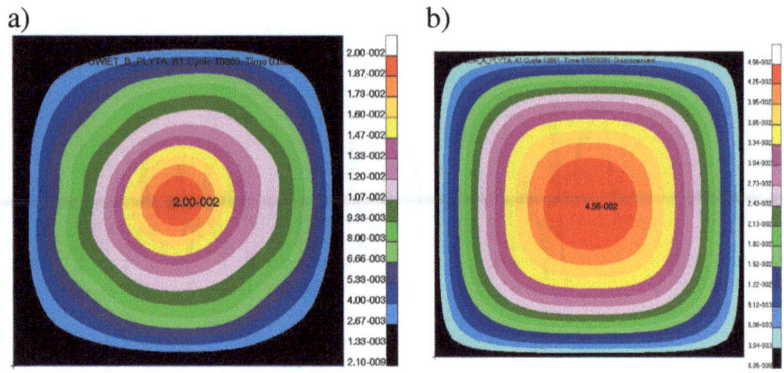

Fig. 10. Displacement map: a) Test 1 - air, b) test 2 - water

The values of displacement permanent have been described previously. An interesting strain of test plate are shown in Figure 12. As in the case of displacement, the maximum values of permanent deformation for each test are respectively equal to: 2.28 E-2, 1.05 E-1.

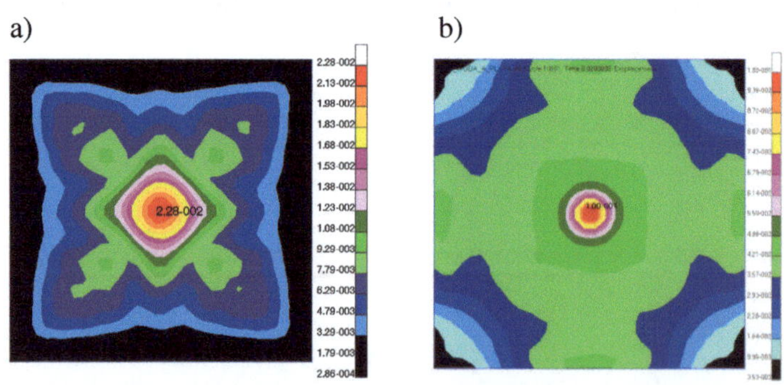

Fig. 11. Strain map: a) Test 1 - air, b) test 2 - water

As in the case of displacements and permanent strain, maximum stress values for each test are respectively equal to: 6.9E8 Pa, 7.35E8 Pa. The values of equivalent tensile stress for each test are shown in Figure 13. They point to the correctness physical quantities previously obtained for test plates. Such high values of equivalent tensile stresses for steel were due to dynamic reinforcement of the material.

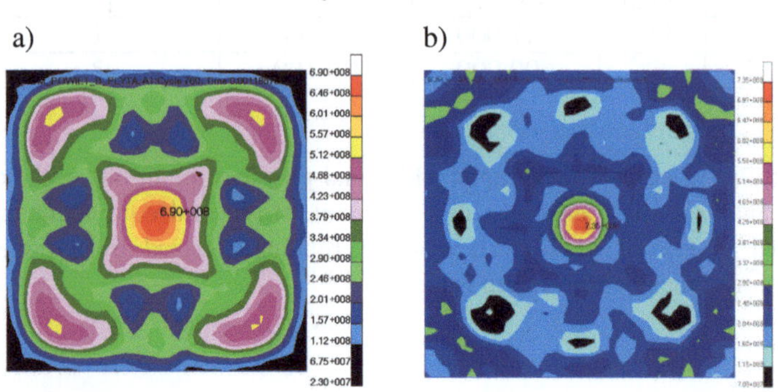

Fig. 12. Map of equivalent tensile stresses: a) Test 1 - air, b) test 2 - water

2.4 Summary

This paper presents the results of experimental and numerical analysis of the dynamic response of plate load by pressure pulse.

Comparing the results of numerical analysis, the charts are the most reliable. In the case of read values especially mechanical quantities from maps there is a concern of poor approximation of results.

Research especially experimental of test plates are extremely important because of the correctness the results. The above example can be used to validate numerical models in the context of the explosive and the constitutive model describing the barrier.

This issue is quite important from the point of view of the protection of vehicles and military facilities. In the numerical analyzes were used coupling to join Euler domain (describing the air) and Lagrangian domain (describing structures).

In case of numerical analysis the gas had parameters of the air, water and soil.

All kind of material data have been selected on the basis of experimental studies and available literature. Plates were described using piecewise linear plastic material model with dynamic strengthening.

During the tests of test plate response for a variety size of charges were used innovative test stand. The aim of the study was to validate numerical models in the aspect of constitutive equations describing the model material and the choice model of explosion and medium in which pressure pulse was spread.

3 Modelling the Under Water Explosion under the Caterpillar Vehicle

A lot of publications about phenomena of detonation concerned the issues the standard detonation, where the wave was propagated in the air. Those publications included also a issues connected with the influence of ground on the character the rising explosion. Sparse phenomenon is the influence the underwater explosion on the military vehicles [1-5]. In this case, there is a large threat of supporting construction of vehicle, because the density of medium is different, the value is 1000 kg/m3, for air the density is 1,2 kg/m3. Due to the denser medium the pressure wave reaches faster to construction with speed similar to sound speed in the medium (the sound speed for water is 1500 m/s and for air it is 340 m/s). Due to denser medium the value of pressure is larger than for explosion in the air.

The sample of the pressure propagation in the water shown in Fig.14.

In fig. 14 presents the way of propagation of the pressure originating from an underwater explosion depending on time in the selected point. When the shock wave arrived the pressure increase in the moment to maximum and decrease with lapse of time. Due to the character of phenomena the gas bubbles which are the products from detonation disperse slower in comparison with blast wave in the medium. The large pressure of the explosive gas causes initially the fast expansion of bubbles.

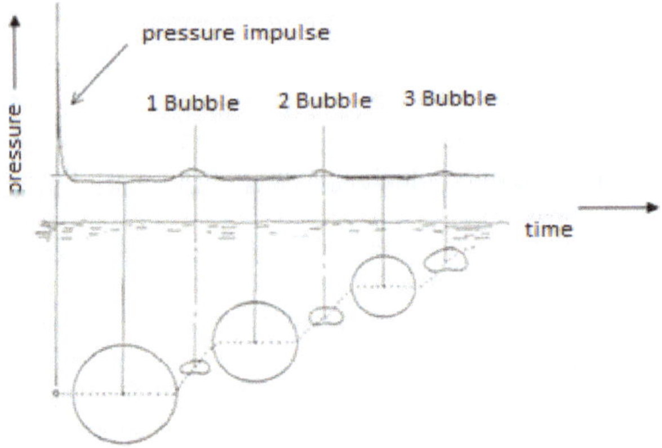

Fig. 13. The sample of propagation the air bubbles creating in underwater explosion [6]

The mathematically the formulation of problem in the case of one dimensional for ideal gas is described by equations expressing the fundamental laws of conservation: mass, momentum and energy.

4 General Description of Numerical Model

The numerical analyses was conducted for an example of numerical model of shell the vehicle placed in Euler environment including the water - underwater explosion. The general view of the numerical model of vehicle and the cross-section of the entire system is shown in Fig. 15.

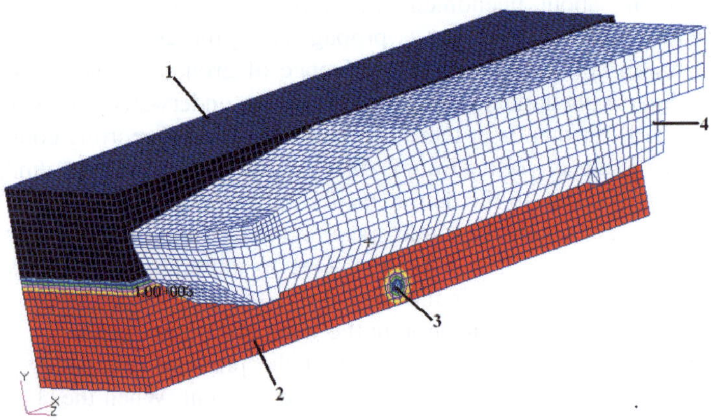

Fig. 14. The scheme of numerical model. 1- air, 2 - water, 3 - explosive charge, 4- vehicle.

The pressure wave originated by detonation (simulated as point detonation) disperse in the area of the cubic shape with suitable boundary conditions corresponding to aquatic environment. The theoretical solution of propagation of strong discontinuities with spherical shape initiated from a point source exists in the form of analytical similarities equations formulated by Taylor. After transformation can be written as [3-6]:

$$p(r) = 0.155E_0r^{-3}, \tag{6}$$

where: E0 - initial internal energy, r - current radius of the sphere.

It allows to computer simulation of process the propagation of the shock wave by suitable initial conditions (density, energy, pressure) for selected elements from Euler domain and solution the laws of conservation: mass, momentum and energy. Typical values for the explosive substances are: density - 1600 kg/m3, and a specific internal energy - 4.2 MJ/kg.

In calculations included the changes made by deformation of hull vehicles (construction). The layer, where the shock wave is propagated, was modeled by Euler elements type Hex 8,which are characterized by ideal gas properties with $\gamma = 1{,}4$ and density corresponding to air density in standard conditions ($\rho = 1{,}2829$ kg/m3). The water is described by the same model as an air with following parameters: $\gamma = 2$ and density corresponding to soil density ($\rho = 2{,}35 \cdot 103$ kg/m3). Assumed that under water is the rock, which is modelled as non-deformable body. Lagrange elements type Shell Quad 4 used to modelling the behaviour of steel plates of vehicle. For this elements given following the mechanical properties: $E = 2{,}1 \cdot 109$ MPa, $\nu = 0{,}31$. To description the behaviour of steel used the bilinear model of material elastoplastic DYMAT 24. The maximum strain failure criterion was adopted [9].

In theoretical description of the coupling procedures included Euler formulations, which are treated as subset. These formulations allows on simulations the phenomena in terms of Euler, in which is a lot of fluids mixing together (multiplate model). Necessary is determinations of parameters for more than one material in single element. This action (increase of number of materials) causes the increase the number of material parameters (density, velocity, energy, etc.), it is characterized by increase the effort of the step.

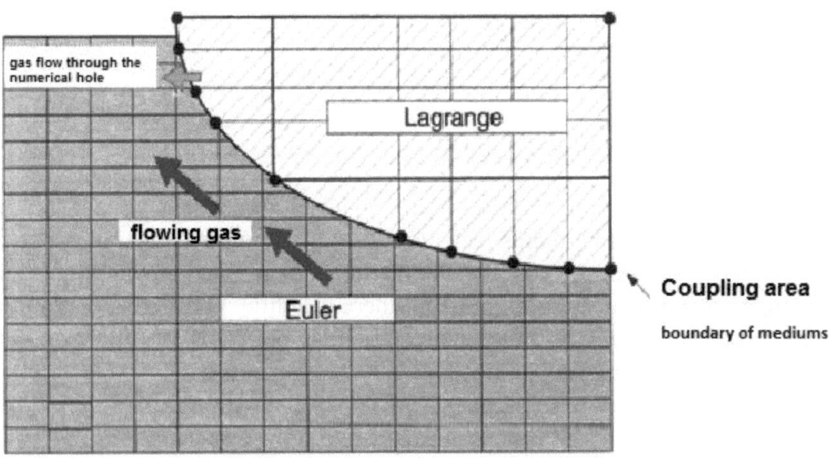

Fig. 15. The sample of general coupling

The range of problems which could be realized with using the general coupling procedure depend on the degree of complexity the algorithms responsible for the smoothing the FEM mesh.

During the realization of Euler step the most of the time occupy the calculations related with transfer the material between adjacent elements but a little part of time is transferred on solution the problem. Used algorithms advection are complicated and time- consuming, but they could eliminate the errors, which appears in algorithms I range of accuracy (false oscillations in the results, failure of stability, reduction of range the parameters, etc.).

In the general coupling process we can distinguish two main steps: carrying out the classics Lagrange steps and execution the checking step. During calculation a checking step the program execute the decision about which nodes should move (displacement the last nodes, displacement insides nodes), conversion of all variables sustained to elements, conversions the value of momentum and the speed update.

Each variable dimension (velocity, density, internal energy, six elements of stress and plastic deformation tensor and kinematic strengthening) should be "transported". Keep in mind that the velocity should be transported separately, because it us referred to nodes, but the other variables - to elements. Moreover, for each element must be respected the laws of conservation of mass, momentum and energy.

Analyzed in the Euler coordinate system the body (gas) moves in the area described by fixed spatial grid. In this case, in that system, described two bodies: air and explosive in the form of stored energy. Physical quantities which described this bodies are defined in appropriate state equations.

EPO elements was described in Lagrange coordinate system. This medium is characterized in that the element mesh covering the test object moves and deforms along the body.

As a simulation results of detonation process obtained a phenomenon a dispersion of pressure the detonation wave by Euler mesh, which meet a moving Lagrange mesh and is a load as a impulse of pressure.

To describe the process of destruction in MES model the Lagrange elements the maximum strain failure criterion was adopted [9].

5 Results Numerical Analyze

As a results the numerical analyzes were obtained the graphs of displacements, accelerations and selected points velocity. In fig. 4 presents the distribution of pressure (fig. 4a) and density (fig. 4b) the mediums (water and air) in initial time t = 0s. Due to the shape of charges (cylindrical) there is a little bit different distribution of pressure and density than was presented in previous points. Is worth noting that the density values of air density to the water density pass smoothly.

a) b)

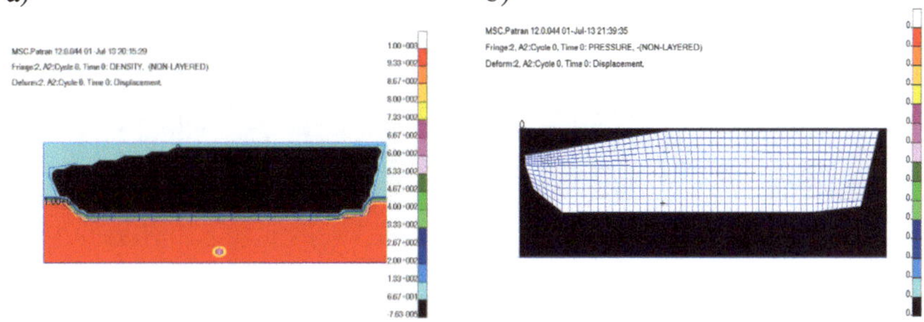

Fig. 16. Distribution the pressure for cross-view of vehicle along plane of symmetry (a) and density of air and water (b) in time moment t=0

In fig. 17 shown the pressure distribution (fig. 17a) and density (fig. 17b) the facilities (water and air) in the time moment t = 0,00018 s. The value of pressure for water is more bigger than for air area. It is caused by differences in the density of mediums. Further pressure and density distribution shown on figures below.

a) b)

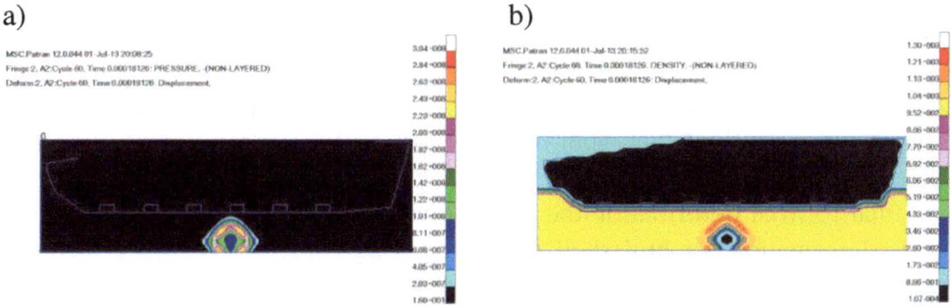

Fig. 17. Distribution the pressure for cross-view of vehicle along plane of symmetry (a) and density of air and water (b) in time moment t = 0,000181 s

a) b)

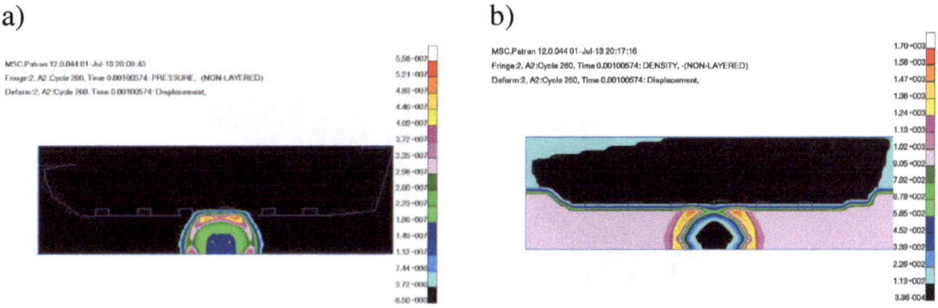

Fig. 18. Distribution the pressure for cross-view of vehicle along plane of symmetry (a) and density of air and water (b) in time moment t = 0,001s

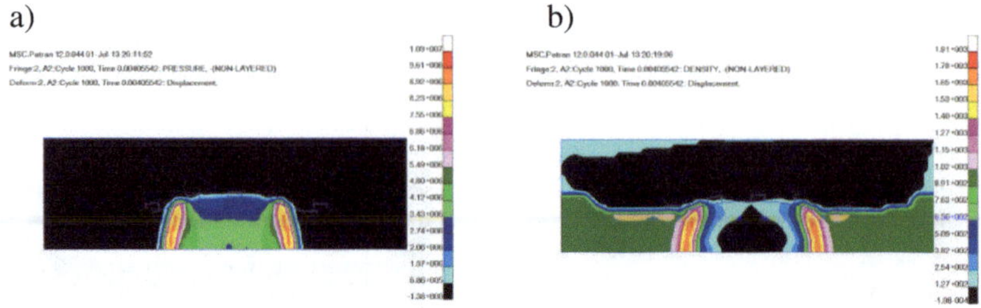

Fig. 19. Distribution the pressure for cross-view of vehicle along plane of symmetry (a) and density of air and water (b) in time moment t = 0,0004 s

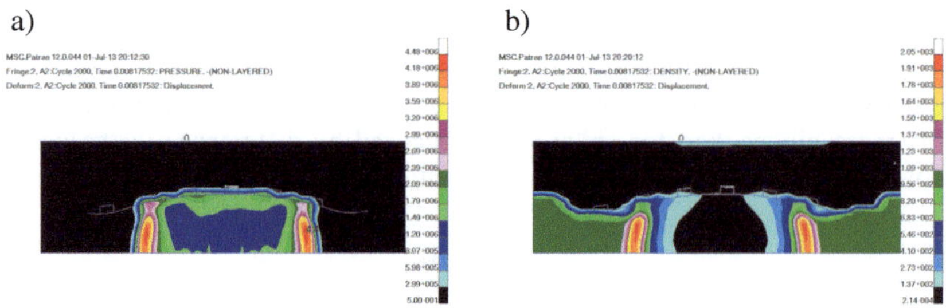

Fig. 20. Distribution the pressure for cross-view of vehicle along plane of symmetry (a) and density of air and water (b) in time moment t = 0,0081s

The final form of deformation the hull shown in Fig. 22 -24. The maximum value of plastic deformation for hull shell loaded by 6 kg charge was 1.46 e-1 for center point, properly for centre the cover of bar torsion tunnel.

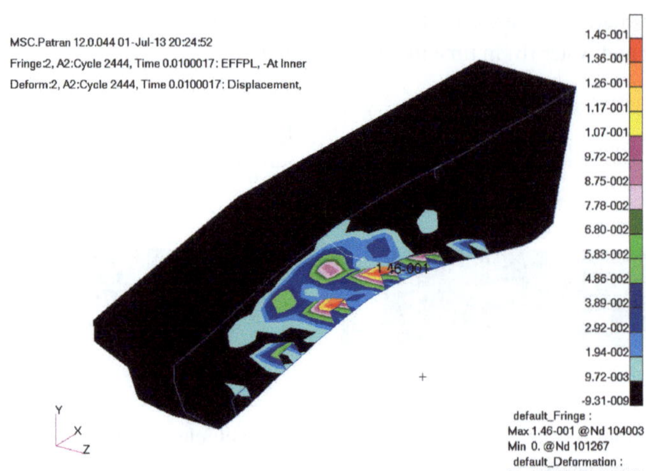

Fig. 21. Distribution of plastic deformation

Regarding the effort of the maximum value of the effort reached 2.16E9 Pa. This value was caused by the increase yield strength by quickly deformation.

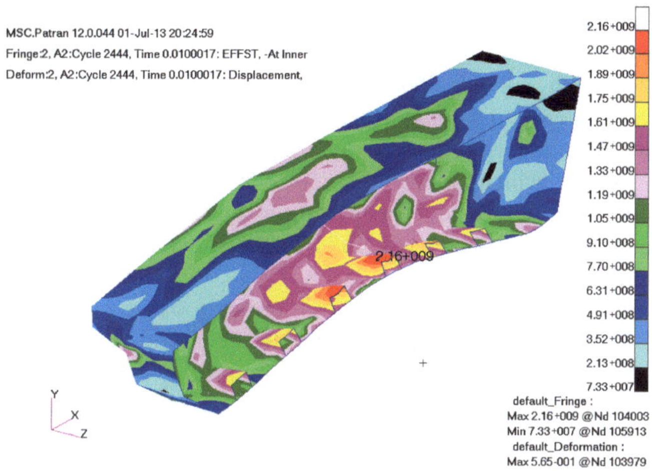

Fig. 22. Distribution the reduction stress

Practically, the bottom of hull deformed permanently. Maximum displacement of bottom was 0,56 m. So large deformation was caused a large density of medium and using material model with dynamic intensification.

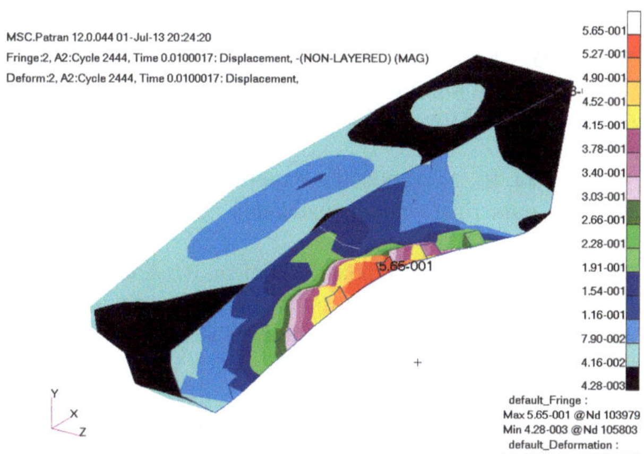

Fig. 23. Displacement map

6 Conclusions

The aim of numerical researches, presented in this article, was to investigate the influence of aquatic environment on results numerical analyze, which was conducted for sample the numerical model of shell the vehicle loaded by pressure wave coming from detonation the explosive. In this article present the issue of dispersion the pressure impulse in Euler area in which was submerge the shell of vehicle. To connection the Euler area with vehicle model using the general coupling.

The pressure wave evoke by detonation (simulated as point detonation) radiate in the cubic area with suitable boundary conditions. The theoretical solution of propagation with strong discontinuity in the spherical shape starting in the point source is as the form the analytical Taylor's similarity equations. The development of the full-size special vehicle model was preceded by additional laboratory test of mechanical properties for armour steel using to built of this type the constructions. The investigations was conducted in Department of Mechanics and Applied Computer Science, MUT.

Cause in the analyze based on cubic, but properly dense mesh of Euler elements the influence of the shape grid on the value the pressure on top the shock wave is relatively slight. The propagation the pressure impulse in denser medium causes that its value and velocity is different from the case of propagation the explosion in thinner medium.

This paper presents part of research conducted on the phenomenon of the pressure wave propagation of.

References

[1] Barnat, W.: Wybrane zagadnienia ochrony życia i zdrowia załóg pojazdów przed wybuchem, Wydawnictwo MilitaryRok (2011)

[2] Barnat, W.: Wstęp do modelowania wybuchu podwodnego pod pojazdem gąsienicowym. Systems 16(3) (2012)

[3] Barnat, W.: Panele kompozytowe do ochrony pojazdów przed skutkami działania prowizorycznych ładunków wybuchowych, Kierownik W. Barnat, Praca badawcza nr R00-O0037/3 (2010)

[4] Barnat, W., Gieleta, R., Niezgoda, T.: Experimental investigation of selected explosion parameters for numerical model validation. Journal of KONES Powertrain and Transport 19(4) (2012)

[5] Barnat, W., Panowicz, R., Niezgoda, T., Gieleta, R.: Analysis of a protective composite panel with energy adsorbent in the form of foamed aluminium. Kones 17(4) (2010)

[6] Babul, W.: Odkształcanie metali wybuchem. WNT, Warszawa (1980)

[7] Dacko M. i inni, Metoda elementów skończonych w mechanice konstrukcji, Warszawa Arkady (1994)

[8] Ding, P., Buijk, A.: Simulation of Under Water Explosion using MSC.Dytran. MSC.Software Corporation

[9] Dytran Theory Manual, 2004; LS-DYNA theoretical manual, FizykaWzrywa (1998)

[10] Ls_Dyna Theory Manual, Livemore, CA (2005)

[11] Włodarczyk, E.: Wstęp do mechaniki wybuchu, PWN, Warszawa (1994)

[12] Zienkiewicz, O.C., Taylor, R.L.: The Finite Element Method, vol. 1-3. Butterworth–Heinemann, Oxford (2000)

The Influence the Location of Large Charge on Behaviour Special Vehicles SHIBA and AV (Armoured Vehicle)

Wiesław Barnat[1] and Aleksander M. Nawrat[2]

[1] Department of Mechanics and Applied Computer Science, Military University of Technology, Gen. S Kaliskiego St. 2, 00-908 Warsaw, Poland
wbarnat@wat.edu.pl
[2] Institute of Automatic Control, Silesian University of Technology, Akademicka 16, 44-100 Gliwice, Poland

Abstract. The problem of impact resistance the vehicles is described in various articles and standardization documents. Mainly concerns on small size of explosive device. In the case of large charges like a 25 to 1000 kg, the literature of subject is very poor. Practically, when the mass of charge is over 1000 kg the level of security of vehicle do not have a sense. The energy emitted by 100kg charge is able to overturn and even lift up vehicle on weight up to 60 t. In the article present the influence of size the large explosive devices on behaviour the special vehicles. Was analyzed two special vehicles. One of them is vehicle to detection and neutralization the mines and improvised explosive devices. The second is armoured personnel carrier. In article presents the results of numerical analyses the mechanical system of vehicle – explosion, which is described by Euler fields and Lagrange elements. The interactions between vehicle and charge obtained by general coupling. Presents the results of numerical analyses for an elements the support structures of combat vehicle, which is loaded by impact generated by large explosive device.

Keywords: large charges, impact resistance, numerical analysis.

1 Introduction

The special vehicles are expose to action with the fire devices, which have an enemy, mainly improvised explosive device (IED). Consequently, this causes, that modern tactic and technical requirements pointing the methods of modeled the shape of hulls of armoured vehicles in direction to ensure high ability to survive on battlefield.

This actions are related to ensuring an appropriate:

- the geometry of vehicles - adequate level of ballistic security
- the secure of the crew and equipment before mine mediums different class (various explosive materials, various masses, various charge placements)

A.M. Nawrat (ed.), *Innovative Control Systems for Tracked Vehicle Platforms*, Studies in Systems, Decision and Control 2,
DOI: 10.1007/978-3-319-04624-2_18, © Springer International Publishing Switzerland 2015

The main problem appears, when is created effective safety of crew and outside equipment before mine devices [1,4], and especially before improvised explosive devices.

Fig. 1. The crater on explosion the large charge, IED (about 500 kg) under tank [photo US ARMY]

Great development of numerical methods and increased the computing capabilities of modern computers allow the modeling of many physical phenomena, also the explosion. Above mentioned development in relation with the increasing care for the passive protection [2-4] cause, that is looking for the better constructional solutions, not only classical experimental methods and numerical methods.

Large difficulty in description phenomena of interactions at bottom of vehicle makes the phenomenon of multiple reflection between flat bottom of vehicle and the ground.

The phenomena of create the Mach wave was described by Ernest Mach in 1887. With some critical angle of incidence do not occur pure reflection (reflection angle equals incidence angle). Ernest Mach shown, that summary incidence wave and reflected would create in new wave, third wave.

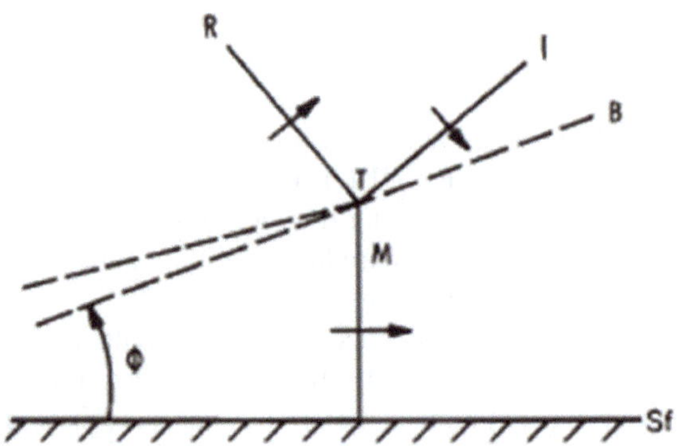

Fig. 2. The way of create Mach wave [9]

The geometry of Mach reflection is presented on fig. 2. Point T is named triple point. On figure shown incidence wave I, reflected wave R and triple point T. Mach wave M moves perpendicular to plane of reflection.

Mach wave creates as a result of intersection of the front of incidence wave and reflected from obstacle (point a on fig. 2). It is worth noting, that this wave is flat and its front is perpendicular to plane of reflected wave. Mach phenomenon is a bit interesting phenomenon and was used during strong nuclear explosions to made stronger destroying force.

The various phenomenon, which made a difficult right evaluation the influence of the charge on the vehicle is intensification the pressure impulse of incidence wave (by reflection). According to stiffness of barrier [6,8] the reinforcement may even be ten times (fig. 3).

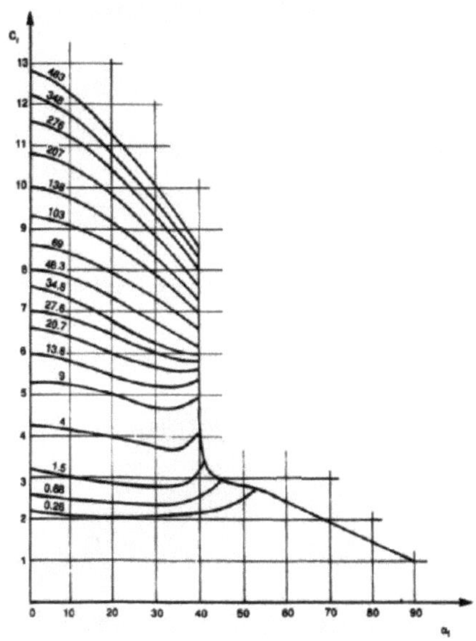

Fig. 3. The diagram of relationship between reinforcement of pressure impulse for incidence angle on stiffness barrier. [8]

Practically, every obstacle, which is on the impulse of pressure way is a reason of increase the value. During numerical analyses should remember, that the standardization documents very clearly define, that security constructions are calculated on value the reflected (reinforced) pressure impulse.

2 General Description of Models the Vehicles SHIBA and AV

Numerical model FEM made in two calculation options:

- option 1: IED explosion in side of vehicle;
- option 2: IED explosion under right front wheel.

In first option 10kg TNT IED charge was placed on height 1m in distance 5m from vehicle. Model was build from: vehicle model, ground, air, TNT charge. Additionally, assumed appropriate for the phenomenon initial and boundary conditions.

Similarly, in second option 10 kg TNT charge was placed under right, front wheel of vehicle. On fig. 4 present scheme of physical model suitable for first option. Second option is present on fig. 5.

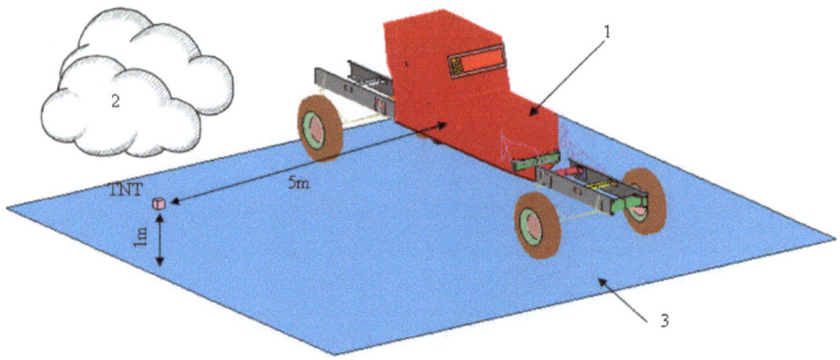

Fig. 4. Physical model of side explosion of simulation (option 1) 1-vehicle, 2- air, 3- ground

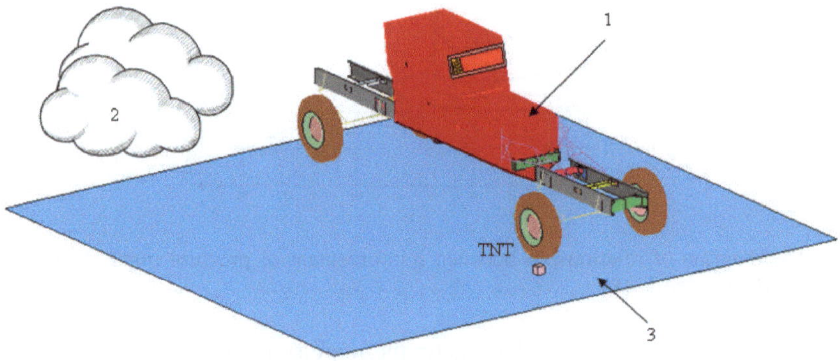

Fig. 5. Physical model of side explosion simulation under right, front wheel (option 2) 1-vehicle, 2- air, 3- ground

To eliminate of reflection effect the shock wave, which arise on the geometrical limits the air model, defined six planes located as shown on fig.6, where the task was dispersion shock wave directed outside the model.

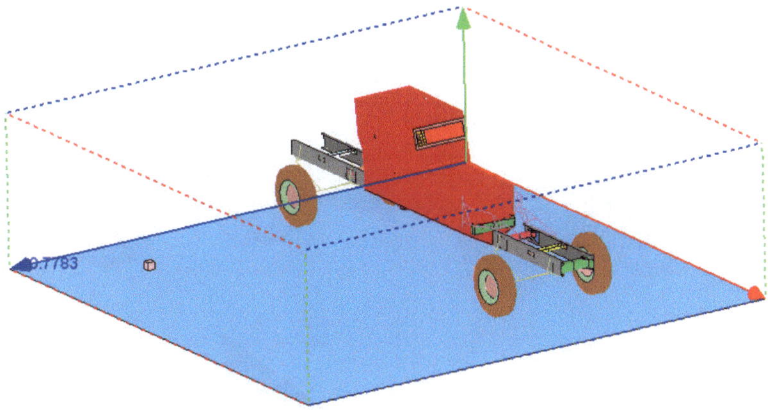

Fig. 6. Six planes (cube) of blast energy dispersion(defined boundary conditions): sides of planes marked by dashed lines

Model of the special vehicle inter alia consisted of: cabin, frame, suspension, chassis. For the model construction defined 36 characteristic (properties) corresponding to the individual structural components. Scheme of model with detailing main components shown on fig. 7.

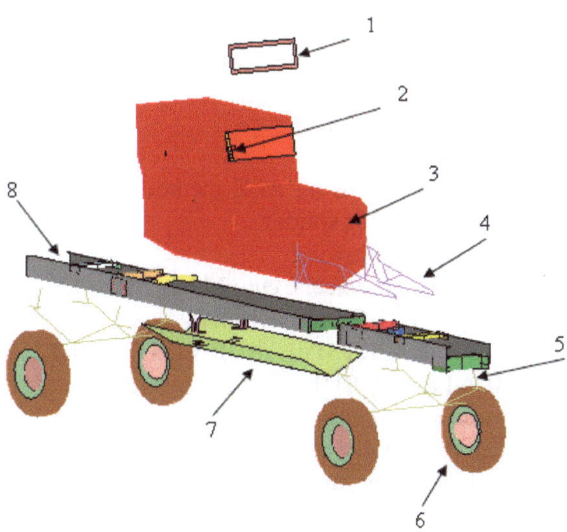

Fig. 7. Main components of special vehicle model: 1- window, 2- seat, 3- cabin, 4 - connecting frame, 5 - suspension, 6 - wheels, 7 - deflector, 8 - frame

Fig. 8. General view the vehicle and frame model

In the numerical model of TNT explosion simulation used following elements:

- SOLID - crew seats, air, TNT,
- BEAM - suspension elements,
- SHELL - the rest of elements.

Process of explosion was modeled using ALE method (using coupling Euler gas description and Lagrange description for solids).

Material properties was defined following:

- cabin, frame, suspension, wheels, etc.: MAT_PLASTIC_KINEMATIC
- gas (air): MAT_NULL (EOS Gruneisena),
- TNT: MAT_HIGH_EXPLOSIVE_BURN (EOS JWL),
- tyres: MAT_BLATZ-KO_RUBBER

Good qualities of explosion proof cabin obtained by application:

- The shape of cab (especially deflector)
- The use of high strength materials (steel Armox, material Dyneema, carbon fibers),
- A large wall thickness.

The explosive material (TNT) was modeled SOLID elements. Using 10 kg of this material. Assuming standard density of TNT ρ= 1.649 g/cm3 corresponding to volume a v = 6064 cm3. The density of designed mesh Lagrange elements (regular cubes with side ~11cm) was chosen so that TNT charge contained in 4 elements.

Material properties of TNT was described by material model MAT_HIGH_EXPLOSIVE_BURN with following parameters:

Table 1. TNT material data for MAT_HIGH_EXPLOSIVE_BURN

parameter	value
density, ρ	1613 kg/m3
speed of detonation, D	6930 m/s
pressure Chapman- Jouget, PCJ	21 GPa

and description of TNT combustion with state equation Jones - Wilkins - Lee (JWL):

$$p = A\left(1 - \frac{\omega}{R_1\bar{p}}\right)\exp(-R_1\bar{p}) + B\left(1 - \frac{\omega}{R_2\bar{p}}\right)\exp(-R_2\bar{p}) + \frac{\omega\bar{e}}{\bar{p}}. \tag{1}$$

Contained in the JWL equation constants are shown in Table2.:

parameter	A [Pa]	B [Pa]	R₁ [-]	R₂ [-]	ω [-]	Eo [Gpa]
value	3.712	3.231	4.150	0.950	0.3	7

To designed the air model used Euler elements, which have material properties for MAT_NULL (Type 9). Constants used in model: density ρ=0,00129 kg/m3, the rest of constants: default.

In state equation (EOS Gruneisen):

- in the case of compression:

$$p = \frac{\rho_0 C^2 \mu\left[1+\left(1-\frac{\gamma_0}{2}\right)\mu-\frac{a}{2}\mu^2\right]}{\left[1-(S_1-1)\mu-S_2\frac{\mu^2}{\mu+1}-S_3\frac{\mu^3}{(\mu+1)^2}\right]^2} + (\gamma_0 + a\mu)E, \tag{2}$$

- in the case of decompression:

$$p = \rho_0 C^2\mu + (\gamma_0 + a\mu)E, \tag{3}$$

where the constants for the state equation assumed following:

Getting the results for structures subjected to influence of burning gas pressure in conditions of sudden increase of load, requiring modeled interaction between solids (Lagrange elements) and gaseous medium (Euler objects). The process of numerical compression between structures described in Lagrange coordinates and pressure wave generated in Euler coordinates, named ALE (Arbitrary Lagrangian-Eulearian), was defined used card CONSTRAINT_LANGRANGE_IN_SOLID.

Table 2. Constants for Gruneisen state equation

parameter	value
C	6000
S1	1.41
S2	1.979
S3	0
Gruneisen gamma, $\gamma 0$	0
a	0.11
Initial energy, E0	0 J
Initial speed V0	0 m/s
ratio	$\mu = \frac{\rho}{\rho_0} - 1.$

General conception of ALE shown on fig. 9.

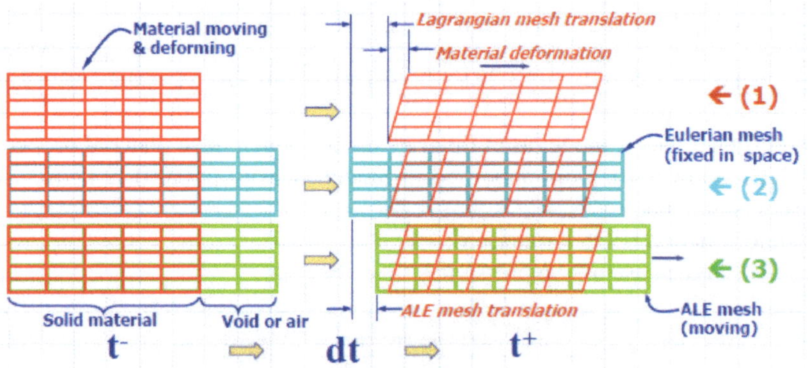

Fig. 9. ALE conception: 1- Lagrange, 2 - Euler, 3 – ALE

3 The Result of Numerical Analyzes for the Special Vehicle SHIBA

Numerical computations were performed based on direct integration process, which is included in LS - Dyna system ("explicit integration"). In accordance with practice of modeled the phenomena of blast, for increase the calculation accuracy, default value time step calculations explicite was decreased to 67%. In the case of SOLID and SHELL elements time step value was:

$$\Delta t_e = 0.67 \frac{V_e}{c \cdot A_e}, \tag{4}$$

where:

Ve - element volume,

Ae- the largest area of the wall element,

c- speed of sound

$$c = \sqrt{\frac{E}{\rho((1-\nu^2)}}, \tag{5}$$

where:

E- Young modulus,

ρ- density,

ν - Possion ratio.

Assumed the run time of simulated phenomena as 2ms. The real time of simulation was correspondingly larger and depends on available computational possibilities and the size of numerical task. It amounted about 12 hours. The calculations were performed on a computing cluster using 18 processors.

After each time steps obtained the results (integration), which constituted basis for the calculation for the next time step. Due to the finite ability to save the results, some results were recorded for further analysis. Assumed, that save of results was realized in 0.02 ms. For assumed time of calculation 2ms corresponded to 100 records. This was sufficient to obtain information about the course of the variability interest parameters. Among obtained the many values of the parameters, a detailed analysis subjected to the time series and extremes:

- reduced stress,
- displacements,
- shock wave pressure.

As a result of explosion of TNT loads produced shock wave (moving to the space of area where is an extremely high pressure). The structure of this region is not uniform. At the moment of the explosion the greatest pressure is at the location of the load, the lowest (but higher than the atmospheric pressure) prevails in the wavefront . The structure of the area shown in the following fig. 10.

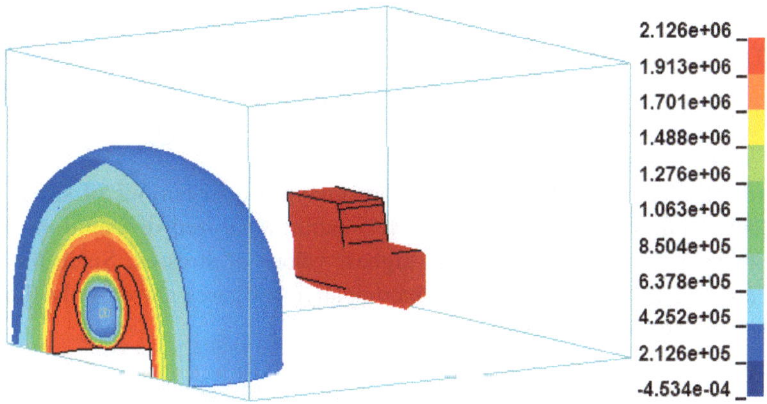

Fig. 10. High pressure area in time t=0.598125 ms(first option)

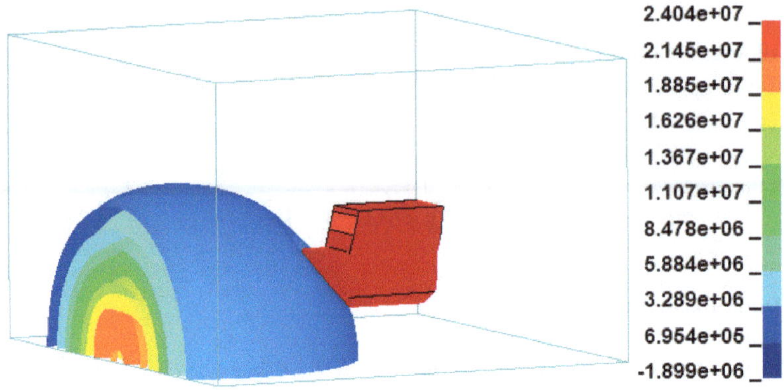

Fig. 11. High pressure area in time t=0.596397 ms (second option)

Energy coming from the combustion load is transmitted through the action of the shock wave, including the parts of the vehicle. Explosive device burning occurs in a highly exothermic. The entire accumulated charge is burned before reaching the shock wave to the vehicle (at the time of to approximately 0.2ms).

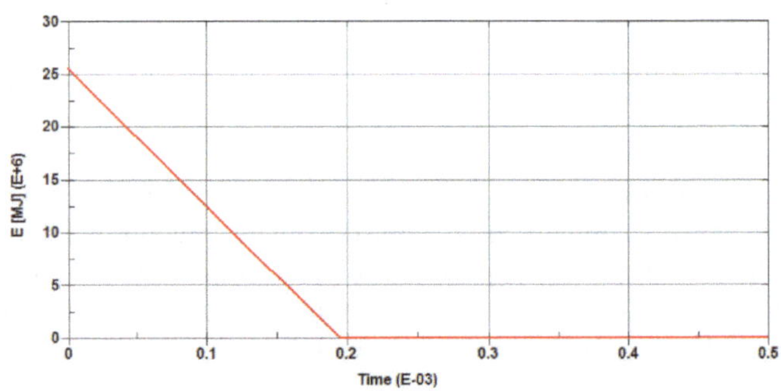

Fig. 12. The diagram of total energy (kinetic and internal) TNT charge - first and second option

The diagram of total energy in time function for second option is the same as for the first option (the same size of charge).

In order to determine the material effort analyzed distributions of stress reduced values (according to the criterion of HMH) cabin. Analyzed distributions for these time moments in which measured the highest values of stresses. Since the reduced stresses are less than 30 MPa, in the accompanying figures shows only those areas where the stress exceeded 10MPa.

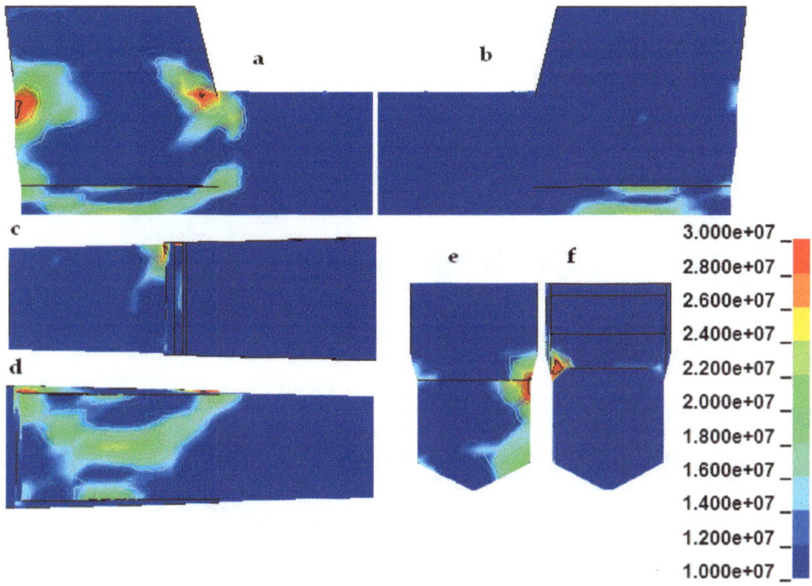

Fig. 13. Maps of reduced stress [Pa] for t = 2ms - first option (side of vehicle explosion), where the area of stress σ > 30MPa marked red color

In the case of first option the highest stress are in side wall from side of explosion, on the floor and on front of vehicle. In second option the maps of stress shown on fig. 14.

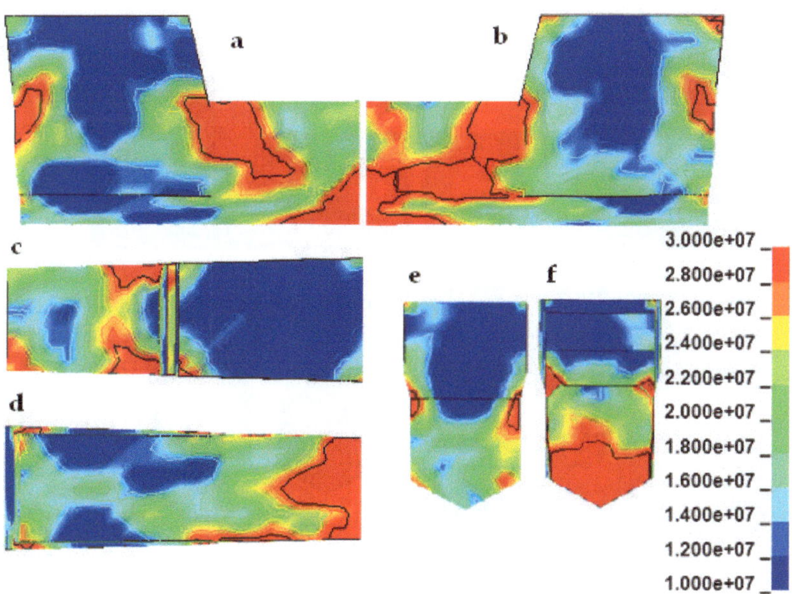

Fig. 14. The maps of reduced stress [Pa] for t= 2ms - second option (explosion under vehicle wheel), red color marked the area where stress σ> 30 MPa; a - left side, b- right side, c - up, d - down, e- back, f - front

In the case of second variant, the highest stress are on the floor of the front of vehicle. In effect of detonation the elements of cabin construction are not destroyed, only deformation. The maps of deformation corresponding in different variants shown on fig. 15.

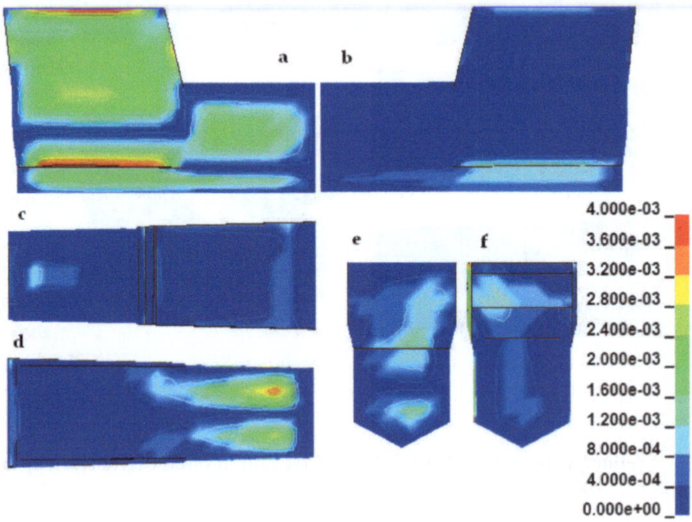

Fig. 15. The maps of summarized deformations [m] for t=2ms - first option. Red color marked the area of deformation T> 4mm. a- left side, b- right side, c - up, d- down, e - back, f - front.

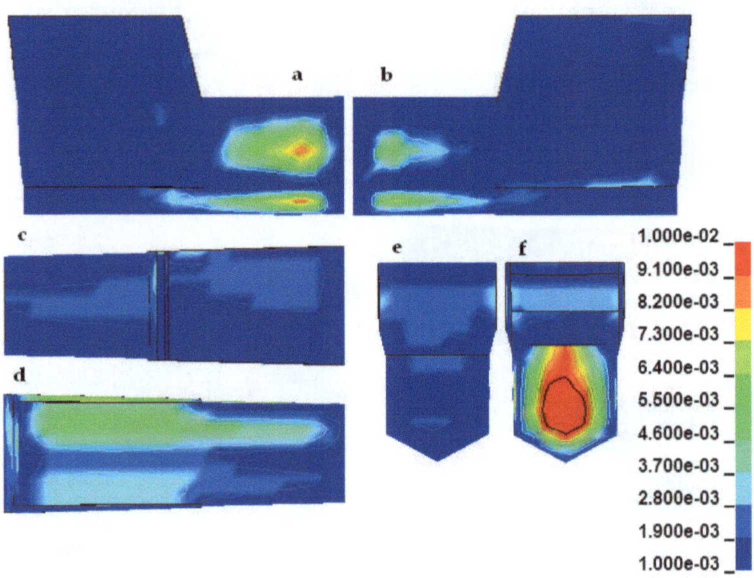

Fig. 16. The maps of summarized deformations [m] for t=2ms - first option. Red color marked the area of deformation T> 1cm. a- left side, b- right side, c - up, d- down, e - back, f - front.

Practically, the size of the loads of 10 kg TNT does not cause damage to the cabin. This means that the cabin has been designed in accordance with applicable standards.

4 The Influence of Side Detonation of IED Charge on Special Vehicle

4.1 The Description of Numerical Models

There is preliminary results of numerical analyses the influence of the side detonation the large IED load on cabin the special vehicle applicable to standardization documents, which are creating. In conducting analyzes by object of the investigations is a wheeled vehicle weighing about 22t.

The load by detonation wave is realized in conducted to accordance with the assumptions included in the newly established standard document. IED charge is located in distance X equal 5m from vehicle and on height H equal 1m above the ground, as shown on fig.17

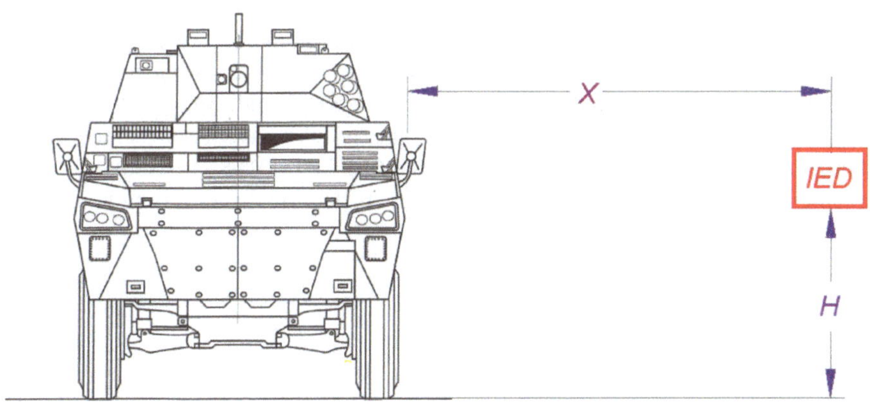

Fig. 17. IED location relative to the special vehicle

The pressure wave induced by detonation (approximately simulated as the point detonation) was distributed in an area of shape of a cube with the conferred the appropriate boundary conditions.

This methodology allows on computer simulation of propagation process the blast wave by giving the appropriate initial conditions (denstity, energry, pressure) for some selected elements from Euler domain, the next solution the conservation the mass, momentum and energy law. Typical values for explosion materials is: density - 1600 kg/m3 and specific internal energy - 4.2 MJ/kg.

The space, where the wave of detonation is propagating, was modeled by Euler elements type Hex 8, which are characterized by ideal gas properties with $\gamma = 1,4$ and the density equivalent to density the air in normal conditions (ρ = 1,2829 kg/m3). The ground was created as an undeformable body. In calculations included the changes made by deformation of vehicle hull (structure) because vehicle was modeled Lagrange elements type Shell Quad 4. Elements have a mechanical properties: E = 2,1 10e9 MPa and ν=0,31. To describe the behaviour of steel used bilinear model of material elastoplastic DYMAT 24, which for destroy criterion adopts maximum deformation [7].

In all models used in this article the nodes of structural elements (nodes of Lagrange domain) do not have included initial conditions. It means, that whole speeds and accelerations for time t=0 was reset.

Numerical analyzed model of special vehicle loaded by wave coming from the blast charges up to 25, 50, 100, 150, 250 and 500kg.

4.2 Results of Numerical Analyses

The results from numerical simulations, for each analyzed cases shown on figures and graphs. On fig. 18 presents the distribution of pressure in Euler domain. On the rest of figures shown the behaviour of the vehicle in various time points after the impact of blast. It should be noted that practically, in all cases the impact of the pressure wave at the vehicle followed the formation of a Mach wave. In addition, the graphs show the displacement, velocity and acceleration of a point on the hull of the vehicle.

During the propagation of pressure impulse for time moment t =0.000012s followed the reflection of pressure impulse from rigid ground, fig. 18a, what made the increase the pressure value. Initial design the Mach wave is presented for time 0.000059s, fig.18b. A full-fledged Mach wave (time 0.000168 s) is shown in fig.18c. Strengthening the pressure pulse through the reflection of the vehicle hull (time 0.000269 s) is shown in fig. 18d.

a)

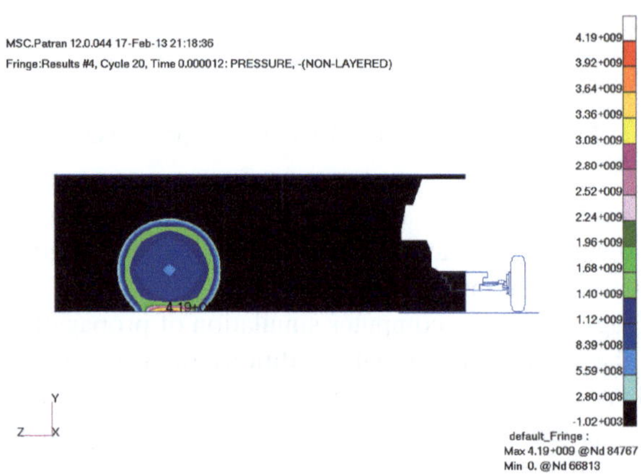

Fig. 18. The distribution of pressure acting to wheel vehicle

b)

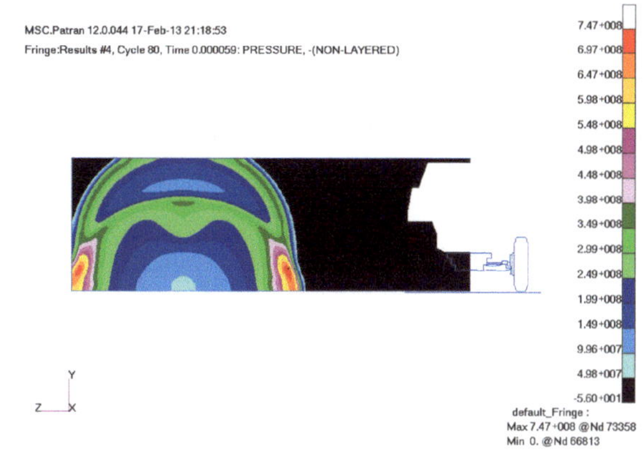

c)

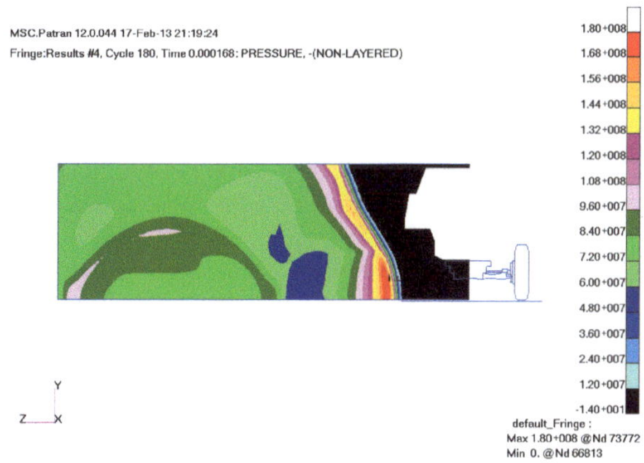

d)

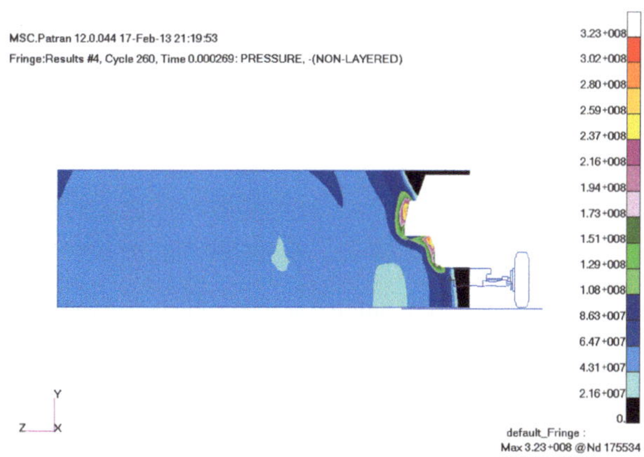

Fig. 18. (*Continued*)

On the following figures presents the displacement of hull the special vehicle coming from various IED charges in different time moments.

As a result of the impact of pressure impulse derived from the 25 kg TNT followed by a light thrown the vehicle at a height of about 1.7 m, fig. 19a. After the collapse on the ground has been shifted by 0.5 m from original position, fig. 19b.

a)

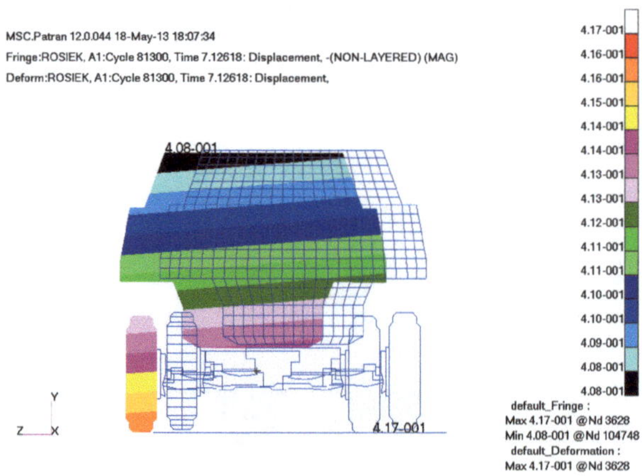

b)

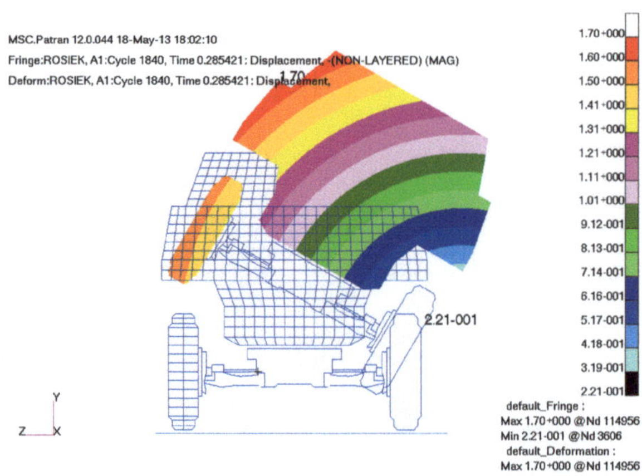

Fig. 19. The displacement of hull the special vehicle derived from 25 kg load

a)

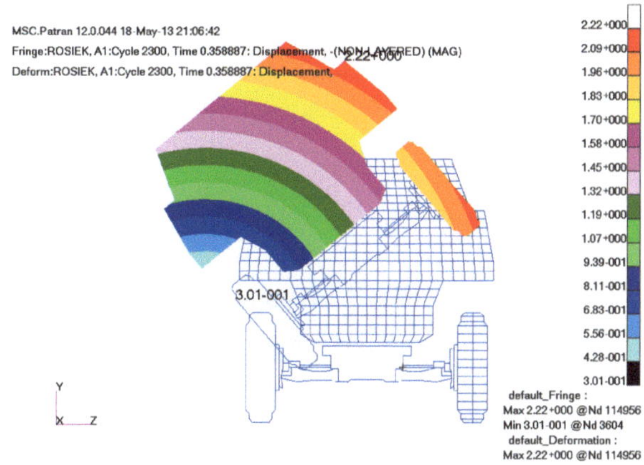

b)

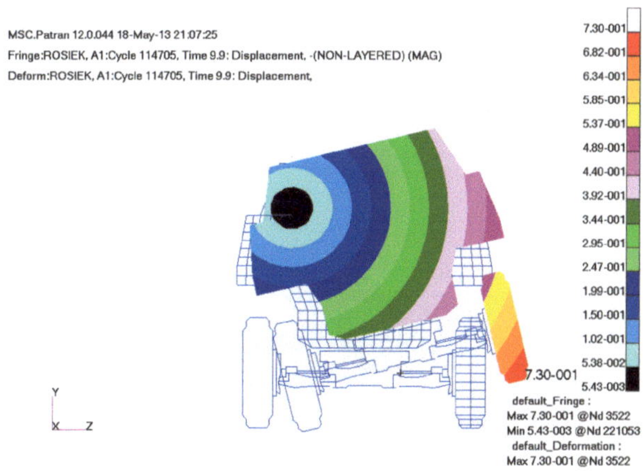

Fig. 20. The displacement of hull special vehicle derived from 50 kg TNT

As a results of impact the pressure impulse derived from 50 kg TNT following the light thrown up the vehicle on height about 2,2 m, fig. 20a. View of vehicle shortly before impact on the ground is shown in fig. 20b. Similarly behaves vehicle laden load of 100 kg (fig. 21).

Practically, the side loading up to 100 kg charge there is a high likelihood of falling vehicle on wheels.

a)

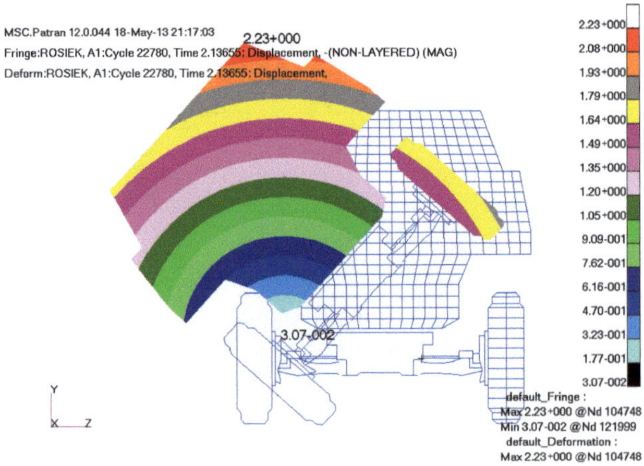

b)

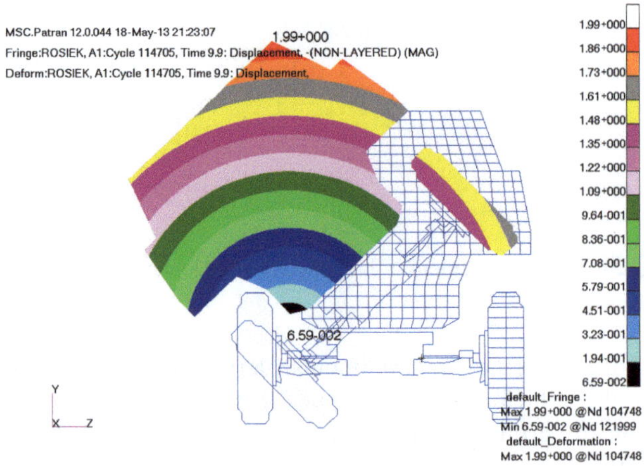

Fig. 21. The displacement of hull of special vehicle derived from 100kg

a)

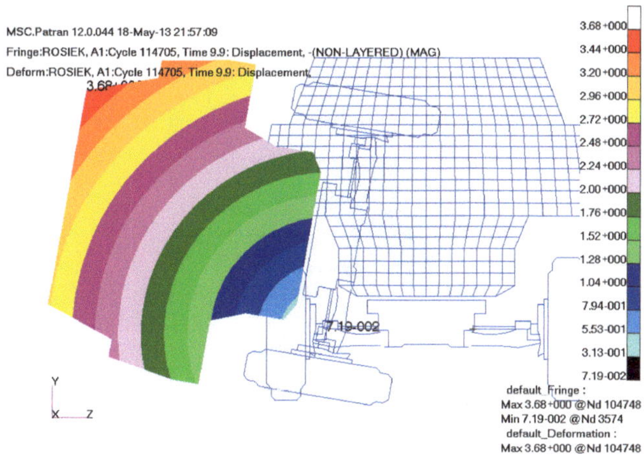

b)

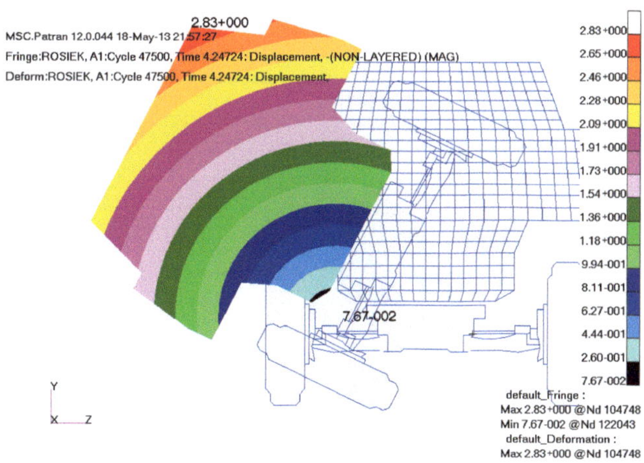

Fig. 22. The displacement of hull the special vehicle derived from 150 kg

a)

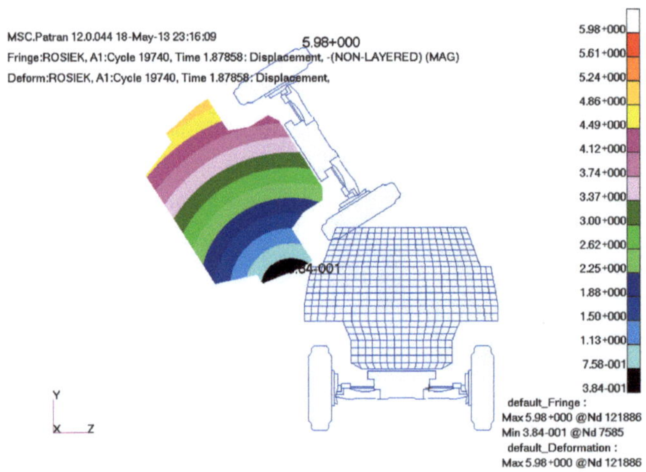

b)

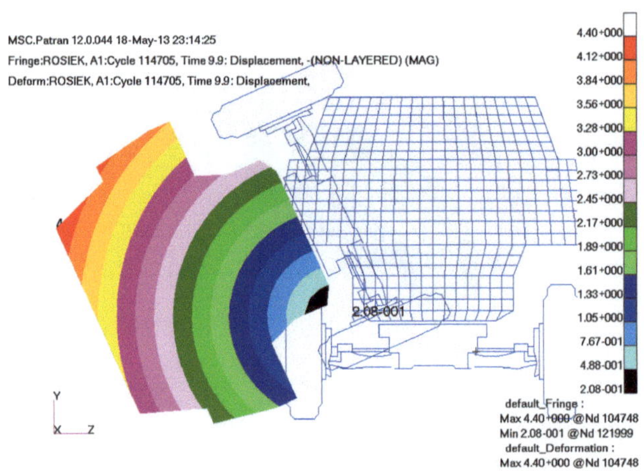

Fig. 23. The displacement of hull the special vehicle derived from 250 kg

a)

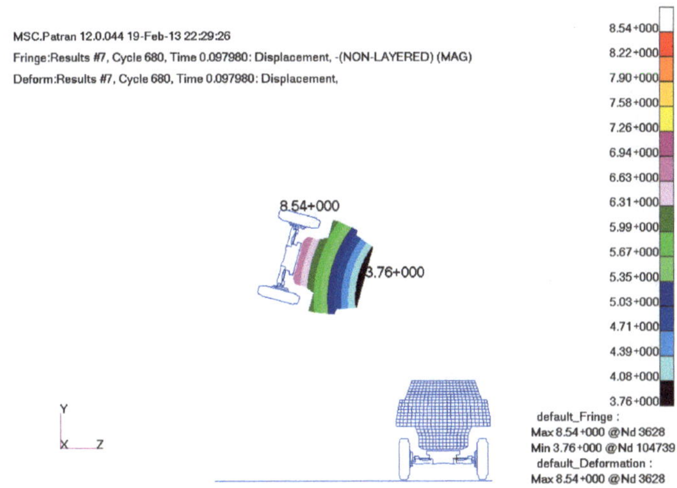

b)

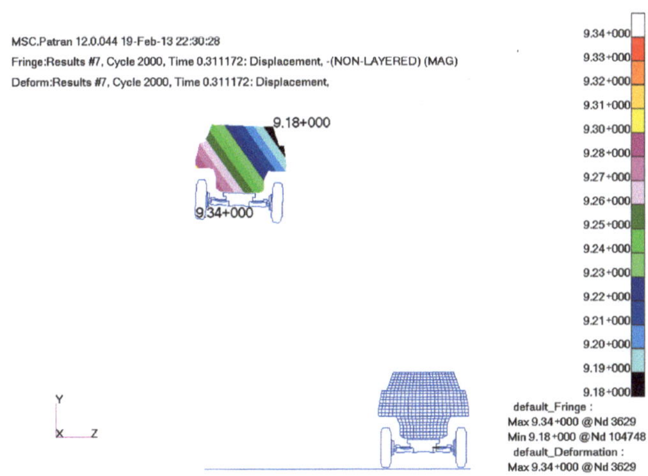

Fig. 24. The displacement of hull the vehicle derived the charge 500 kg

In other consequences than the previous is load the vehicle by charge 150 kg. As a result of the impact of pulse pressure obtained from 150 kg TNT takes the vehicle to overturn on its side.

For larger loads (from 250 kg TNT) the vehicle behaves differently than in previous cases. (Figure 21, 22 and 23). In the first phase the impact load to the vehicle following the separation of the wheels on the side of the explosion. As a result of such separation occurs eccentric hull of the vehicle from the ground. In subsequent time points the vehicle is lifted up a rotating motion. This movement is caused by the eccentric impact on the vehicle.

Due to sparse a record caused by length the time of analyze and the time of phenomenon (about 10s) the graphs have a approximately character. It is worth noting that in the abovementioned graphs are shown accident of magnitude as in the case of rotational movement of the vehicle may cause certain misrepresentation read results. In the following articles will focus on selected fragments of analysis noted that the maximum value of the resultant velocity vector and the graphs presented was for a load 250 kg and the total was about 0.4 m / s. Furthermore, the maximum acceleration value us 0.06 m/s2. In the case of analyze the model loaded by 500 kg charge the vehicle was take off on height 10m. The most interesting is an example for load 150 kg. Practically, the vehicle slightly tilt, leap and its trajectory of flight was plate and low.

It should be noted, that in results shown a magnitude for speeds, accelerations and displacements. Properly, for larger loads from 150 kg should consider the individual components.

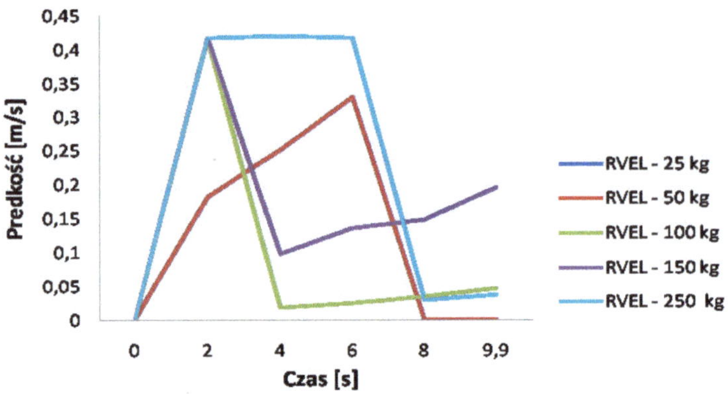

Fig. 25. Graph the speed of point located on hull the special vehicle

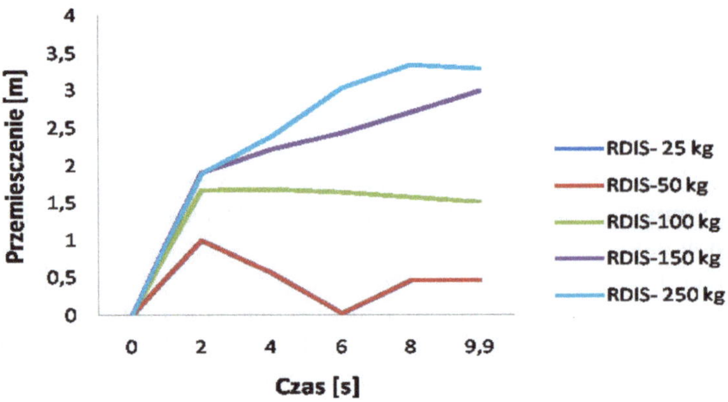

Fig. 26. Graph of displacement the point located on hull the special vehicle

The large acceleration acting on human body made a injuries, which also effectively exclude located inside the crew from combat operations causing permanent disability excluding the soldier from further service and even with a normal existence in society.

It follows that the determination of the acceleration acting on the vehicle, and thus the soldiers who are inside, it can also prevent the dangerous effects of explosion. The summarized of accelerations in the characteristic point on hull for each cases of size the IED charge presents on fig. 27.

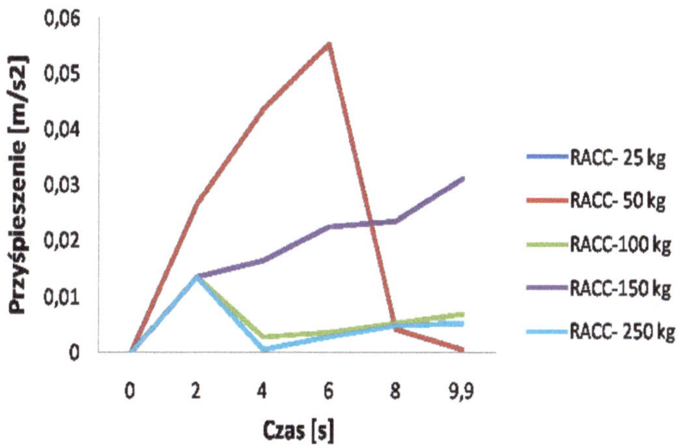

Fig. 27. Graph of acceleration the point located on hull the special vehicle

The analysis of the received courses can see that the most dangerous for the crew inside the vehicle which caused loads are loads of weights over 100 kg of explosive device. When the shock wave reaches the hull acts on the operating pressure peak in a relatively short period of time. As known this acting, it is a large pressure value in short period of time made the most dangerous injuries on human. When the smaller value of acceleration and longer duration than is the less probability of permanent injuries of body the crew in vehicle subjected to this type of load.

5 Conclusions

Previous articles did not included the studies about influence large charges on hull of vehicle. Suitable located the charge to the vehicle may causes many of adverse from the point of view the protecting the vehicle phenomena. A restriction in the form of soil increased the pressure impulse transmission on the side of the vehicle.

Because the influence the shape of grid on pressure value on front the shock wave is relatively slight. In analyze based on cubic, but properly thick, the grid of Euler elements.

In this article used the coupling between the medium described by continuum mechanics terms of Euler and the medium described by Lagrange equations to describe the interaction between the construction and loading by pressure wave coming from detonation. In terms of the Euler equations describe the usual liquid - in this case, it is the air, which comes to the detonation of the explosive and the propagation of the shock wave. However, the equations expressing the laws of conservation of mass, momentum and energy in terms of the Lagrangian describes the behavior of the structure (the structure). For the coupling fluid and structure interaction is selected general coupling which is included in implementation the standard software Dytran MSC.

The pressure wave induced of detonation (approximately simulated by the point detonation) was spread in an area of cube shape from conferred the appropriate boundary conditions. The development of the full-size model was preceded by a of special vehicle with additional laboratory testing of mechanical properties of armor steel used for the construction of this type of vehicle. These tests were conducted in Department of Mechanics and Applied Computer Science, MUT.

References

[1] Barnat, W.: Wybrane zagadnienia ochrony życia i zdrowia załóg pojazdów przed wybuchem, Military Rok (2011)
[2] Trzciński, W., Trębiński, R., Cudziło, S.: Investigation of the Behaviour of Steel and Lami-nated Fabric Plates Under Blast Wave Load, Part I: Experimental Approach. In: V International Armament Conference, Waplewo (2005)
[3] Krzewiński, R., Rekrucki, R.: Roboty budowlane przy użyciu materiałów wybuchowych, Polcen (2005)
[4] Krzystała, E., Kciuk, S., Mężyk, A.: Identyfikacja zagrożeń załogi pojazdów specjalnych podczas wybuchu, Politechnika Śląska, Gliwice (2013r)
[5] Babul, W.: Odkształcanie metali wybuchem. WNT, Warszawa (1980)
[6] Włodarczyk, E.: Wstęp do mechaniki wybuchu. PWN, Warszawa (1994)
[7] Dytran Theory Manual, 2004; LS-DYNA theoretical manual (1998)
[8] Smith, P.D., Hetherington, J.G.: Blast and Ballistic Loading of Structures. Butterworth-Heinemann (1994)
[9] Baker, W.E.: Explosions in Air. University of Texas Press, Austin (1973)
[10] Instrukcja "Prace inżynierskie i niszczenia". Sztab Generalny WP, Szefostwo Wojsk Inżynieryjnych, Warszawa (1995)
[11] Barnat, W.: Numeryczno doświadczalna analiza złożonych warstw ochronnych obciążonych falą uderzeniową wybuchu. Bell Studio, Warszawa (2010)

About the Book

Nowadays the unmanned vehicles are becoming more and more common. Therefore there is a need for innovative mechanical constructions capable of adapting to various applications regardless the ground, air or water/underwater environment. Military operations in order to complete successfully require a vast amount of information and high quality equipment and vehicles capable of movement both in a city and terrain. An example of such are tracked vehicles. As well as in today's and tomorrow's army, tracked vehicles will be one of the most important equipment. Those objects combines with innovative algorithms become autonomous objects capable of patrolling chosen areas by themselves and automatically inform the operator of the system about potential place of detection of a specific incident. This is achieved by using sophisticated methods of generation of non-collision trajectory for those types of objects and enabling automatic integration of both ground and aerial unmanned vehicles. Many research groups works on design and implementation of control systems for unmanned vehicles. One of the most challenging tasks is to design robust control system that allows human operator stay away from operation zone which can be very dangerous. There are few approaches to this problem. First proposition of control system is remote control. Human operator is away from vehicle and has full control on the vehicle. Operator controls signals which are send directly to vehicle actuators. Control signals are transferred from ground station to vehicle by wireless link. Another approach is tele-operation where human operator has visual feedback from cameras mounted on vehicle. Final approach is to implement autonomy functions on vehicle which is able to do tasks without human intervention. That option requires full sensor system with control processor unit on the vehicle.

This book is a valuable source of information for constructors and developers of tracked vehicles or accompanying systems. Scientists and researchers involved in mechanics, control algorithms, image processing, computer vision, data fusion, or IC can find many valuable suggestions and solutions. We took an effort to present multiple solutions and challenges in the described fields.

Author Index

MIX
Papier aus verantwortungsvollen Quellen
Paper from responsible sources
FSC® C105338

If you have any concerns about our products,
you can contact us on
ProductSafety@springernature.com

In case Publisher is established outside the EU,
the EU authorized representative is:
Springer Nature Customer Service Center GmbH
Europaplatz 3, 69115 Heidelberg, Germany

Printed by Libri Plureos GmbH
in Hamburg, Germany